SYSTEMS ENGINEERING PRINCIPLES AND PRACTICE

SYSTEMS ENGINEERING PRINCIPLES AND PRACTICE

SECOND EDITION

Alexander Kossiakoff
William N. Sweet
Samuel J. Seymour
Steven M. Biemer

A JOHN WILEY & SONS, INC. PUBLICATION

Published by John Wiley & Sons, Inc., Hoboken, New Jersey
Published simultaneously in Canada

For general information on our other products and services or for technical support, please contact our Customer Care Department within the United States at (800) 762-2974, outside the United States at (317) 572-3993 or fax (317) 572-4002.

Wiley also publishes its books in a variety of electronic formats. Some content that appears in print may not be available in electronic formats. For more information about Wiley products, visit our web site at www.wiley.com.

Library of Congress Cataloging-in-Publication Data:

Systems engineering : principles and practice/Alexander Kossiakoff ... [et al.].—2nd ed.
    p. cm.—(Wiley series in systems engineering and management; 67)
  Rev. ed. of: Systems engineering: principles and practices/Alexander Kossiakoff, William N. Sweet. 2003.
  ISBN 978-0-470-40548-2 (hardback)
  1. Systems engineering.   I. Kossiakoff, Alexander, 1945–   II. Title.
  TA168.K68 2010
  620.001'171–dc22

                                                        2010036856

Printed in the United States of America

oBook ISBN: 9781118001028
ePDF ISBN: 9781118001011
ePub ISBN: 9781118009031

V10008069_020619

To Alexander Kossiakoff,

who never took "no" for an answer and refused to believe that anything was impossible. He was an extraordinary problem solver, instructor, mentor, and friend.

Samuel J. Seymour

Steven M. Biemer

CONTENTS

LIST OF ILLUSTRATIONS

LIST OF TABLES

PREFACE TO THE SECOND EDITION

It is an incredible honor and privilege to follow in the footsteps of an individual who had a profound influence on the course of history and the field of systems engineering. Since publication of the first edition of this book, the field of systems engineering has seen significant advances, including a significant increase in recognition of the discipline, as measured by the number of conferences, symposia, journals, articles, and books available on this crucial subject. Clearly, the field has reached a high level of maturity and is destined for continued growth. Unfortunately, the field has also seen some sorrowful losses, including one of the original authors, Alexander Kossiakoff, who passed away just 2 years after the publication of the book. His vision, innovation, excitement, and perseverance were contagious to all who worked with him and he is missed by the community. Fortunately, his vision remains and continues to be the driving force behind this book. It is with great pride that we dedicate this second edition to the enduring legacy of Alexander Ivanovitch Kossiakoff.

ALEXANDER KOSSIAKOFF, 1914–2005

Alexander Kossiakoff, known to so many as "Kossy," gave shape and direction to the Johns Hopkins University Applied Physics Laboratory as its director from 1969 to 1980. His work helped defend our nation, enhance the capabilities of our military, pushed technology in new and exciting directions, and bring successive new generations to an understanding of the unique challenges and opportunities of systems engineering. In 1980, recognizing the need to improve the training and education of technical professionals, he started the master of science degree program at Johns Hopkins University in Technical Management and later expanded it to Systems Engineering, one of the first programs of its kind.

Today, the systems engineering program he founded is the largest part-time graduate program in the United States, with students enrolled from around the world in classroom, distance, and organizational partnership venues; it continues to evolve as the field expands and teaching venues embrace new technologies, setting the standard for graduate programs in systems engineering. The first edition of the book is the foundational systems engineering textbook for colleges and universities worldwide.

OBJECTIVES OF THE SECOND EDITION

Traditional engineering disciplines do not provide the training, education, and experience necessary to ensure the successful development of a large, complex system program from inception to operational use. The advocacy of the systems engineering viewpoint and the goal for the practitioners to think like a systems engineer are still the major premises of this book.

This second edition of *Systems Engineering Principles and Practice* continues to be intended as a graduate-level textbook for courses introducing the field and practice of systems engineering. We continue the tradition of utilizing models to assist students in grasping abstract concepts presented in the book. The five basic models of the first edition are retained, with only minor refinements to reflect current thinking. Additionally, the emphasis on application and practice is retained throughout and focuses on students pursuing their educational careers in parallel with their professional careers. Detailed mathematics and other technical fields are not explored in depth, providing the greatest range of students who may benefit, nor are traditional engineering disciplines provided in detail, which would violate the book's intended scope.

The updates and additions to the first edition revolve around the changes occurring in the field of systems engineering since the original publication. Special attention was made in the following areas:

- *The Systems Engineer's Career.* An expanded discussion is presented on the career of the systems engineer. In recent years, systems engineering has been recognized by many companies and organizations as a separate field, and the position of "systems engineer" has been formalized. Therefore, we present a model of the systems engineer's career to help guide prospective professionals.

- *The Systems Engineering Landscape.* The only new chapter introduced in the second edition is titled by the same name and reinforces the concept of the systems engineering viewpoint. Expanded discussions of the implications of this viewpoint have been offered.

- *System Boundaries.* Supplemental material has been introduced defining and expanding our discussion on the concept of the system boundary. Through the use of the book in graduate-level education, the authors recognized an inherent misunderstanding of this concept—students in general have been unable to recognize the boundary between the system and its environment. This area has been strengthened throughout the book.

- *System Complexity.* Significant research in the area of system complexity is now available and has been addressed. Concepts such as system of systems engineering, complex systems management, and enterprise systems engineering are introduced to the student as a hierarchy of complexity, of which systems engineering forms the foundation.

- *Systems Architecting.* Since the original publication, the field of systems architecting has expanded significantly, and the tools, techniques, and practices of this field have been incorporated into the concept exploration and definition chapters. New models and frameworks for both traditional structured analysis and object-oriented analysis techniques are described and examples are provided, including an expanded description of the Unified Modeling Language and the Systems Modeling Language. Finally, the extension of these new methodologies, model-based systems engineering, is introduced.

- *Decision Making and Support.* The chapter on systems engineering decision tools has been updated and expanded to introduce the systems engineering student to the variety of decisions required in this field, and the modern processes, tools, and techniques that are available for use. The chapter has also been moved from the original special topics part of the book.

- *Software Systems Engineering.* The chapter on software systems engineering has been extensively revised to incorporate modern software engineering techniques, principles, and concepts. Descriptions of modern software development life cycle models, such as the agile development model, have been expanded to reflect current practices. Moreover, the section on capability maturity models has been updated to reflect the current integrated model. This chapter has also been moved out of the special topics part and introduced as a full partner of advanced development and engineering design.

In addition to the topics mentioned above, the chapter summaries have been reformatted for easier understanding, and the lists of problems and references have been updated and expanded. Lastly, feedback, opinions, and recommendations from graduate students have been incorporated where the wording or presentation was awkward or unclear.

CONTENT DESCRIPTION

This book continues to be used to support the core courses of the Johns Hopkins University Master of Science in Systems Engineering program and is now a primary textbook used throughout the United States and in several other countries. Many programs have transitioned to online or distance instruction; the second edition was written with distance teaching in mind, and offers additional examples.

The length of the book has grown, with the updates and new material reflecting the expansion of the field itself.

The second edition now has four parts:

- *Part I.* The Foundation of Systems Engineering, consisting of Chapters 1–5, describes the origins and structure of modern systems, the current field of systems engineering, the structured development process of complex systems, and the organization of system development projects.
- *Part II.* Concept Development, consisting of Chapters 6–9, describes the early stages of the system life cycle in which a need for a new system is demonstrated,

its requirements identified, alternative implementations developed, and key program and technical decisions made.

- *Part III.* Engineering Development, consisting of Chapters 10–13, describes the later stages of the system life cycle, in which the system building blocks are engineered (to include both software and hardware subsystems) and the total system is integrated and evaluated in an operational environment.
- *Part IV.* Postdevelopment, consisting of Chapters 14 and 15, describes the roles of systems in the production, operation, and support phases of the system life cycle and what domain knowledge of these phases a systems engineer should acquire.

Each chapter contains a summary, homework problems, and bibliography.

ACKNOWLEDGMENTS

The authors of the second edition gratefully acknowledge the family of Dr. Kossiakoff and Mr. William Sweet for their encouragement and support of a second edition to the original book. As with the first edition, the authors gratefully acknowledge the many contributions made by the present and past faculties of the Johns Hopkins University Systems Engineering graduate program. Their sharp insight and recommendations on improvements to the first edition have been invaluable in framing this publication. Particular thanks are due to E. A. Smyth for his insightful review of the manuscript.

Finally, we are exceedingly grateful to our families—Judy Seymour and Michele and August Biemer—for their encouragement, patience, and unfailing support, even when they were continually asked to sacrifice, and the end never seemed to be within reach.

Much of the work in preparing this book was supported as part of the educational mission of the Johns Hopkins University Applied Physics Laboratory.

Samuel J. Seymour
Steven M. Biemer
2010

PREFACE TO THE FIRST EDITION

Learning how to be a successful systems engineer is entirely different from learning how to excel at a traditional engineering discipline. It requires developing the ability to think in a special way, to acquire the "systems engineering viewpoint," and to make the central objective the system as a whole and the success of its mission. The systems engineer faces three directions: the system user's needs and concerns, the project manager's financial and schedule constraints, and the capabilities and ambitions of the engineering specialists who have to develop and build the elements of the system. This requires learning enough of the language and basic principles of each of the three constituencies to understand their requirements and to negotiate balanced solutions acceptable to all. The role of interdisciplinary leadership is the key contribution and principal challenge of systems engineering and it is absolutely indispensable to the successful development of modern complex systems.

1.1 OBJECTIVES

Systems Engineering Principles and Practice is a textbook designed to help students learn to think like systems engineers. Students seeking to learn systems engineering after mastering a traditional engineering discipline often find the subject highly abstract and ambiguous. To help make systems engineering more tangible and easier to grasp, the book provides several models: (1) a hierarchical model of complex systems, showing them to be composed of a set of commonly occurring building blocks or components; (2) a system life cycle model derived from existing models but more explicitly related to evolving engineering activities and participants; (3) a model of the steps in the systems engineering method and their iterative application to each phase of the life cycle; (4) a concept of "materialization" that represents the stepwise evolution of an abstract concept to an engineered, integrated, and validated system; and (5) repeated references to the specific responsibilities of systems engineers as they evolve during the system life cycle and to the scope of what a systems engineer must know to perform these effectively. The book's significantly different approach is intended to complement the several excellent existing textbooks that concentrate on the quantitative and analytical aspects of systems engineering.

Particular attention is devoted to systems engineers as professionals, their responsibilities as part of a major system development project, and the knowledge, skills, and mind-set they must acquire to be successful. The book stresses that they must be innovative and resourceful, as well as systematic and disciplined. It describes the special functions and responsibilities of systems engineers in comparison with those of system analysts, design specialists, test engineers, project managers, and other members of the system development team. While the book describes the necessary processes that systems engineers must know and execute, it stresses the leadership, problem-solving, and innovative skills necessary for success.

The function of systems engineering as defined here is to "guide the engineering of complex systems." To learn how to be a good guide requires years of practice and the help and advice of a more experienced guide who knows "the way." The purpose of this book is to provide a significant measure of such help and advice through the organized collective experience of the authors and other contributors.

This book is intended for graduate engineers or scientists who aspire to or are already engaged in careers in systems engineering, project management, or engineering management. Its main audience is expected to be engineers educated in a single discipline, either hardware or software, who wish to broaden their knowledge so as to deal with systems problems. It is written with a minimum of mathematics and specialized jargon so that it should also be useful to managers of technical projects or organizations, as well as to senior undergraduates.

1.2 ORIGIN AND CONTENTS

The main portion of the book has been used for the past 5 years to support the five core courses of the Johns Hopkins University Master of Science in Systems Engineering program and is thoroughly class tested. It has also been used successfully as a text for distance course offerings. In addition, the book is well suited to support short courses and in-house training.

The book consists of 14 chapters grouped into five parts:

- *Part I.* The Foundations of Systems Engineering, consisting of Chapters 1–4, describes the origin and structure of modern systems, the stepwise development process of complex systems, and the organization of system development projects.

- *Part II.* Concept Development, consisting of Chapters 5–7, describes the first stage of the system life cycle in which a need for a new system is demonstrated, its requirements are developed, and a specific preferred implementation concept is selected.

- *Part III.* Engineering Development, consisting of Chapters 8–10, describes the second stage of the system life cycle, in which the system building blocks are engineered and the total system is integrated and evaluated in an operational environment.

- *Part IV.* Postdevelopment, consisting of Chapters 11 and 12, describes the role of systems engineering in the production, operation, and support phases of the system life cycle, and what domain knowledge of these phases in the system life cycle a systems engineer should acquire.
- *Part V.* Special Topics consists of Chapters 13 and 14. Chapter 13 describes the pervasive role of software throughout system development, and Chapter 14 addresses the application of modeling, simulation, and trade-off analysis as systems engineering decision tools.

Each chapter also contains a summary, homework problems, and a bibliography. A glossary of important terms is also included. The chapter summaries are formatted to facilitate their use in lecture viewgraphs.

ACKNOWLEDGMENTS

The authors gratefully acknowledge the many contributions made by the present and past faculties of the Johns Hopkins University Systems Engineering Masters program. Particular thanks are due to S. M. Biemer, J. B. Chism, R. S. Grossman, D. C. Mitchell, J. W. Schneider, R. M. Schulmeyer, T. P. Sleight, G. D. Smith, R. J. Thompson, and S. P. Yanek, for their astute criticism of passages that may have been dear to our hearts but are in need of repairs.

An even larger debt is owed to Ben E. Amster, who was one of the originators and the initial faculty of the Johns Hopkins University Systems Engineering program. Though not directly involved in the original writing, he enhanced the text and diagrams by adding many of his own insights and fine-tuned the entire text for meaning and clarity, applying his 30 years' experience as a systems engineer to great advantage.

We especially want to thank H. J. Gravagna for her outstanding expertise and inexhaustible patience in typing and editing the innumerable rewrites of the drafts of the manuscript. These were issued to successive classes of systems engineering students as the book evolved over the past 3 years. It was she who kept the focus on the final product and provided invaluable assistance with the production of this work.

Finally, we are eternally grateful to our wives, Arabelle and Kathleen, for their encouragement, patience, and unfailing support, especially when the written words came hard and the end seemed beyond our reach.

Much of the work in preparing this book was supported as part of the educational mission of the Johns Hopkins Applied Physics Laboratory.

ALEXANDER KOSSIAKOFF
WILLIAM N. SWEET
2002

PART I

FOUNDATIONS OF SYSTEMS ENGINEERING

Part I provides a multidimensional framework that interrelates the basic principles of systems engineering, and helps to organize the areas of knowledge that are required to master this subject. The dimensions of this framework include

1. a hierarchical model of the structure of complex systems;
2. a set of commonly occurring functional and physical system building blocks;
3. a systems engineering life cycle, integrating the features of the U.S Department of Defense, ISO/IEC, IEEE, and NSPE models;
4. four basic steps of the systems engineering method that are iterated during each phase of the life cycle;
5. three capabilities differentiating project management, design specialization, and systems engineering;
6. three different technical orientations of a scientist, a mathematician, and an engineer and how they combine in the orientation of a systems engineer; and
7. a concept of "materialization" that measures the degree of transformation of a system element from a requirement to a fully implemented part of a real system.

Systems Engineering Principles and Practice, Second Edition. Alexander Kossiakoff, William N. Sweet, Samuel J. Seymour, and Steven M. Biemer

Chapter 1 describes the origins and characteristics of modern complex systems and systems engineering as a profession.

Chapter 2 defines the "systems engineering viewpoint" and how it differs from the viewpoints of technical specialists and project managers. This concept of a systems viewpoint is expanded to describe the domain, fields, and approaches of the systems engineering discipline.

Chapter 3 develops the hierarchical model of a complex system and the key building blocks from which it is constituted. This framework is used to define the breadth and depth of the knowledge domain of systems engineers in terms of the system hierarchy.

Chapter 4 derives the concept of the systems engineering life cycle, which sets the framework for the evolution of a complex system from a perceived need to operation and disposal. This framework is systematically applied throughout Parts II–IV of the book, each part addressing the key responsibilities of systems engineering in the corresponding phase of the life cycle.

Finally, Chapter 5 describes the key parts that systems engineering plays in the management of system development projects. It defines the basic organization and the planning documents of a system development project, with a major emphasis on the management of program risks.

1

SYSTEMS ENGINEERING AND THE WORLD OF MODERN SYSTEMS

1.1 WHAT IS SYSTEMS ENGINEERING?

There are many ways in which to define systems engineering. For the purposes of this book, we will use the following definition:

The function of systems engineering is to *guide* the *engineering* of *complex systems*.

The words in this definition are used in their conventional meanings, as described further below.

To guide is defined as "to lead, manage, or direct, usually based on the superior experience in pursuing a given course" and "to show the way." This characterization emphasizes the process of selecting the path for others to follow from among many possible courses—a primary function of systems engineering. A dictionary definition of engineering is "the application of scientific principles to practical ends; as the design, construction and operation of efficient and economical structures, equipment, and systems." In this definition, the terms "efficient" and "economical" are particular contributions of good systems engineering.

The word "system," as is the case with most common English words, has a very broad meaning. A frequently used definition of a system is "*a set of interrelated*

Systems Engineering Principles and Practice, Second Edition. Alexander Kossiakoff, William N. Sweet, Samuel J. Seymour, and Steven M. Biemer
© 2011 by John Wiley & Sons, Inc. Published 2011 by John Wiley & Sons, Inc.

components working *together* toward some *common* objective." This definition implies a multiplicity of interacting parts that collectively perform a significant function. The term *complex* restricts this definition to systems in which the elements are diverse and have intricate relationships with one another. Thus, a home appliance such as a washing machine would not be considered sufficiently diverse and complex to require systems engineering, even though it may have some modern automated attachments. On the other hand, the context of an *engineered* system excludes such complex systems as living organisms and ecosystems. The restriction of the term "system" to one that is complex and engineered makes it more clearly applicable to the function of systems engineering as it is commonly understood. Examples of systems requiring systems engineering for their development are listed in a subsequent section.

The above definitions of "systems engineering" and "system" are not represented as being unique or superior to those used in other textbooks, each of which defines them somewhat differently. In order to avoid any potential misunderstanding, the meaning of these terms *as used in this book* is defined at the very outset, before going on to the more important subjects of the responsibilities, problems, activities, and tools of systems engineering.

Systems Engineering and Traditional Engineering Disciplines

From the above definition, it can be seen that systems engineering differs from mechanical, electrical, and other engineering disciplines in several important ways:

1. Systems engineering is focused on the system as a whole; it emphasizes its total operation. It looks at the system from the outside, that is, at its interactions with other systems and the environment, as well as from the inside. It is concerned not only with the engineering design of the system but also with external factors, which can significantly constrain the design. These include the identification of customer needs, the system operational environment, interfacing systems, logistics support requirements, the capabilities of operating personnel, and such other factors as must be correctly reflected in system requirements documents and accommodated in the system design.

2. While the primary purpose of systems engineering is to guide, this does not mean that systems engineers do not themselves play a key role in system design. On the contrary, they are responsible for leading the formative (concept development) stage of a new system development, which culminates in the functional design of the system reflecting the needs of the user. Important design decisions at this stage cannot be based entirely on quantitative knowledge, as they are for the traditional engineering disciplines, but rather must often rely on qualitative judgments balancing a variety of incommensurate quantities and utilizing experience in a variety of disciplines, especially when dealing with new technology.

3. Systems engineering *bridges* the traditional engineering disciplines. The diversity of the elements in a complex system requires different engineering disci-

plines to be involved in their design and development. For the system to perform correctly, each system element must function properly in combination with one or more other system elements. Implementation of these interrelated functions is dependent on a complex set of physical and functional interactions between separately designed elements. Thus, the various elements cannot be engineered independently of one another and then simply assembled to produce a working system. Rather, systems engineers must guide and coordinate the design of each individual element as necessary to assure that the interactions and interfaces between system elements are compatible and mutually supporting. Such coordination is especially important when individual system elements are designed, tested, and supplied by different organizations.

Systems Engineering and Project Management

The engineering of a new complex system usually begins with an exploratory stage in which a new system concept is evolved to meet a recognized need or to exploit a technological opportunity. When the decision is made to engineer the new concept into an operational system, the resulting effort is inherently a major enterprise, which typically requires many people, with diverse skills, to devote years of effort to bring the system from concept to operational use.

The magnitude and complexity of the effort to engineer a new system requires a dedicated team to lead and coordinate its execution. Such an enterprise is called a "project" and is directed by a project manager aided by a staff. Systems engineering is an inherent part of project management—the part that is concerned with guiding the engineering effort itself—setting its objectives, guiding its execution, evaluating its results, and prescribing necessary corrective actions to keep it on course. The management of the planning and control aspects of the project fiscal, contractual, and customer relations is supported by systems engineering but is usually not considered to be part of the systems engineering function. This subject is described in more detail in Chapter 5.

Recognition of the importance of systems engineering by every participant in a system development project is essential for its effective implementation. To accomplish this, it is often useful to formally assign the leader of the systems engineering team to a recognized position of technical responsibility and authority within the project.

1.2 ORIGINS OF SYSTEMS ENGINEERING

No particular date can be associated with the origins of systems engineering. Systems engineering principles have been practiced at some level since the building of the pyramids and probably before. (The Bible records that Noah's Ark was built to a system specification.)

The recognition of systems engineering as a distinct activity is often associated with the effects of World War II, and especially the 1950s and 1960s when a number of textbooks were published that first identified systems engineering as a distinct discipline and defined its place in the engineering of systems. More generally, the recognition of systems engineering as a unique activity evolved as a necessary corollary to the rapid growth of technology, and its application to major military and commercial operations during the second half of the twentieth century.

The global conflagration of World War II provided a tremendous spur to the advancement of technology in order to gain a military advantage for one side or the other. The development of high-performance aircraft, military radar, the proximity fuse, the German VI and V2 missiles, and especially the atomic bomb required revolutionary advances in the application of energy, materials, and information. These systems were complex, combining multiple technical disciplines, and their development posed engineering challenges significantly beyond those that had been presented by their more conventional predecessors. Moreover, the compressed development time schedules imposed by wartime imperatives necessitated a level of organization and efficiency that required new approaches in program planning, technical coordination, and engineering management. Systems engineering, as we know it today, developed to meet these challenges.

During the Cold War of the 1950s, 1960s, and 1970s, military requirements continued to drive the growth of technology in jet propulsion, control systems, and materials. However, another development, that of solid-state electronics, has had perhaps a more profound effect on technological growth. This, to a large extent, made possible the still evolving "information age," in which computing, networks, and communications are extending the power and reach of systems far beyond their previous limits. Particularly significant in this connection is the development of the digital computer and the associated software technology driving it, which increasingly is leading to the replacement of human control of systems by automation. Computer control is qualitatively increasing the complexity of systems and is a particularly important concern of systems engineering.

The relation of modern systems engineering to its origins can be best understood in terms of three basic factors:

1. *Advancing Technology,* which provide opportunities for increasing system capabilities, but introduces development risks that require systems engineering management; nowhere is this more evident than in the world of automation. Technology advances in human–system interfaces, robotics, and software make this particular area one of the fastest growing technologies affecting system design.
2. *Competition,* whose various forms require seeking superior (and more advanced) system solutions through the use of system-level trade-offs among alternative approaches.
3. *Specialization,* which requires the partitioning of the system into building blocks corresponding to specific product types that can be designed and built by specialists, and strict management of their interfaces and interactions.

These factors are discussed in the following paragraphs.

Advancing Technology: Risks

The explosive growth of technology in the latter half of the twentieth century and into this century has been the single largest factor in the emergence of systems engineering as an essential ingredient in the engineering of complex systems. Advancing technology has not only greatly extended the capabilities of earlier systems, such as aircraft, telecommunications, and power plants, but has also created entirely new systems such as those based on jet propulsion, satellite communications and navigation, and a host of computer-based systems for manufacturing, finance, transportation, entertainment, health care, and other products and services. Advances in technology have not only affected the nature of products but have also fundamentally changed the way they are engineered, produced, and operated. These are particularly important in early phases of system development, as described in Conceptual Exploration, in Chapter 7.

Modern technology has had a profound effect on the very approach to engineering. Traditionally, engineering applies known principles to practical ends. Innovation, however, produces new materials, devices, and processes, whose characteristics are not yet fully measured or understood. The application of these to the engineering of new systems thus increases the risk of encountering unexpected properties and effects that might impact system performance and might require costly changes and program delays.

However, failure to apply the latest technology to system development also carries risks. These are the risks of producing an inferior system, one that could become prematurely obsolete. If a competitor succeeds in overcoming such problems as may be encountered in using advanced technology, the competing approach is likely to be superior. The successful entrepreneurial organization will thus assume carefully selected technological risks and surmount them by skillful design, systems engineering, and program management.

The systems engineering approach to the early application of new technology is embodied in the practice of "risk management." Risk management is a process of dealing with calculated risks through a process of analysis, development, test, and engineering oversight. It is described more fully in Chapters 5 and 9.

Dealing with risks is one of the essential tasks of systems engineering, requiring a broad knowledge of the total system and its critical elements. In particular, systems engineering is central to the decision of how to achieve the best balance of risks, that is, which system elements should best take advantage of new technology and which should be based on proven components, and how the risks incurred should be reduced by development and testing.

The development of the digital computer and software technology noted earlier deserves special mention. This development has led to an enormous increase in the automation of a wide array of control functions for use in factories, offices, hospitals, and throughout society. Automation, most of it being concerned with information processing hardware and software, and its sister technology, autonomy, which adds in capability of command and control, is the fastest growing and most powerful single influence on the engineering of modern systems.

The increase in automation has had an enormous impact on people who operate systems, decreasing their number but often requiring higher skills and therefore special training. Human–machine interfaces and other people–system interactions are particular concerns of systems engineering.

Software continues to be a growing engineering medium whose power and versatility has resulted in its use in preference to hardware for the implementation of a growing fraction of system functions. Thus, the performance of modern systems increasingly depends on the proper design and maintenance of software components. As a result, more and more of the systems engineering effort has had to be directed to the control of software design and its application.

Competition: Trade-offs

Competitive pressures on the system development process occur at several different levels. In the case of defense systems, a primary drive comes from the increasing military capabilities of potential adversaries, which correspondingly decrease the effectiveness of systems designed to defeat them. Such pressures eventually force a development program to redress the military balance with a new and more capable system or a major upgrade of an existing one.

Another source of competition comes with the use of competitive contracting for the development of new system capabilities. Throughout the competitive period, which may last through the initial engineering of a new system, each contractor seeks to devise the most cost-effective program to provide a superior product.

In developing a commercial product, there are nearly always other companies that compete in the same market. In this case, the objective is to develop a new market or to obtain an increased market share by producing a superior product ahead of the competition, with an edge that will maintain a lead for a number of years. The above approaches nearly always apply the most recent technology in an effort to gain a competitive advantage.

Securing the large sums of money needed to fund the development of a new complex system also involves competition on quite a different level. In particular, both government agencies and industrial companies have many more calls on their resources than they can accommodate and hence must carefully weigh the relative payoff of proposed programs. This is a primary reason for requiring a phased approach in new system development efforts, through the requirement for justification and formal approval to proceed with the increasingly expensive later phases. The results of each phase of a major development must convince decision makers that the end objectives are highly likely to be attained within the projected cost and schedule.

On a still different basis, the competition among the essential characteristics of the system is always a major consideration in its development. For example, there is always competition between performance, cost, and schedule, and it is impossible to optimize all three at once. Many programs have failed by striving to achieve levels of performance that proved unaffordable. Similarly, the various performance parameters of a vehicle, such as speed and range, are not independent of one another; the efficiency of most vehicles, and hence their operating range, decreases at higher speeds.

Thus, it is necessary to examine alternatives in which these characteristics are allowed to vary and to select the combination that best balances their values for the benefit of the user.

All of the forms of competition exert pressure on the system development process to produce the best performing, most affordable system, in the least possible time. The process of selecting the most desirable approach requires the examination of numerous potential alternatives and the exercise of a breadth of technical knowledge and judgment that only experienced systems engineers possess. This is often referred to as "trade-off analysis" and forms one of the basic practices of systems engineering.

Specialization: Interfaces

A complex system that performs a number of different functions must of necessity be configured in such a way that each major function is embodied in a separate component capable of being specified, developed, built, and tested as an individual entity. Such a subdivision takes advantage of the expertise of organizations specializing in particular types of products, and hence is capable of engineering and producing components of the highest quality at the lowest cost. Chapter 3 describes the kind of functional and physical building blocks that make up most modern systems.

The immensity and diversity of engineering knowledge, which is still growing, has made it necessary to divide the education and practice of engineering into a number of specialties, such as mechanical, electrical, aeronautical, and so on. To acquire the necessary depth of knowledge in any one of these fields, further specialization is needed, into such subfields as robotics, digital design, and fluid dynamics. Thus, engineering specialization is a predominant condition in the field of engineering and manufacturing and must be recognized as a basic condition in the system development process.

Each engineering specialty has developed a set of specialized tools and facilities to aid in the design and manufacture of its associated products. Large and small companies have organized around one or several engineering groups to develop and manufacture devices to meet the needs of the commercial market or of the system-oriented industry. The development of interchangeable parts and automated assembly has been one of the triumphs of the U.S. industry.

The convenience of subdividing complex systems into individual building blocks has a price: that of integrating these disparate parts into an efficient, smoothly operating system. Integration means that each building block fits perfectly with its neighbors and with the external environment with which it comes into contact. The "fit" must be not only physical but also functional; that is, its design will both affect the design characteristics and behavior of other elements, and will be affected by them, to produce the exact response that the overall system is required to make to inputs from its environment. The physical fit is accomplished at intercomponent boundaries called *interfaces*. The functional relationships are called *interactions*.

The task of analyzing, specifying, and validating the component interfaces with each other and with the external environment is beyond the expertise of the individual design specialists and is the province of the systems engineer. Chapter 3 discusses further the importance and nature of this responsibility.

A direct consequence of the subdivision of systems into their building blocks is the concept of modularity. Modularity is a measure of the degree of mutual independence of the individual system components. An essential goal of systems engineering is to achieve a high degree of modularity to make interfaces and interactions as simple as possible for efficient manufacture, system integration, test, operational maintenance, reliability, and ease of in-service upgrading. The process of subdividing a system into modular building blocks is called "functional allocation" and is another basic tool of systems engineering.

1.3 EXAMPLES OF SYSTEMS REQUIRING SYSTEMS ENGINEERING

As noted at the beginning of this chapter, the generic definition of a system as a *set* of *interrelated components* working *together* as an integrated whole to achieve some common objective would fit most familiar home appliances. A washing machine consists of a main clothes tub, an electric motor, an agitator, a pump, a timer, an inner spinning tub, and various valves, sensors, and controls. It performs a sequence of timed operations and auxiliary functions based on a schedule and operation mode set by the operator. A refrigerator, microwave oven, dishwasher, vacuum cleaner, and radio all perform a number of useful operations in a systematic manner. However, these appliances involve only one or two engineering disciplines, and their design is based on well-established technology. Thus, they fail the criterion of being *complex*, and we would not consider the development of a new washer or refrigerator to involve much systems engineering as we understand the term, although it would certainly require a high order of reliability and cost engineering. Of course, home appliances increasingly include clever automatic devices that use newly available microchips, but these are usually self-contained add-ons and are not necessary to the main function of the appliance.

Since the development of new modern systems is strongly driven by technological change, we shall add one more characteristic to a system requiring systems engineering, namely, that some of its key elements use advanced technology. The characteristics of a system whose development, test, and application require the practice of systems engineering are that the system

- is an engineered product and hence satisfies a specified need,
- consists of diverse components that have intricate relationships with one another and hence is multidisciplinary and relatively complex, and
- uses advanced technology in ways that are central to the performance of its primary functions and hence involves development risk and often a relatively high cost.

Henceforth, references in this text to an *engineered* or *complex* system (or in the proper context, just *system*) will mean the type that has the three attributes noted above, that is, is an engineered product, contains diverse components, and uses advanced technology. These attributes are, of course, in addition to the generic definition stated earlier and serve to identify the systems of concern to the systems engineer as those that require system design, development, integration, test, and evaluation. In Chapter 2, we explore the full spectrum of systems complexity and why the systems engineering landscape presents a challenge for systems engineers.

Examples of Complex Engineered Systems

To illustrate the types of systems that fit within the above definition, Tables 1.1 and 1.2 list 10 modern systems and their principal inputs, processes, and outputs.

TABLE 1.1. Examples of Engineered Complex Systems: Signal and Data Systems

System	Inputs	Process	Outputs
Weather satellite	Images	• Data storage • Transmission	Encoded images
Terminal air traffic control system	Aircraft beacon responses	• Identification • Tracking	• Identity • Air tracks • Communications
Track location system	Cargo routing requests	• Map tracing • Communication	• Routing information • Delivered cargo
Airline reservation system	Travel requests	Data management	• Reservations • Tickets
Clinical information system	• Patient ID • Test records • Diagnosis	Information management	• Patient status • History • Treatment

TABLE 1.2. Examples of Engineered Complex Systems: Material and Energy Systems

System	Inputs	Process	Outputs
Passenger aircraft	• Passengers • Fuel	• Combustion • Thrust • Lift	Transported passengers
Modern harvester combine	• Grain field • Fuel	• Cutting • Threshing	Harvested grain
Oil refinery	• Crude oil • Catalysts • Energy	• Cracking • Separation • Blending	• Gasoline • Oil products • Chemicals
Auto assembly plant	• Auto parts • Energy	• Manipulation • Joining • Finishing	Assembled auto
Electric power plant	• Fuel • Air	• Power generation • Regulation	• Electric AC power • Waste products

It has been noted that a system consists of a multiplicity of elements, some of which may well themselves be complex and deserve to be considered a system in their own right. For example, a telephone-switching substation can well be considered as a system, with the telephone network considered as a "system of systems." Such issues will be discussed more fully in Chapters 2 and 4, to the extent necessary for the understanding of systems engineering.

Example: A Modern Automobile. A more simple and familiar system, which still meets the criteria for an engineered system, is a fully equipped passenger automobile. It can be considered as a lower limit to more complex vehicular systems. It is made up of a large number of diverse components requiring the combination of several different disciplines. To operate properly, the components must work together accurately and efficiently. Whereas the operating principles of automobiles are well established, modern autos must be designed to operate efficiently while at the same time maintaining very close control of engine emissions, which requires sophisticated sensors and computer-controlled mechanisms for injecting fuel and air. Antilock brakes are another example of a finely tuned automatic automobile subsystem. Advanced materials and computer technology are used to an increasing degree in passenger protection, cruise control, automated navigation and autonomous driving and parking. The stringent requirements on cost, reliability, performance, comfort, safety, and a dozen other parameters present a number of substantive systems engineering problems. Accordingly, an automobile meets the definition established earlier for a system requiring the application of systems engineering, and hence can serve as a useful example.

An automobile is also an example of a large class of systems that require active interaction (control) by a human operator. To some degree, all systems require such interaction, but in this case, continuous control is required. In a very real sense, the operator (driver) functions as an integral part of the overall automobile system, serving as the steering feedback element that detects and corrects deviations of the car's path on the road. The design must therefore address as a critical constraint the inherent sensing and reaction capabilities of the operator, in addition to a range of associated human–machine interfaces such as the design and placement of controls and displays, seat position, and so on. Also, while the passengers may not function as integral elements of the auto steering system, their associated interfaces (e.g., weight, seating and viewing comfort, and safety) must be carefully addressed as part of the design process. Nevertheless, since automobiles are developed and delivered without the human element, for purposes of systems engineering, they may be addressed as systems in their own right.

1.4   SYSTEMS ENGINEERING AS A PROFESSION

With the increasing prevalence of complex systems in modern society, and the essential role of systems engineering in the development of systems, systems engineering as a profession has become widely recognized. Its primary recognition has come in companies specializing in the development of large systems. A number of these have established departments of systems engineering and have classified those engaging in the process as systems engineers. In addition, global challenges in health care, communications, environment, and many other complex areas require engineering systems methods to develop viable solutions.

To date, the slowness of recognition of systems engineering as a career is the fact that it does not correspond to the traditional academic engineering disciplines. Engineering disciplines are built on quantitative relationships, obeying established physical laws, and measured properties of materials, energy, or information. Systems engineering, on the other hand, deals mainly with problems for which there is incomplete knowledge, whose variables do not obey known equations, and where a balance must be made among conflicting objectives involving incommensurate attributes. The absence of a quantitative knowledge base previously inhibited the establishment of systems engineering as a unique discipline.

Despite those obstacles, the recognized need for systems engineering in industry and government has spurred the establishment of a number of academic programs offering master's degrees and doctoral degrees in systems engineering. An increasing number of universities are offering undergraduate degrees in systems engineering as well.

The recognition of systems engineering as a profession has led to the formation of a professional society, the International Council on Systems Engineering (INCOSE), one of whose primary objectives is the promotion of systems engineering, and the recognition of systems engineering as a professional career.

Career Choices

Systems engineers are highly sought after because their skills complement those in other fields and often serve as the "glue" to bring new ideas to fruition. However, career choices and the related educational needs for those choices is complex, especially when the role and responsibilities of a systems engineer is poorly understood.

Four potential career directions are shown in Figure 1.1: financial, management, technical, and systems engineering. There are varying degrees of overlap between them despite the symmetry shown in the figure. The systems engineer focuses on the whole system product, leading and working with many diverse technical team members, following the systems engineering development cycle, conducting studies of alternatives, and managing the system interfaces. The systems engineer generally matures in the field after a technical undergraduate degree with work experience and a master of science degree in systems engineering, with an increasing responsibility of successively larger projects, eventually serving as the chief or lead systems engineer for a major systems, or systems-of-systems development. Note the overlap and need to understand the content and roles of the technical specialists and to coordinate with the program manager (PM).

The project manager or PM, often with a technical or business background, is responsible for interfacing with the customer and for defining the work, developing the plans, monitoring and controlling the project progress, and delivering the finished output to the customer. The PM often learns from on the job training (OJT) with

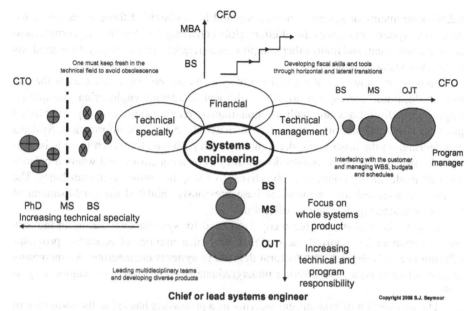

Figure 1.1. Career opportunities and growth.

projects of increasing size and importance, enhancing the toolset available with a master of science degree in technical/program management. While not exclusively true, the chief executive officer (CEO) frequently originates from the ranks of the organization's PMs.

The financial or business career path that ultimately could lead to a chief financial officer (CFO) position usually includes business undergraduate and master of business administration (MBA) degrees. Individuals progress through their careers with various horizontal and vertical moves, often with specialization in the field. There is an overlap in skill and knowledge with the PM in areas of contract and finance management.

Many early careers start with a technical undergraduate degree in engineering, science or information technology. The technical specialist makes contributions as part of a team in the area of their primary knowledge, honing skills and experience to develop and test individual components or algorithms that are part of a larger system. Contributions are made project to project over time, and recognition is gained from innovative, timely, and quality workmanship. Technical specialists need to continue to learn about their field and to stay current in order to be employable compared to the next generation of college graduates. Often advanced degrees (MS and PhDs) are acquired to enhance knowledge, capability, and recognition, and job responsibilities can lead to positions such as lead engineer, lead scientist, or chief technology officer (CTO) in an organization. The broader minded or experienced specialist often considers a career in systems engineering.

Orientation of Technical Professionals

The special relationship of systems engineers with respect to technical disciplines can be better understood when it is realized that technical people not only engage in widely different professional specialties, but their intellectual objectives, interests, and attitudes, which represent their technical orientations, can also be widely divergent. The typical scientist is dedicated to understanding the nature and behavior of the physical world. The scientist asks the questions "Why?" and "How?" The mathematician is usually primarily concerned with deriving the logical consequences of a set of assumptions, which may be quite unrelated to the real world. The mathematician develops the proposition "If A, then B." Usually, the engineer is mainly concerned with creating a useful product. The engineer exclaims "Voila!"

These orientations are quite different from one another, which accounts for why technical specialists are focused on their own aspects of science and technology. However, in most professionals, those orientations are not absolute; in many cases, the scientist may need some engineering to construct an apparatus, and the engineer may need some mathematics to solve a control problem. So, in the general case, the orientation of a technical professional might be modeled by a sum of three orthogonal vectors, each representing the extent of the individual's orientation being in science, mathematics, or engineering.

To represent the above model, it is convenient to use a diagram designed to show the composition of a mixture of three components. Figure 1.2a is such a diagram in which the components are science, mathematics, and engineering. A point at each vertex represents a mixture with 100% of the corresponding component. The composition of the mixture marked by the small triangle in the figure is obtained by finding the percentage of each component by projecting a line parallel to the baseline opposite each vertex to the scale radiating from the vertex. This process gives intercepts of 70% science, 20% mathematics, and 10% engineering for the orientation marked by the triangle.

Because the curricula of technical disciplines tend to be concentrated in specialized subjects, most students graduate with limited general knowledge. In Figure 1.2b, the circles representing the orientation of individual graduates are seen to be concentrated in the corners, reflecting their high degree of specialization.

The tendency of professional people to polarize into diverse specialties and interests tends to be accentuated after graduation, as they seek to become recognized in their respective fields. Most technical people resist becoming generalists for fear they will lose or fail to achieve positions of professional leadership and the accompanying recognition. This specialization of professionals inhibits technical communication between them; the language barrier is bad enough, but the differences in basic objectives and methods of thought are even more serious. The solution of complex interdisciplinary problems has had to depend on the relatively rare individuals who, for one reason or another, after establishing themselves in their principal profession, have become interested and involved in solving system problems and have learned to work jointly with specialists in various other fields.

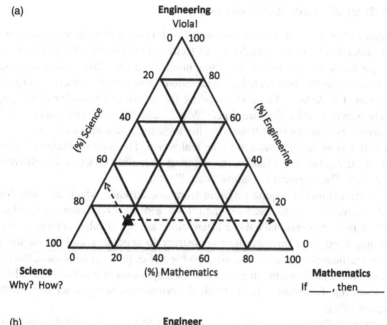

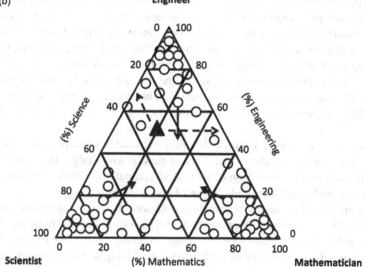

Figure 1.2. (a) Technical orientation phase diagram. (b) Technical orientation population density distribution.

The occasional evolution of technical specialists into systems engineers is symbolized in Figure 1.2b by the arrows directed from the vertices toward the center. The small black triangle corresponds to such an evolved individual whose orientation is 30% science, 50% engineering, and 20% mathematics, a balance that would be effective in the type of problem solving with which a systems engineer is typically involved. It is the few individuals who evolve into systems engineers or system architects who become the technical leaders of system development programs.

The Challenge of Systems Engineering

An inhibiting factor in becoming a professional systems engineer is that it represents a deviation from a chosen established discipline to a more diverse, complicated professional practice. It requires the investment of time and effort to gain experience and an extensive broadening of the engineering base, as well as learning communication and management skills, a much different orientation from the individual's original professional choice.

For the above reasons, an engineer considering a career in systems engineering may come to the conclusion that the road is difficult. It is clear that a great deal must be learned; that the educational experience in a traditional engineering discipline is necessary; and that there are few tools and few quantitative relationships to help make decisions. Instead, the issues are ambiguous and abstract, defying definitive solutions. There may appear to be little opportunity for individual accomplishment and even less for individual recognition. For a systems engineer, success is measured by the accomplishment of the development team, not necessarily the system team leader.

What Then Is the Attraction of Systems Engineering?

The answer may lie in the challenges of systems engineering rather than its direct rewards. Systems engineers deal with the most important issues in the system development process. They design the overall system architecture and the technical approach and lead others in designing the components. They prioritize the system requirements in conjunction with the customer to ensure that the different system attributes are appropriately weighted when balancing the various technical efforts. They decide which risks are worth undertaking and which are not, and how the former should be hedged to ensure program success.

It is the systems engineers who map out the course of the development program that prescribes the type and timing of tests and simulations to be performed along the way. They are the ultimate authorities on how the system performance and system affordability goals may be achieved at the same time.

When unanticipated problems arise in the development program, as they always do, it is the systems engineers who decide how they may be solved. They determine whether an entirely new approach to the problem is necessary, whether more intense effort will accomplish the purpose, whether an entirely different part of the system can be modified to compensate for the problem, or whether the requirement at issue can best be scaled back to relieve the problem.

Systems engineers derive their ability to guide the system development not from their position in the organization but from their superior knowledge of the system as a whole, its operational objectives, how all its parts work together, and all the technical factors that go into its development, as well as from their proven experience in steering complex programs through a maze of difficulties to a successful conclusion.

Attributes and Motivations of Systems Engineers

In order to identify candidates for systems engineering careers, it is useful to examine the characteristics that may be useful to distinguish people with a talent for systems engineering from those who are not likely to be interested or successful in that discipline. Those likely to become talented systems engineers would be expected to have done well in mathematics and science in college.

A systems engineer will be required to work in a multidisciplinary environment and to grasp the essentials of related disciplines. It is here that an aptitude for science and engineering helps a great deal because it makes it much easier and less threatening for individuals to learn the essentials of new disciplines. It is not so much that they require in depth knowledge of higher mathematics, but rather, those who have a limited mathematical background tend to lack confidence in their ability to grasp subjects that inherently contain mathematical concepts.

A systems engineer should have a creative bent and must like to solve practical problems. An interest in the job should be greater than an interest in career advancement. Systems engineering is more of a challenge than a quick way to the top.

The following characteristics are commonly found in successful systems engineers. They

1. enjoy learning new things and solving problems,
2. like challenges,
3. are skeptical of unproven assertions,
4. are open-minded to new ideas,
5. have a solid background in science and engineering,
6. have demonstrated technical achievement in a specialty area,
7. are knowledgeable in several engineering areas,
8. pick up new ideas and information quickly, and
9. have good interpersonal and communication skills.

1.5 SYSTEMS ENGINEER CAREER DEVELOPMENT MODEL

When one has the characteristics noted above and is attracted to become a systems engineer, there are four more elements that need to be present in the work environment. As shown in Figure 1.3a, one should seek assignments to problems and tasks that are

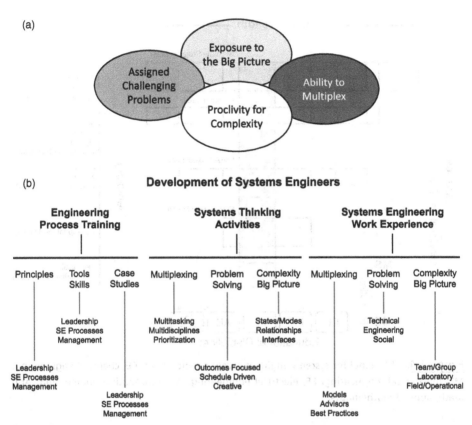

Figure 1.3. (a) Systems engineering (SE) career elements derived from quality work experiences. (b) Components of employer development of systems engineers.

very challenging and are likely to expand technical domain knowledge and creative juices. Whatever the work assignment, understanding the context of the work and understanding the big picture is also essential. Systems engineers are expected to manage many activities at the same time, being able to have broad perspectives but able to delve deeply into to many subjects at once. This ability to multiplex is one that takes time to develop. Finally, the systems engineer should not be intimidated by complex problems since this is the expected work environment. It is clear these elements are not part of an educational program and must be gained through extended professional work experience. This becomes the foundation for the systems engineering career growth model.

Employers seeking to develop systems engineers to competitively address more challenging problems should provide key staff with relevant systems engineering work experience, activities that require mature systems thinking, and opportunities for systems engineering education and training. In Figure 1.3b, it can be seen that the experience can be achieved not only with challenging problems but also with

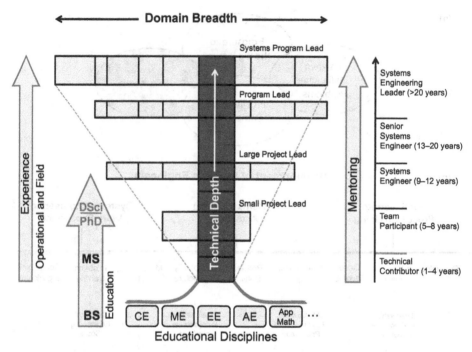

Figure 1.4. "T" model for systems engineer career development. CE, chemical engineering; ME, mechanical engineering; EE, electrical engineering; AE, aeronautical engineering; App Math, applied mathematics.

experienced mentors and real, practical exercises. While using systems thinking to explore complex problem domains, staff should be encouraged to think creatively and out of the box. Often, technically trained people rigidly follow the same processes and tired ineffective solutions. Using lessons learned from past programs and case studies creates opportunities for improvements. Formal training and use of systems engineering tools further enhance employee preparation for tackling complex issues.

Interests, attributes, and training, along with an appropriate environment, provide the opportunity for individuals to mature into successful systems engineers. The combination of these factors is captured in the "T" model for systems engineer career development illustrated in Figure 1.4. In the vertical, from bottom to top is the time progression in a professional's career path. After completion of a technical undergraduate degree, shown along the bottom of the chart, an individual generally enters professional life as a technical contributor to a larger effort. The effort is part of a project or program that falls in a particular domain such as aerodynamics, biomedicine, combat systems, information systems, or space exploration. Within a domain, there are several technical competencies that are fundamental for systems to operate or to be developed.

The T is formed by snapshots during a professional's career that illustrates in the horizontal part of the T the technical competencies at the time that were learned and used to meet the responsibilities assigned at that point in their career. After an initial experience in one or two technical domains as technical contributor, one progresses to increasing responsibilities in a team setting and eventually to leading small technical groups. After eight or more years, the professional has acquired both sufficient technical depth and technical domain depth to be considered a systems engineer. Additional assignments lead to project and program systems engineering leadership and eventually to being the senior systems engineer for a major development program that exercises the full range of the technical competencies for the domain.

In parallel with broadening and deepening technical experience and competencies, the successful career path is augmented by assignments that involve operational field experiences, advanced education and training, and a strong mentoring program. In order to obtain a good understanding of the environment where the system under development will operate and to obtain firsthand knowledge of the system requirements, it is essential for the early systems engineer professional to visit the "field site" and operational location. This approach is important to continue throughout one's career. A wide variety of systems engineering educational opportunities are available in both classroom and online formats. As in most engineering disciplines where the student is not planning on an academic career, the master of science is the terminal degree. Courses are usually a combination of systems engineering and domain or concentration centric focused with a thesis or capstone project for the students to demonstrate their knowledge and skills on a practical systems problem. Large commercial companies also provide training in systems engineering and systems architecting with examples and tools that are specific to their organization and products. Finally, the pairing of a young professional with an experienced systems engineer will enhance the learning process.

1.6 THE POWER OF SYSTEMS ENGINEERING

If power is measured by authority over people or money, then systems engineers would appear to have little power as members of the system development team. However, if power is measured by the influence over the design of the system and its major characteristics, and over the success or failure of the system development, then systems engineers can be more powerful than project managers. The sources of this power come from their knowledge, skills, and attitude. Each of these is discussed in the following paragraphs.

The Power of Multidisciplinary Knowledge

A major system development project is a veritable "Tower of Babel." There are literally dozens of specialists in different disciplines whose collective efforts are necessary to develop and produce a successful new system. Each group of specialists has its own language, making up for the imprecision of the English language with a rich set of acronyms, which convey a very specific meaning but are unintelligible to those outside the specialty. The languages, in turn, are backed up by knowledge bases, which the specialists use to ply their trade. These knowledge bases contain descriptions of the different materials peculiar to each discipline, as well as bodies of relationships, many of them expressed in mathematical terms, that enable the specialists to compute various characteristics of their components on the basis of design assumptions. These knowledge bases are also foreign to those outside the discipline.

Such a collection of multi-tongued participants could never succeed in collectively developing a new system by themselves, just as the citizens of Babylon could never build their tower. It is the systems engineers who provide the linkages that enable these disparate groups to function as a team. The systems engineers accomplish this feat through the power of multidisciplinary knowledge. This means that they are sufficiently literate in the different disciplines involved in their system that they can understand the languages of the specialists, appreciate their problems, and are able to interpret the necessary communications for their collective endeavor. Thus, they are in the same position as a linguist in the midst of a multinational conference, with people speaking in their native tongues. Through the ability to understand different languages comes the capability to obtain cooperative effort from people who would otherwise never be able to achieve a common objective. This capability enables systems engineers to operate as leaders and troubleshooters, solving problems that no one else is capable of solving. It truly amounts to a power that gives systems engineers a central and decisive role to play in the development of a system.

It is important to note that the depth of interdisciplinary knowledge, which is required to interact effectively with specialists in a given field, is a very small fraction of the depth necessary to work effectively in that field. The number of new acronyms that one has to learn in a given technical area is nearer to a dozen of the more frequently used ones than to a hundred. It also turns out that once one gets past the differences in semantics, there are many common principles in different disciplines and many similar relationships. For instance, the equation used in communications, connecting signal, noise, antenna gain, receiver sensitivity, and other factors, is directly analogous to a similar relationship in acoustics.

These facts mean that a systems engineer does not need to spend a lifetime becoming expert in associated disciplines, but rather can accumulate a working knowledge of related fields through selected readings, and more particularly, discussion with colleagues knowledgeable in each field. The important thing is to know which principles, relationships, acronyms, and the like are important at the system level and which are details. The power of multidisciplinary knowledge is so great that, to a systems engineer, the effort required to accumulate it is well worth the learning time.

The Power of Approximate Calculation

The practice of systems engineering requires another talent besides multidisciplinary knowledge. The ability to carry out "back of the envelope" calculations to obtain a "sanity check" on the result of a complex calculation or test is of inestimable value to the systems engineer. In a few cases, this can be done intuitively on the basis of past experience, but more frequently, it is necessary to make a rough estimate to ensure that a gross omission or error has not been committed. Most successful systems engineers have the ability, using first principles, to apply basic relationships, such as the communications equation or other simple calculation, to derive an order of magnitude result to serve as a check. This is particularly important if the results of the calculation or experiment turn out very differently from what had been originally expected.

When the sanity check does not confirm the results of a simulation or experiment, it is appropriate to go back to make a careful examination of the assumptions and conditions on which the latter were based. As a matter of general experience, more often than not, such examinations reveal an error in the conditions or assumptions under which the simulation or experiment was conducted.

The Power of Skeptical Positive Thinking

The above seemingly contradictory title is meant to capture an important characteristic of successful systems engineering. The skeptical part is important to temper the traditional optimism of the design specialist regarding the probability of success of a chosen design approach. It is the driving force for the insistence of validation of the approach selected at the earliest possible opportunity.

The other dimension of skepticism, which is directly related to the characteristic of positive thinking, refers to the reaction in the face of failure or apparent failure of a selected technique or design approach. Many design specialists who encounter an unexpected failure are plunged into despair. The systems engineer, on the other hand, cannot afford the luxury of hand wringing but must have, first of all, a healthy skepticism of the conditions under which the unexpected failure occurred. Often, it is found that these conditions did not properly test the system. When the test conditions are shown to be valid, the systems engineer must set about finding ways to circumvent the cause of failure. The conventional answer that the failure must require a new start along a different path, which in turn will lead to major delays and increases in program cost, is simply not acceptable unless heroic efforts to find an alternative solution do not succeed. This is where the power of multidisciplinary knowledge permits the systems engineer to look for alternative solutions in other parts of the system, which may take the stress off the particular component whose design proved to be faulty.

The characteristic of positive thinking is absolutely necessary in both the systems engineer and the project manager so that they are able to generate and sustain the confidence of the customer and of company management, as well as the members of the design team. Without the "can-do" attitude, the esprit de corps and productivity of the project organization is bound to suffer.

1.7 SUMMARY

What Is Systems Engineering?

The function of systems engineering is to guide the engineering of complex systems. And a system is defined as a set of interrelated components working together toward a common objective. Furthermore, a complex engineered system (as defined in this book) is (1) composed of a multiplicity of intricately interrelated diverse elements and (2) requires systems engineering to lead its development.

Systems engineering differs from traditional disciplines in that (1) it is focused on the system as a whole; (2) it is concerned with customer needs and operational environment; (3) it leads system conceptual design; and (4) it bridges traditional engineering disciplines and gaps between specialties. Moreover, systems engineering is an integral part of project management in that it plans and guides the engineering effort.

Origins of Systems Engineering

Modern systems engineering originated because advancing technology brought risks and complexity with the growth of automation; competition required expert risk taking; and specialization required bridging disciplines and interfaces.

Examples of Systems Requiring Systems Engineering

Examples of engineered complex systems include

- weather satellites,
- terminal air traffic control,
- truck location systems,
- airline navigation systems,
- clinical information systems,
- passenger aircraft,
- modern harvester combines,
- oil refineries,
- auto assembly plants, and
- electric power plants.

Systems Engineering as a Profession

Systems engineering is now recognized as a profession and has an increasing role in government and industry. In fact, numerous graduate (and some undergraduate) degree programs are now available across the country. And a formal, recognized organization exists for systems engineering professionals: the INCOSE.

Technical professionals have specific technical orientations—technical graduates tend to be highly specialized. Only a few become interested in interdisciplinary problems—it is these individuals who often become systems engineers.

Systems Engineer Career Development Model

The systems engineering profession is difficult but rewarding. A career in systems engineering typically features technical satisfaction—finding the solution of abstract and ambiguous problems—and recognition in the form of a pivotal program role. Consequently, a successful systems engineer has the following traits and attributes:

- a good problem solver and should welcome challenges;
- well grounded technically, with broad interests;
- analytical and systematic, but also creative; and
- a superior communicator, with leadership skills.

The "T" model represents the proper convergence of experience, education, mentoring, and technical depth necessary to become a successful and influential systems engineer.

The Power of Systems Engineering

Overall, systems engineering is a powerful discipline, requiring a multidisciplinary knowledge, integrating diverse system elements. Systems engineers need to possess the ability to perform approximate calculations of complex phenomena, thereby providing sanity checks. And finally, they must have skeptical positive thinking as a prerequisite to prudent risk taking.

PROBLEMS

1.1 Write a paragraph explaining what is meant by the statement "Systems engineering is focused on the system as a whole." State what characteristics of a system you think this statement implies and how they apply to systems engineering.

1.2 Discuss the difference between engineered complex systems and complex systems that are not engineered. Give three examples of the latter. Can you think of systems engineering principles that can also be applied to nonengineered complex systems?

1.3 For each of the following areas, list and explain how at least two major technological advances/breakthroughs occurring since 1990 have radically changed them. In each case, explain how the change was effected in
  (a) transportation,
  (b) communication,
  (c) financial management,
  (d) manufacturing,
  (e) distribution and sales,
  (f) entertainment, and
  (g) medical care.

1.4 What characteristics of an airplane would you attribute to the system as a whole rather than to a collection of its parts? Explain why.

1.5 List four pros and cons (two of each) of incorporating some of the latest technology into the development of a new complex system. Give a specific example of each.

1.6 What is meant by the term "modularity?" What characteristics does a modular system possess? Give a specific example of a modular system and identify the modules.

1.7 The section Orientation of Technical Professionals uses three components to describe this characteristic: science, mathematics, and engineering. Using this model, describe what you think your orientation is in terms of $x\%$ science, $y\%$ mathematics, and $z\%$ engineering. Note that your "orientation" does not measure your knowledge or expertise, but rather your interest and method of thought. Consider your relative interest in discovering new truths, finding new relationships, or building new things and making them work. Also, try to remember what your orientation was when you graduated from college, and explain how and why it has changed.

1.8 Systems engineers have been described as being an advocate for the whole system. Given this statement, which stakeholders should the systems engineer advocate the most? Obviously, there are many stakeholders and the systems engineer must be concerned with most, if not all, of them. Therefore, rank your answer in priority order—which stakeholder is the most important to the systems engineer; which is second; which is third?

FURTHER READING

B. Blanchard. *Systems Engineering Management*, Third Edition. John Wiley & Sons, 2004.

B. Blanchard and W. Fabrycky. *Systems Engineering and Analysis*, Fourth Edition. Prentice Hall, 2006, Chapter 1.

W. P. Chase. *Management of System Engineering*. John Wiley, 1974, Chapter 1.

H. Chesnut. *System Engineering Methods*. John Wiley, 1967.

H. Eisner. *Essentials of Project and Systems Engineering Management*, Second Edition. Wiley, 2002, Chapter 1.

C. D. Flagle, W. H. Huggins, and R. R. Roy. *Operations Research and Systems Engineering*. Johns Hopkins Press, 1960, Part I.

A. D. A. Hall. *Methodology for Systems Engineering*. Van Nostrand, 1962, Chapters 1–3; *Systems Engineering Handbook*. International Council on Systems Engineering, *A Guide for System Life Cycle Processes and Activities*, Version 3.2, July 2010.

E. Rechtin. *Systems Architecting: Creating and Building Complex Systems*. Prentice Hall, 1991, Chapters 1 and 11.

E. Rechtin and M. W. Maier. *The Art of Systems Architecting*. CRC Press, 1997.

A. P. Sage. *Systems Engineering*. McGraw Hill, 1992, Chapter 1.

A. P. Sage and J. E. Armstrong, Jr. *Introduction to Systems Engineering*. Wiley, 2000, Chapter 1.

R. Stevens, P. Brook, K. Jackson, and S. Arnold. *Systems Engineering, Coping with Complexity*. Prentice Hall, 1988.

SYSTEMS ENGINEERING LANDSCAPE

2.1 SYSTEMS ENGINEERING VIEWPOINT

The origins of the systems engineering section in Chapter 1 described how the emergence of complex systems and the prevailing conditions of advancing technology, competitive pressures, and specialization of engineering disciplines and organizations required the development of a new profession: systems engineering. This profession did not, until much later, bring with it a new academic discipline, but rather, it was initially filled by engineers and scientists who acquired through experience the ability to lead successfully complex system development programs. To do so, they had to acquire a greater breadth of technical knowledge and, more importantly, to develop a different way of thinking about engineering, which has been called "the systems engineering viewpoint."

The essence of the systems engineering viewpoint is exactly what it implies—making the central objective the system as a whole and the success of its mission. This, in turn, means the subordination of individual goals and attributes in favor of those of the overall system. The systems engineer is always the advocate of the total system in any contest with a subordinate objective.

Systems Engineering Principles and Practice, Second Edition. Alexander Kossiakoff, William N. Sweet, Samuel J. Seymour, and Steven M. Biemer
© 2011 by John Wiley & Sons, Inc. Published 2011 by John Wiley & Sons, Inc.

Successful Systems

The principal focus of systems engineering, from the very start of a system development, is the success of the system—in meeting its requirements and development objectives, its successful operation in the field, and a long, useful operating life. The systems engineering viewpoint encompasses all of these objectives. It seeks to look beyond the obvious and the immediate, to understand the user's problems, and the environmental conditions that the system will be subjected to during its operation. It aims at the establishment of a technical approach that will both facilitate the system's operational maintenance and accommodate the eventual upgrading that will likely be required at some point in the future. It attempts to anticipate developmental problems and to resolve them as early as possible in the development cycle; where this is not practicable, it establishes contingency plans for later implementation as required.

Successful system development requires the use of a consistent, well-understood systems engineering approach within the organization, which involves the exercise of systematic and disciplined direction, with extensive planning, analysis, reviews, and documentation. Just as important, however, is a side of systems engineering that is often overlooked, namely, innovation. For a new complex system to compete successfully in a climate of rapid technological change and to retain its edge for many years of useful life, its key components must use some of the latest technological advances. These will inevitably introduce risks, some known and others as yet unknown, which in turn will entail a significant development effort to bring each new design approach to maturity and later to validate the use of these designs in system components. Selecting the most promising technological approaches, assessing the associated risks, rejecting those for which the risks outweigh the potential payoff, planning critical experiments, and deciding on potential fallbacks are all primary responsibilities of systems engineering. Thus, the systems engineering viewpoint includes a combination of risk taking and risk mitigation.

The "Best" System

In characterizing the systems engineering viewpoint, two oft-stated maxims are "the best is the enemy of the good enough" and "systems engineering is the art of the good enough." These statements may be misleading if they are interpreted to imply that systems engineering means settling for second best. On the contrary, systems engineering does seek the best possible system, which, however, is often not the one that provides the best performance. The seeming inconsistency comes from what is referred to by best. The popular maxims use the terms "best" and "good enough" to refer to system performance, whereas systems engineering views performance as only one of several critical attributes; equally important ones are affordability, timely availability to the user, ease of maintenance, and adherence to an agreed-upon development completion schedule. Thus, the systems engineer seeks the *best balance* of the critical system attributes from the standpoint of the success of the development program and of the value of the system to the user.

The interdependence of performance and cost can be understood in terms of the law of diminishing returns. Assuming a particular technical approach to the achieve-

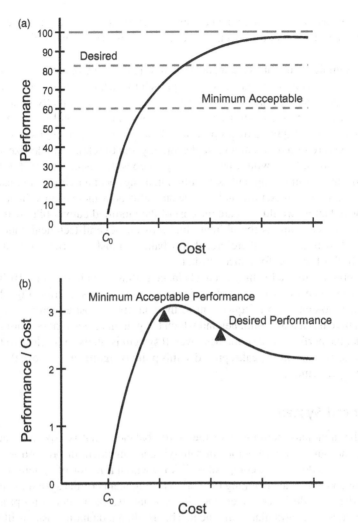

Figure 2.1. (a) Performance versus cost. (b) Performance/cost versus cost.

ment of a given performance attribute of a system under development, Figure 2.1a is a plot of a typical variation in the level of performance of a hypothetical system component as a function of the cost of the expended development effort. The upper horizontal line represents the theoretical limit in performance inherent in the selected technical approach. A more sophisticated approach might produce a higher limit, but at a higher cost. The dashed horizontal lines represent the minimum acceptable and desirable performance levels.

The curve of Figure 2.1a originates at C_0, which represents the cost of just achieving any significant performance. The slope is steep at first, becoming less steep as the performance asymptotically approaches the theoretical limit. This decreasing slope, which is a measure of the incremental gain in performance with an increment of added cost, illustrates the law of diminishing returns that applies to virtually all developmental activities.

An example of the above general principle is the development of an automobile with a higher maximum speed. A direct approach to such a change would be to use an engine that generates greater power. Such an engine would normally be larger, weigh more, and use gas less efficiently. Also, an increase in speed will result in greater air drag, which would require a disproportionately large increase in engine power to overcome. If it was required to maintain fuel economy and to retain vehicle size and weight as nearly as possible, it would be necessary to consider using or developing a more advanced engine, improving body streamlining, using special lightweight materials, and otherwise seeking to offset the undesirable side effects of increasing vehicle speed. All of the above factors would escalate the cost of the modified automobile, with the incremental costs increasing as the ultimate limits of the several technical approaches are approached. It is obvious, therefore, that a balance must be struck well short of the ultimate limit of any performance attribute.

An approach to establishing such a balance is illustrated in Figure 2.1b. This figure plots performance divided by cost against cost (i.e., y/x vs. x from Fig. 2.1a). This performance-to-cost ratio is equivalent to the concept of cost-effectiveness. It is seen that this curve has a maximum, beyond which the gain in effectiveness diminishes. This shows that the performance of the best overall system is likely to be close to that where the performance/cost ratio peaks, provided this point is significantly above the minimum acceptable performance.

A Balanced System

One of the dictionary definitions of the word "balance" that is especially appropriate to system design is "a harmonious or satisfying arrangement or proportion of parts or elements, as in a design or a composition." An essential function of systems engineering is to bring about a balance among the various components of the system, which, it was noted earlier, are designed by engineering specialists, each intent on optimizing the characteristics of a particular component. This is often a daunting task, as illustrated in Figure 2.2. The figure is an artist's conception of what a guided missile might look like if it were designed by a specialist in one or another guided missile component technology. While the cartoons may seem fanciful, they reflect a basic truth, that is, that design specialists will seek to optimize the particular aspect of a system that they best understand and appreciate. In general, it is to be expected that, while the design specialist does understand that the system is a group of components that in combination provide a specific set of capabilities, during system development, the specialist's attention is necessarily focused on those issues that most directly affect his or her own area of technical expertise and assigned responsibilities.

Conversely, the systems engineer must always focus on the system as a whole, while addressing design specialty issues only in so far as they may affect overall system performance, developmental risk, cost, or long-term system viability. In short, it is the responsibility of the systems engineer to guide the development so that each of the

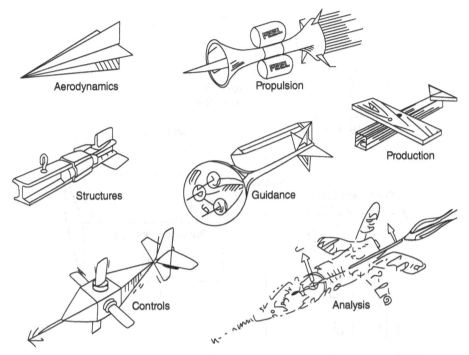

Aerodynamics   Propulsion

Production

Structures   Guidance

Controls   Analysis

Figure 2.2. The ideal missile design from the viewpoint of various specialists.

components receives the proper balance of attention and resources while achieving the capabilities that are optimal for the best overall system behavior. This often involves serving as an "honest technical broker" who guides the establishment of technical design compromises in order to achieve a workable interface between key system elements.

A Balanced Viewpoint

A system view thus connotes a focus on balance, ensuring that no system attribute is allowed to grow at the expense of an equally important or more important attribute, for example, greater performance at the expense of acceptable cost, high speed at the expense of adequate range, or high throughput at the expense of excessive errors. Since virtually all critical attributes are interdependent, a proper balance must be struck in essentially all system design decisions. These characteristics are typically incommensurable, as in the above examples, so that the judgment of how they should be balanced must come from a deep understanding of how the system works. It is such judgment that systems engineers have to exercise every day, and they must be able to think at a level that encompasses all of the system characteristics.

The viewpoint of the systems engineer calls for a different combination of skills and areas of knowledge than those of a design specialist or a manager. Figure 2.3 is

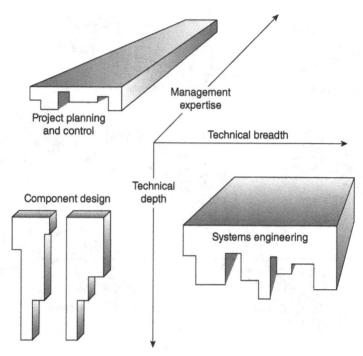

Figure 2.3. The dimensions of design, systems engineering, and project planning and control.

intended to illustrate the general nature of these differences. Using the three dimensions to represent technical depth, technical breadth, and management depth, respectively, it is seen that the design specialist may have limited managerial skills but has a deep understanding in one or a few related areas of technology. Similarly, a project manager needs to have little depth in any particular technical discipline but must have consider-able breadth and capability to manage people and technical effort. A systems engineer, on the other hand, requires significant capabilities in all three components, representing the balance needed to span the needs of a total system effort. In that sense, the systems engineer operates in more dimensions than do his or her coworkers.

2.2 PERSPECTIVES OF SYSTEMS ENGINEERING

While the field of systems engineering has matured rapidly in the past few decades, there will continue to exist a variety of differing perspectives as more is learned about the potential and the utility of systems approaches to solve the increasing complex problems around the world. The growth of systems engineering is evidenced in the number of academic programs and graduates in the area. Some surveys note that systems engineering is a favored and potentially excellent career path. Employers in all sectors, private and government, seek experienced systems engineering candidates. Experts in workforce development look for ways to encourage more secondary school

TABLE 2.1. Comparison of Systems Perspectives

Systems thinking	Systems engineering	Engineering systems
Focus on process	Focus on whole product	Focus on both process and product
Consideration of issues	Solve complex technical problems	Solve complex interdisciplinary technical, social, and management issues
Evaluation of multiple factors and influences	Develop and test tangible system solutions	Influence policy, processes and use systems engineering to develop system solutions
Inclusion of patterns relationships, and common understanding	Need to meet requirements, measure outcomes and solve problems	Integrate human and technical domain dynamics and approaches

and college students to pursue degrees in science, technology, engineering, and mathematics (STEM). With experience and additional knowledge, these students would mature into capable systems engineers.

Since it often requires professional experience in addition to education to tackle the most complex and challenging problems, developing a systems mindset—to "think like a systems engineer"—is a high priority at any stage of life. A perspective that relates a progression in the maturity of thinking includes concepts of systems thinking, systems engineering, and engineering systems (see Table 2.1) An approach to understanding the environment, process, and policies of a systems problem requires one to use systems thinking. This approach to a problem examines the domain and scope of the problem and defines it in quantitative terms. One looks at the parameters that help define the problem and then, through research and surveys, develops observations about the environment the problem exists in and finally generates options that could address the problem. This approach would be appropriate for use in secondary schools to have young students gain an appreciation of the "big picture" as they learn fundamental science and engineering skills.

The systems engineering approach discussed in this book and introduced in Chapter 1 focuses on the products and solutions of a problem, with the intent to develop or build a system to address the problem. The approach tends to be more technical, seeking from potential future users and developers of the solution system, what are the top level needs, requirements, and concepts of operations, before conducting a functional and physical design, development of design specifications, production, and testing of a system solution for the problem. Attention is given to the subsystem interfaces and the need for viable and tangible results. The approach and practical end could be applied to many degrees of complexity, but there is an expectation of a successful field operation of a product. The proven reliability of the systems engineering approach for product development is evident in many commercial and military sectors.

A broader and robust perspective to systems approaches to solve very extensive complex engineering problems by integrating engineering, management, and social science approaches using advanced modeling methodologies is termed "engineering systems." The intent is to tackle some of society's grandest challenges with significant global impact by investigating ways in which engineering systems behave and interact with one another including social, economic, and environmental factors. This approach encompasses engineering, social science, and management processes without the implied rigidity of systems engineering. Hence, applications to critical infrastructure, health care, energy, environment, information security, and other global issues are likely areas of attention.

Much like the proverbial blind men examining the elephant, the field of systems engineering can be considered in terms of various domains and application areas where it is applied. Based on the background of the individuals and on the needs of the systems problems to be solved, the systems environment can be discussed in terms of the fields and technologies that are used in the solution sets. Another perspective can be taken from the methodologies and approaches taken to solve problems and to develop complex systems. In any mature discipline, there exist for systems engineering a number of processes, standards, guidelines, and software tools to organize and enhance the effectiveness of the systems engineering professional. The International Council of Systems Engineering maintains current information and reviews in these areas. These perspectives will be discussed in the following sections.

2.3 SYSTEMS DOMAINS

With a broad view of system development, it can be seen that the traditional approach to systems now encompasses a growing domain breadth. And much like a Rubik's Cube, the domain faces are now completely integrated into the systems engineer's perspective of the "big (but complex) picture." The systems domain faces shown in Figure 2.4 include not only the engineering, technical, and management domains but

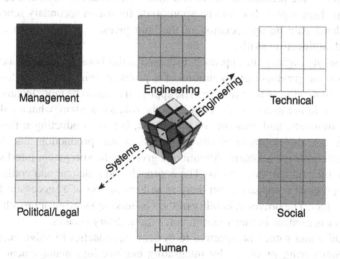

Figure 2.4. Systems engineering domains.

also social, political/legal, and human domains. These latter softer dimensions require additional attention and research to fully understand their impact and utility in system development, especially as we move to areas at the enterprise and global family of systems levels of complexity.

Particularly interesting domains are those that involve scale, such as nano- and microsystems, or systems that operate (often autonomously) in extreme environments, such as deep undersea or outer space. Much like physical laws change with scale, does the systems engineering approach need to change? Should systems engineering practices evolve to address the needs for submersibles, planetary explorers, or intravascular robotic systems?

2.4 SYSTEMS ENGINEERING FIELDS

Since systems engineering has a strong connection bridging the traditional engineering disciplines like electrical, mechanical, aerodynamic, and civil engineering among others, it should be expected that engineering specialists look at systems engineering with a perspective more strongly from their engineering discipline. Similarly, since systems engineering is a guide to design of systems often exercised in the context of a project or program, then functional, project, and senior managers will consider the management elements of planning and control to be key aspects of system development. The management support functions that are vital to systems engineering success such as quality management, human resource management, and financial management can all claim an integral role and perspective to the system development.

These perceptions are illustrated in Figure 2.5, and additional fields that represent a few of the traditional areas associated with systems engineering methods and practices are also shown. An example is the area of operations research whose view of systems engineering includes provision of a structure that will lead to a quantitative analysis of

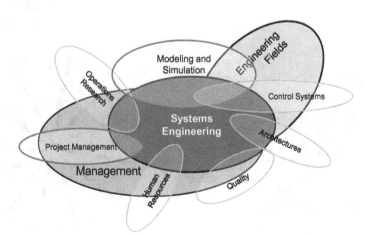

Figure 2.5. Examples of systems engineering fields.

alternatives and optimal decisions. The design of systems also has a contingency of professionals who focus on the structures and architectures. In diverse areas such as manufacturing to autonomous systems, another interpretation of systems engineering comes from engineers who develop control systems, who lean heavily on the systems engineering principles that focus on management of interfaces and feedback systems. Finally, the overlap of elements of modeling and simulation with systems engineering provides a perspective that is integral to a cost-effective examination of systems options to meet the requirements and needs of the users. As systems engineering matures, there will be an increasing number of perspectives from varying fields that adopt it as their own.

2.5 SYSTEMS ENGINEERNG APPROACHES

Systems engineering can also be viewed in terms of the depictions of the sequence of processes and methodologies used in the execution of the design, development, integration, and testing of a system (see Figure 2.6 for examples). Early graphics were linear

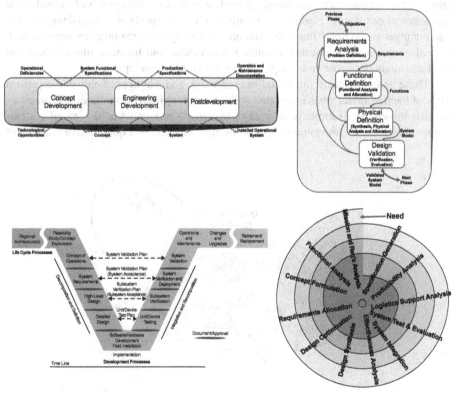

Figure 2.6. Examples of systems engineering approaches.

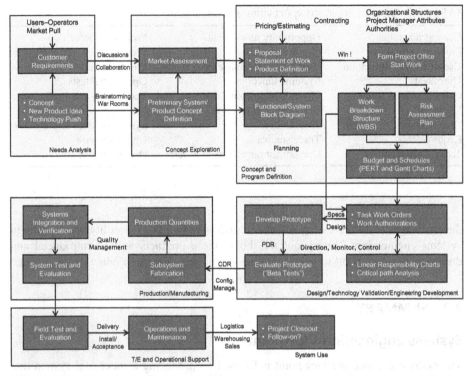

Figure 2.7. Life cycle systems engineering view. PERT, Program Evaluation and Review Technique; PDR, Preliminary Design Review; CDR, Critical Design Review.

in the process flow with sequences of steps that are often iterative to show the logical means to achieve consistency and viability. Small variations are shown in the waterfall charts that provide added means to illustrate interfaces and broader interactions. Many of the steps that are repeated and dependent on each other lead to the spiral or loop conceptual diagrams. The popular systems engineering "V" diagram provides a view of life cycle development with explicit relationships shown between requirements and systems definition and the developed and validated product.

A broader perspective shown in Figure 2.7 provides a full life cycle view and includes the management activities in each phase of development. This perspective illustrates the close relationship between management planning and control and the systems engineering process.

2.6   SYSTEMS ENGINEERING ACTIVITIES AND PRODUCTS

Sometimes followed as a road map, the life cycle development of a system can be associated with a number of systems engineering and project management products or outputs that are listed in Table 2.2. The variety and breadth of these products reflect

TABLE 2.2. Systems Engineering Activities and Documents

Context diagrams	Opportunity assessments	Prototype integration
Problem definition	Candidate concepts	Prototype test and evaluation
User/owner identification	Risk analysis/management plan	Production/operations plan
User needs	Systems functions	Operational tests
Concept of operations	Physical allocation	Verification and validation
Scenarios	Component interfaces	Field support/maintenance
Use cases	Traceability	System/product effectiveness
Requirements	Trade studies	Upgrade/revise
Technology readiness	Component development & test	Disposal/reuse

the challenges early professionals have in understanding the full utility of engaging in systems engineering. Throughout this book, these products will be introduced and discussed in some detail to help guide the systems engineer in product development.

2.7 SUMMARY

Systems Engineering Viewpoint

The systems engineering viewpoint is focused on producing a successful system that meets requirements and development objectives, is successful in its operation in the field, and achieves its desired operating life. In order to achieve this definition of success, the systems engineer must balance superior performance with affordability and schedule constraints. In fact, many aspects of systems engineering involve achieving a balance among conflicting objectives. For example, the systems engineering typically must apply new technology to the development of a new system while managing the inherent risks that new technology poses.

Throughout the development period, the systems engineer focuses his or her perspective on the total system, making decisions based on the impacts and capabilities of the system as a whole. Often, this is accomplished by bridging multiple disciplines and components to ensure a total solution. Specialized design is one dimensional in that it has great technical depth, but little technical breadth and little management expertise. Planning and control is two dimensional: it has great management expertise, but moderate technical breadth and small technical depth. But systems engineering is three dimensional: it has great technical breadth, as well as moderate technical depth and management expertise.

Perspectives of Systems Engineering

A spectrum of views exist in understanding systems engineering, from a general systems thinking approach to problems, to the developmental process approach for systems engineering, to the broad perspective of engineering systems.

Systems Domains

The engineering systems view encompasses not only traditional engineering disciplines but also technical and management domains and social, political/legal, and human domains. Scales at the extremes are of particular interest due to their complexity.

Systems Engineering Fields

Systems engineering encompasses or overlaps with many related fields including engineering, management, operations analysis, architectures, modeling and simulation, and many more.

Systems Engineering Approaches

As the field of systems engineering matures and is used for many applications, several process models have been developed including the linear, V, spiral, and waterfall models.

Systems Engineering Activities and Products

A full systems life cycle view illustrated the close relationship with management process and leads to a large, diverse set of activities and products.

PROBLEMS

2.1 Figure 2.1 illustrates the law of diminishing returns in seeking the optimum system (or component) performance and hence the need to balance the performance against the cost. Give examples of two pairs of characteristics other than performance versus cost where optimizing one frequently competes with the other, and briefly explain why they do.

2.2 Explain the advantages and disadvantages of introducing system concepts to secondary students in order to encourage them to pursue STEM careers.

2.3 Select a very large complex system of system example and explain how the engineering systems approach could provide useful solutions that would have wide acceptance across many communities.

2.4 Referring to Figure 2.5, identify and justify other disciplines that overlap with systems engineering and give examples how those disciplines contribute to solving complex systems problems.

2.5 Discuss the use of different systems engineering process models in terms of their optimal use for various system developments. Is one model significantly better than another?

FURTHER READING

B. Blanchard. *Systems Engineering Management*, Third Edition. John Wiley & Sons, 2004.

H. Eisner. *Essentials of Project and Systems Engineering Management*, Second Edition. John Wiley & Sons, 2002.

3

STRUCTURE OF COMPLEX SYSTEMS

3.1 SYSTEM BUILDING BLOCKS AND INTERFACES

The need for a systems engineer to attain a broad knowledge of the several interacting disciplines involved in the development of a complex system raises the question of how deep that understanding needs to be. Clearly, it cannot be as deep as the knowledge possessed by the specialists in these areas. Yet it must be sufficient to recognize such factors as program risks, technological performance limits, and interfacing requirements, and to make trade-off analyses among design alternatives.

Obviously, the answers depend on specific cases. However, it is possible to provide an important insight by examining the structural hierarchy of modern systems. Such an examination reveals the existence of identifiable types of the building blocks that make up the large majority of systems and represent the lower working level of technical understanding that the systems engineer must have in order to do the job. This is the level at which technical trade-offs affecting system capabilities must be worked out and at which interface conflicts must be resolved in order to achieve a balanced design across the entire system. The nature of these building blocks in their context as fundamental system elements and their interfaces and interactions are discussed in the ensuing sections.

Systems Engineering Principles and Practice, Second Edition. Alexander Kossiakoff, William N. Sweet, Samuel J. Seymour, and Steven M. Biemer
© 2011 by John Wiley & Sons, Inc. Published 2011 by John Wiley & Sons, Inc.

3.2 HIERARCHY OF COMPLEX SYSTEMS

In order to understand the scope of systems engineering and what a systems engineer must learn to carry out the responsibilities involved in guiding the engineering of a complex system, it is necessary to define the general scope and structure of that system. Yet, the definition of a "system" is inherently applicable to different levels of aggregation of complex interacting elements. For example, a telephone substation, with its distributed lines to the area that it serves, can be properly called a system. Hotel and office building switchboards, with their local lines, may be called "subsystems," and the telephone instruments may be called "components" of the system. At the same time, the substation may be regarded as a subsystem of the city telephone system and that, in turn, to be a subsystem of the national telephone system.

In another example, a commercial airliner certainly qualifies to be called a system, with its airframe, engines, controls, and so on, being subsystems. The airliner may also be called a subsystem of the air transportation system, which consists of the air terminal, air traffic control, and other elements of the infrastructure in which the airliner operates. Thus, it is often said that every system is a subsystem of a higher-level system, and every subsystem may itself be regarded as a system.

The above relationships have given rise to terms such as "supersystems" to refer to overarching systems like the wide-area telephone system and the air transportation system. In networked military systems, the term "system of systems" (SoS) has been coined to describe integrated distributed sensor and weapon systems. This nomenclature has migrated to the commercial world as well; however, the use and definition of the term varies by area and specialty.

Model of a Complex System

While learning the fundamentals of systems engineering, this ambiguity of the scope of a system may be confusing to some students. Therefore, for the purpose of illustrating the typical scope of a systems engineer's responsibilities, it is useful to create a more specific model of a typical system. As will be described later, the technique of modeling is one of the basic tools of systems engineering, especially in circumstances where unambiguous and quantitative facts are not readily available. In the present instance, this technique will be used to construct a model of a typical complex system in terms of its constituent parts. The purpose of this model is to define a relatively simple and readily understood system architecture, which can serve as a point of reference for discussing the process of developing a new system and the role of systems engineering throughout the process. While the scope of this model does not extend to that of supersystems or an SoS, it is representative of the majority of systems that are developed by an integrated acquisition process, such as a new aircraft or a terminal air traffic control system.

By their nature, complex systems have a hierarchical structure in that they consist of a number of major interacting elements, generally called *subsystems*, which themselves are composed of more simple functional entities, and so on down to primitive elements such as gears, transformers, or light bulbs, usually referred to as *parts*.

TABLE 3.1. System Design Hierarchy

Systems			
Communications systems	Information systems	Material processing systems	Aerospace systems

Subsystems			
Signal networks	Databases	Material preparation	Engines

Components					
Signal receivers	Data displays	Database programs	Power transfer	Material reactors	Thrust generators

Subcomponents					
Signal amplifiers	Cathode ray tubes	Library utilities	Gear trains	Reactive valves	Rocket nozzles

Parts						
Transformer	LED	Algorithms	Gears		Couplings	Seals

Commonly used terminology for the various architectural levels in the structure of systems is confined to the generic system and subsystem designation for the uppermost levels and parts for the lowest.

For reasons that will become evident later in this section, the system model as defined in this book will utilize two additional intermediate levels, which will be called *components* and *subcomponents*. While some models use one or two more intermediate levels in their representation of systems, these five have proven to be sufficient for the intended purpose.

Definition of System Levels. Table 3.1 illustrates the above characterization of the hierarchical structure of the system model. In this table, four representative system types employing advanced technology are listed horizontally, and successive levels of subdivisions within each system are arranged vertically.

In describing the various levels in the system hierarchy depicted in the figure, it was noted previously that the term *system* as commonly used does not correspond to a specific level of aggregation or complexity, it being understood that systems may serve as parts of more complex aggregates or supersystems, and subsystems may themselves be thought of as systems. For the purpose of the ensuing discussion, this ambiguity will be avoided by limiting the use of the term system to those entities that

1. possess the properties of an engineered system and
2. perform a significant useful service with only the aid of human operators and standard infrastructures (e.g., power grid, highways, fueling stations, and

communication lines). According to the above conditions, a passenger aircraft would fit the definition of a system, as would a personal computer with its normal peripherals of input and output keyboard, display, and so on.

The first subordinate level in the system hierarchy defined in Table 3.1 is appropriately called a subsystem and has the conventional connotation of being a major portion of the system that performs a closely related subset of the overall system functions. Each subsystem may in itself be quite complex, having many of the properties of a system except the ability to perform a useful function in the absence of its companion subsystems. Each subsystem typically involves several technical disciplines (e.g., electronic and mechanical).

The term component is commonly used to refer to a range of mostly lower-level entities, but in this book, the term component will be reserved to refer to the middle level of system elements described above. Components will often be found to correspond to configuration items (CIs) in government system acquisition notation.

The level below the component building blocks is composed of entities, referred to as subcomponents, which perform elementary functions and are composed of several parts. The lowest level, composed of parts, represents elements that perform no significant function except in combination with other parts. The great majority of parts come in standard sizes and types and can usually be obtained commercially.

Domains of the Systems Engineer and Design Specialist

From the above discussion, the hierarchical structure of engineered systems can be used to define the respective knowledge domains of both the systems engineer and the design specialist. The intermediate system components occupy a central position in the system development process, representing elements that are, for the most part, products fitting within the domain of industrial design specialists, who can adapt them to a particular application based on a given set of specifications. The proper specification of components, especially to define performance and to ensure compatible interfaces, is the particular task of systems engineering. This means that the systems engineer's knowledge must extend to the understanding of the key characteristics of components from which the system may be constituted, largely through dialogue and interaction with the design specialists, so that he or she may select the most appropriate types and specify their performance and interfaces with other components.

The respective knowledge domains of the systems engineer and the design specialist are shown in Figure 3.1 using the system hierarchy defined above. It shows that the systems engineer's knowledge needs to extend from the highest level, the system and its environment, down through the middle level of primary system building blocks or components. At the same time, the design specialist's knowledge needs to extend from the lowest level of parts up through the components level, at which point their two knowledge domains "overlap." This is the level at which the systems engineer and the design specialist must communicate effectively, identify and discuss technical problems, and negotiate workable solutions that will not jeopardize either the system design process or the capabilities of the system as a whole.

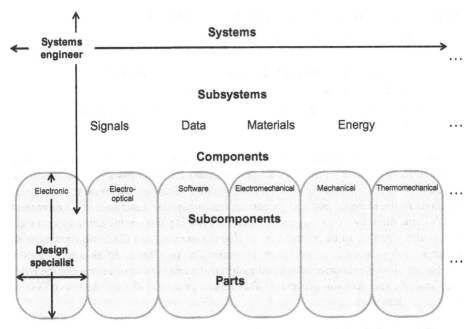

Figure 3.1. Knowledge domains of the systems engineer and the design specialist.

The horizontal boundaries of these domains are deliberately shown as continuity lines in the figure to indicate that they should be extended as necessary to reflect the composition of the particular system. When a subcomponent or part happens to be critical to the system's operation (e.g., the ill-fated seal in the space shuttle *Challenger*'s booster rocket), the systems engineer should be prepared to learn enough about its behavior to identify its potential impact on the system as a whole. This is frequently the case in high-performance mechanical and thermomechanical devices, such as turbines and compressors. Conversely, when the specified function of a particular component imposes unusual demands on its design, the design specialist should call on the systems engineer to reexamine the system-level assumptions underlying this particular requirement.

3.3 SYSTEM BUILDING BLOCKS

Using this system model provides systems engineers with a simple method of partitioning a system along a functional and physical dimension: understanding the functional aspects of the system, then partitioning the system into a physical hierarchy. Each dimensional description of the system can then be decomposed into elements. Below is the description of these two categories of building blocks and a recommended set of elements used in defining the components of each.

Functional Building Blocks: Functional Elements

The three basic entities that constitute the media on which systems operate are

1. *Information:* the content of all knowledge and communication,
2. *Material:* the substance of all physical objects, and
3. *Energy:* energizes the operation and movement of all active system components.

Because all system functions involve a purposeful alteration in some characteristic of one or more of these entities, the latter constitutes a natural basis for classifying the principal system functional units. Since information elements are more than twice as populous as the material and energy entities among system functions, it is convenient to subdivide them into two classes: (1) elements dealing with propagating information (e.g., radio signals), to be referred to as *signal elements*, and (2) those dealing with stationary information (e.g., computer programs), to be referred to as *data elements*. The former class is primarily associated with sensing and communications and the latter with analysis and decision processes. This results in a total of four classes of system functional elements:

1. *Signal Elements,* which sense and communicate information;
2. *Data Elements,* which interpret, organize, and manipulate information;
3. *Material Elements,* which provide structure and transformation of materials; and
4. *Energy Elements,* which provide energy and motive power.

To provide a context for acquainting the student with significant design knowledge peculiar to each of the four broad classes of functional elements, a set of generic functional elements has been defined that represents the majority of important types for each class.

To make the selected elements self-consistent and representative, three criteria may be used to ensure that each element is neither trivially simple nor inordinately complex and has wide application:

1. *Significance.* Each functional element must perform a distinct and significant function, typically involving several elementary functions.
2. *Singularity.* Each functional element should fall largely within the technical scope of a single engineering discipline.
3. *Commonality.* The function performed by each element can be found in a wide variety of system types.

In configuring the individual functional elements, it is noted that regardless of their primary function and classification, their physical embodiments are necessarily built of material usually controlled by external information and powered by electricity or some

TABLE 3.2. System Functional Elements

Class function	Element function	Applications
Signal—generate, transmit, distribute, and receive signals used in passive or active sensing and in communications	Input signal	TV camera
	Transmit signal	FM radio transmitter
	Transduce signal	Radar antenna
	Receive signal	Radio receiver
	Process signal	Image processor
	Output signal	
Data—analyze, interpret, organize, query, and/or convert data and information into forms desired by the user or other systems	Input data	Keyboard
	Process data	Computer CPU
	Control data	Operating system
	Control processing	Word processor
	Store data	Printer
	Output data	
	Display data	
Material—provide system structural support or enclosure, or transform the shape, composition, or location of material substances	Support material	Airframe
	Store material	Shipping container
	React material	Autoclave
	Form material	Milling machine
	Join material	Welding machine
	Control position	Servo actuator
Energy—provide and convert energy or propulsive power to the system	Generate thrust	Turbojet engine
	Generate torque	Reciprocating engine
	Generate electricity	Solar cell array
	Control temperature	Refrigerator
	Control motion	Auto transmission

other source of energy. Thus, a television set, whose main function is to process information in the form of a radio frequency signal into information in the form of a TV picture and sound, is built of materials, powered by electricity, and controlled by user-generated information inputs. Accordingly, it should be expected that most elements in all classes would have information and energy inputs in addition to their principal processing inputs and outputs.

The above process converges on a set of 23 functional elements, five or six in each class. These are listed in the middle column of Table 3.2. The function of the class as a whole is shown in the left column, and typical applications that might embody the individual elements are listed in the right column. It should be noted that the above classification is not meant to be absolute, but is established solely to provide a systematic and logical framework for discussing the properties of systems at the levels of importance to systems engineers.

Fundamentally, the functional design of any system may be defined by conceptually combining and interconnecting the identified functional elements along with perhaps one or two very specialized elements that might perform a unique function in certain system applications so as to logically derive the desired system capabilities from the available system inputs. In effect, the system inputs are transformed and processed through the interconnected functions to provide the desired system outputs.

Physical Building Blocks: Components

System physical building blocks are the physical embodiments of the functional elements consisting of hardware and software. Consequently, they have the same distinguishing characteristics of significance, singularity, and commonality and are at the same level in the system hierarchy, generally one level below a typical subsystem and two levels above a part. They will be referred to as *component elements* or simply as components.

The classes into which the component building blocks have been categorized are based on the different design disciplines and technologies that they represent. In total, 31 different component types were identified and grouped into six categories, as shown in Table 3.3. The table lists the category, component name, and the functional element(s) with which it is associated. As in the case of functional elements, the component names are indicative of their primary function but, in this case, represent things rather than processes. Many of these represent devices that are in widespread use.

The systems engineer's concern with the implementation of the functional elements within components is related to a different set of factors than those associated with the initial functional design itself. Here, the predominant issues are reliability, form and fit, compatibility with the operational environment, maintainability, producibility, testability, safety, and cost, along with the requirement that product design does not violate the integrity of the functional design. The depth of the systems engineer's understanding of the design of individual components needs to extend to the place where the system-level significance of these factors may be understood, and any risks, conflicts, and other potential problems addressed.

The required extent and nature of such knowledge varies widely according to the type of system and its constitution. A systems engineer dealing with an information system can expect to concentrate largely on the details of the software and user aspects of the system while considering mainly the external aspects of the hardware components, which are usually standard (always paying special attention to component interfaces). At another extreme, an aerospace system such as an airplane consists of a complex and typically nonstandard assemblage of hardware and software operating in a highly dynamic and often adverse environment. Accordingly, an aerospace systems engineer needs to be knowledgeable about the design of system components to a considerably more detailed level so as to be aware of the potentially critical design features before they create reliability, producibility, or other problems during the product engineering, test, and operational stages.

Common Building Blocks

An important and generally unrecognized observation resulting from an examination of the hierarchical structure of a large variety of systems is the existence of an intermediate level of elements of types that recur in a variety of systems. Devices such as

TABLE 3.3. Component Design Elements

Category	Component	Functional element(s)
Electronic	Receiver	Receive signal
	Transmitter	Transmit signal
	Data processor	Process data
	Signal processor	Process signal
	Communications processors	Process signal/data
	Special electronic equipment	Various
Electro-optical	Optical sensing device	Input signal
	Optical storage device	Store data
	Display device	Output signal/data
	High-energy optics device	Form material
	Optical power generator	Generate electricity
Electromechanical	Inertial instrument	Input data
	Electric generator	Generate electricity
	Data storage device	Store data
	Transducer	Transduce signal
	Data input/output device	Input/output data
Mechanical	Framework	Support material
	Container	Store material
	Material processing machine	Form/join material
	Material reactor	React material
	Power transfer device	Control motion
Thermomechanical	Rotary engine	Generate torque
	Jet engine	Generate thrust
	Heating unit	Control temperature
	Cooling unit	Control temperature
	Special energy source	Generate electricity
Software	Operating system	Control system
	Application	Control processing
	Support software	Control processing
	Firmware	Control system

signal receivers, data displays, torque generators, containers, and numerous others perform significant functions used in many systems. Such elements typically constitute product lines of commercial organizations, which may configure them for the open market or customize them to specifications to fit a complex system. In Table 3.1, the above elements are situated at the third or middle level and are referred to by the generic name component.

The existence of a distinctive set of middle-level system building blocks can be seen as a natural result of the conditions discussed in Chapter 1 for the origin of complex systems, namely, (1) advancing technology, (2) competition, and (3) special-ization. Technological advances are generally made at basic levels, such as the develop-ment of semiconductors, composite materials, light-emitting devices, graphic user interfaces, and so on. The fact of specialization tends to apply such advances primarily to devices that can be designed and manufactured by people and organizations specialized in certain types of products. Competition, which drives technology advances, also favors specialization in a variety of specific product lines. A predictable result is the proliferation of advanced and versatile products that can find a large market (and hence achieve a low cost) in a variety of system applications. The current emphasis in defense system development on adapting commercial off-the-shelf (COTS) components, wherever practicable, attempts to capitalize on economies of scale found in the commercial component market.

Referring back to Table 3.1, it is noted that as one moves up through the hierarchy of system element levels, the functions performed by those in the middle or component level are the first that provide a significant functional capability, as well as being found in a variety of different systems. For this reason, the types of elements identified as components in the figure were identified as basic system building blocks. Effective systems engineering therefore requires a fundamental understanding of both the functional and physical attributes of these ubiquitous system constituents. To provide a framework for gaining an elementary knowledge base of system building blocks, a set of models has been defined to represent commonly occurring system components. This section is devoted to the derivation, classification, interrelationships, and common examples of the defined system building blocks.

Applications of System Building Blocks

The system building block model described above may be useful in several ways:

1. The categorization of functional elements into the four classes of signal, data, material, and energy elements can help suggest what kind of actions may be appropriate to achieve required operational outcomes.

2. Identifying the classes of functions that need to be performed by the system may help group the appropriate functional elements into subsystems and thus may facilitate functional partitioning and definition.

3. Identifying the individual functional building blocks may help define the nature of the interfaces within and between subsystems.

4. The interrelation between the functional elements and the corresponding one or more physical implementations can help visualize the physical architecture of the system.

5. The commonly occurring examples of the system building blocks may suggest the kinds of technology appropriate to their implementation, including possible alternatives.

6. For those specialized in software and unfamiliar with hardware technology, the relatively simple framework of four classes of functional elements and six classes of physical components should provide an easily understood organization of hardware domain knowledge.

3.4 THE SYSTEM ENVIRONMENT

The system environment may be broadly defined as everything outside of the system that interacts with the system. The interactions of the system with its environment form the main substance of system requirements. Accordingly, it is important at the outset of system development to identify and specify in detail all of the ways in which the system and its environment interact. It is the particular responsibility of the systems engineer to understand not only what these interactions are but also their physical basis, to make sure that the system requirements accurately reflect the full range of operating conditions.

System Boundaries

To identify the environment in which a new system operates, it is necessary to identify the system's boundaries precisely, that is, to define what is inside the system and what is outside. Since we are treating systems engineering in the context of a system development project, the totality of the system will be taken as that of the product to be developed.

Although defining the system boundary seems almost trivial at first glance, in practice, it is very difficult to identify what is part of the system and what is part of the environment. Many systems have failed due to miscalculations and assumptions about what is internal and what is external. Moreover, different organizations tend to define boundaries differently, even with similar systems.

Fortunately, several criteria are available to assist in determining whether an entity should be defined as part of a system:

- *Developmental Control.* Does the system developer have control over the entity's development? Can the developer influence the requirements of the entity, or are requirements defined outside of the developer's sphere of influence? Is funding part of the developer's budget, or is it controlled by another organization?
- *Operational Control.* Once fielded, will the entity be under the operational control of the organization that controls the system? Will the tasks and missions performed by the entity be directed by the owner of the system? Will another organization have operational control at times?
- *Functional Allocation.* In the functional definition of the system, is the systems engineer "allowed" to allocate functions to the entity?
- *Unity of Purpose.* Is the entity dedicated to the system's success? Once fielded, can the entity be removed without objection by another entity?

Systems engineers have made mistakes by defining entities as part of the system when, in fact, the span of control (as understood by the above criteria) was indeed small. And typically, either during development or operations, the entity was not available to perform its assigned functions or tasks.

One of the basic choices required early is to determine whether human users or operators of a system are considered part of the system or are external entities. In a majority of cases, the user or operator should be considered external to the system. The system developer and owner rarely have sufficient control over operators to justify their inclusion in the system. When operators are considered external to the system, the systems engineer and the developer will focus on the operator interface, which is critical to complex systems.

From another perspective, most systems cannot operate without the active participation of human operators exercising decision and control functions. In a functional sense, the operators may well be considered to be integral parts of the system. However, to the systems engineer, the operators constitute elements of the system environment and impose interface requirements that the system must be engineered to accommodate. Accordingly, in our definition, the operators will be considered to be external to the system.

As noted earlier, many, if not most, complex systems can be considered as parts of larger systems. An automobile operates on a network of roads and is supported by an infrastructure of service stations. However, these are not changed to suit a new automobile. A spacecraft must be launched from a complex gantry, which performs the fueling and flight preparation functions. The gantry, however, is usually a part of the launch complex and not a part of the spacecraft's development. In the same manner, the electrical power grid is a standard source of electricity, which a data processing system may utilize. Thus, the supersystems identified in the above examples need not be considered in the engineering process as part of the system being developed but as an essential element in its operational environment, and to the extent required to assure that all interfacing requirements are correctly and adequately defined.

Systems engineers must also become involved in interface decisions affecting designs both of their own and of an interfacing system. In the example of a spacecraft launched from a gantry, some changes to the information handling and perhaps other functions of the gantry may well be required. In such instances, the definition of common interfaces and any associated design issues would need to be worked out with engineers responsible for the launch complex.

System Boundaries: The Context Diagram

An important communications tool available to the systems engineer is the context diagram. This tool effectively displays the external entities and their interactions with the system and instantly allows the reader to identify those external entities. Figure 3.2 shows a generic context diagram. This type of diagram is known as a black box diagram in that the system is represented by a single geographic figure in the center, without any detail. Internal composition or functionality is hidden from the reader. The diagram consists of three components:

1. *External Entities.* These constitute all entities in which the system will interact. Many of these entities can be considered as sources for inputs into the system and destinations of outputs from the system.

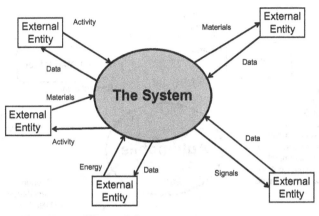

Figure 3.2. Context diagram.

2. *Interactions.* These represent the interactions between the external entities and the system and are represented by arrows. Arrowheads represent the direction or flow of a particular interaction. While double-headed arrows are allowed, single-headed arrows communicate clearer information to the reader. Thus, the engineer should be careful when using two-directional interactions—make sure the meanings of your interactions are clear. Regardless, each interaction (arrow) is labeled to identify what is being passed across the interface.

The diagram depicts the common types of interactions that a context diagram typically contains. In an actual context diagram, these interactions would be labeled with the specific interactions, not the notional words used above. The labels need to be sufficiently detailed to communicate meaning, but abstract enough to fit into the diagram. Thus, words such as "data" or "communications" are to be avoided in the actual diagram since they convey little meaning.

3. *The System.* This is the single geographic figure mentioned already. Typically, this is an oval, circle, or rectangle in the middle of the figure with only the name of the system within. No other information should be present.

We can categorize what can be passed across these external interfaces by utilizing our definitions of the four basic elements above. Using these elements and adding one additional element, we can form five categories:

- data,
- signals,
- materials,
- energy, and
- activities.

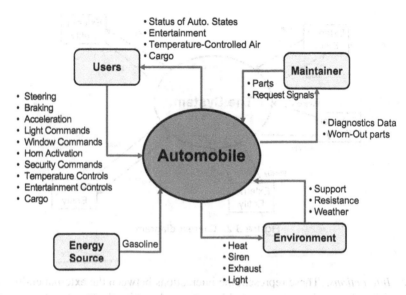

Figure 3.3. Context diagram for an automobile.

Thus, a system interacts with its environment (and specifically, the external entities) by accepting and providing either one of the first four elements or by performing an activity that influences the system or the environment in some manner.

Constructing a diagram such as the system context diagram can be invaluable in communicating the boundary of the system. The picture clearly and easily identifies the external interfaces needed and provides a short description of what is being passed into and out of the system—providing a good pictorial of the system's inputs and outputs.

Figure 3.3 provides a simple example using a typical automobile as the system. Although the system is rather simple, it nicely illustrates all five types of interfaces. Four external entities are identified: users (to include the driver and passengers), the maintainer (which could be a user, but, because of his specialized interactions with the system, is listed separately), an energy source, and the environment. Most systems will interact with these four external entity types. Of course, many other entities may interact with a system as well.

The user provides a multitude of inputs to the system, including various commands and controls as well as actions, such as steering and braking. Materials are also passed to the system: cargo. In return, several outputs are passed from the automobile back to the user, including various status indications on the state of the system. Additionally, an activity is performed: entertainment, representing the various forms of entertainment available in today's automobile. Finally, cargo is returned to the users when desired.

Other entities also interact with the system. The maintainer must provide a request for diagnostics data, typically in the form of signals passed to the auto via an interface. Diagnostics data are returned along with the exchange of parts.

The last two external entities represent somewhat specialized entities: an energy source and the ubiquitous environment. In the automobile case, the energy source provides gasoline to the automobile. This energy source can be one of many types: a gasoline pump at a station or a small container with a simple nozzle. The environment requires some special consideration, if for no other reason than it includes everything not specifically contained in the other external entities. So, in some respects, the environment entity represents "other." In our example, the automobile will generate heat and exhaust in its typical operation. Additionally, a siren and light from various light bulbs, horns, and signals will also radiate from the auto. The environment is also a source of many inputs, such as physical support, air resistance, and weather.

It takes some thought to identify the inputs, outputs, and activities that are part of the system–environment interaction. The creator of this diagram could have really gone "overboard" and specified temperature, pressure, light, humidity, and a number of other factors in this interaction. This brings up an interesting question: what do we include in listing the interactions between the system and the external entity? For that matter, how do we know whether an external entity should be included in our diagram? Fortunately, there is a simple answer to this: if the interaction is important for the design of the system, then it should be included.

In our automobile case, physical support is important for our design and will influence the type of transmission, steering, and tires. So we include "support" in our diagram. Temperature, humidity, pressure, and so on, will be a factor, but we are not sure about their importance to design, so we group these characteristics under "weather." This does not mean that the automobile will be designed for all environmental conditions, only that we are not considering all conditions in our design. We should have an idea of the environmental conditions from the requirements, and therefore, we can determine whether they should be in our context diagram.

Output from the system to the environment also depends on whether it will influence the design. The automobile will in fact output many things into the environment: heat, smells, texture, colors … and especially carbon dioxide as part of the exhaust! But which of these influence our design? Four will be major influences: heat, noise from the siren, exhaust, and light. Therefore, we include only those for now and omit the others. We can always go back and update the context diagram (in fact, we should, as we progress through both the systems engineering process and the system development life cycle).

The system context diagram is a very simple yet powerful tool to identify, evaluate, and communicate the boundaries of our system. Therefore, it becomes the first tool we introduce in this book. More will follow that will eventually provide the systems engineer with the collection needed to adequately develop his system.

Types of Environmental Interactions

To understand the nature of the interactions of a system with its surroundings, it is convenient to distinguish between primary and secondary interactions. The former involves elements that interact with the system's primary functions, that is, represent functional inputs, outputs, and controls; the latter relates to elements that interact with the system in an indirect nonfunctional manner, such as physical supports, ambient temperature, and so on. Thus, the functional interactions of a system with its environment include its inputs and outputs and human control interfaces. Operational maintenance may be considered a quasi-functional interface. Threats to the system are those entities that deny or disrupt the system's ability to perform its activities. The physical environment includes support systems, system housing, and shipping, handling, and storage. Each of these is briefly described below.

Inputs and Outputs. The primary purpose of most systems is to operate on external stimuli and/or materials in such a manner as to process these inputs in a useful way. For a passenger aircraft, the materials are the passengers, their luggage, and fuel, and the aircraft's function is to transport the passengers and their belongings to a distant destination rapidly, safely, and comfortably. Figure 3.4 illustrates some of the large

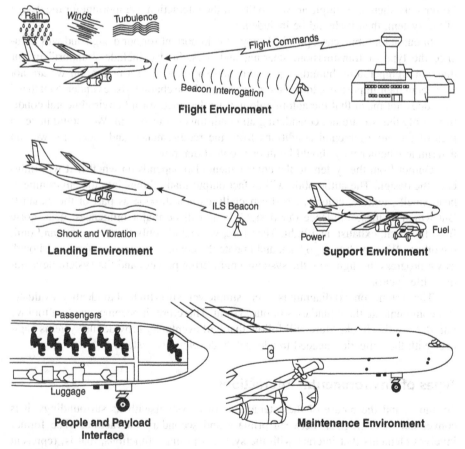

Figure 3.4. Environments of a passenger airliner. ILS, instrument landing system.

variety of interactions that a complex system has with its operating environment for the case of a passenger aircraft.

System Operators. As noted previously, virtually all systems, including automated systems, do not operate autonomously but are controlled to some degree by human operators in performing their function. For the purposes of defining the systems engineer's task, the operator is part of the system's environment. The interface between the operator and the system (human–machine interface) is one of the most critical of all because of the intimate relationship between the control exercised by the operator and the performance of the system. It is also one of the most complex to define and test.

Operational Maintenance. The requirements for system readiness and operational reliability relate directly to the manner in which it is to be maintained during its operating life. This requires that the system be designed to provide access for monitoring, testing, and repair requirements that are frequently not obvious at the outset, but nevertheless must be addressed early in the development process. Thus, it is necessary to recognize and explicitly provide for the maintenance environment.

Threats. This class of external entities can be man-made or natural. Clearly, weather could be considered a threat to a system exposed to the elements. For example, when engineering naval systems, the salt water environment becomes a corrosive element that must be taken into consideration. Threats can also be man-made. For example, a major threat to an automatic teller machine (ATM) would be the thief, whose goal might be access to the stored cash. System threats need to be identified early to design countermeasures into the system.

Support Systems. Support systems are that part of the infrastructure on which the system depends for carrying out its mission. As illustrated in Figure 3.4, the airport, the air traffic control system, and their associated facilities constitute the infrastructure in which an individual aircraft operates, but which is also available to other aircraft. These are parts of the SoS represented by the air transportation system, but for an airplane, they represent standard available resources with which it rousts interface harmoniously.

Two examples of common support systems that have been mentioned previously are the electric power grids, which distribute usable electric power throughout the civilized world, and the network of automobile filling stations and their suppliers. In building a new airplane, automobile, or other systems, it is necessary to provide interfaces that are compatible with and capable of utilizing these support facilities.

System Housing. Most stationary systems are installed in an operating site, which itself imposes compatibility constraints on the system. In some cases, the installation site provides protection for the system from the elements, such as variations in temperature, humidity, and other external factors. In other cases, such as installations on board ship, these platforms provide the system's mechanical mounting but, otherwise, may expose the system to the elements, as well as subject it to shock, vibration, and other rigors.

Shipping and Handling Environment. Many systems require transport from the manufacturing site to the operating site, which imposes special conditions for which the system must be designed. Typical of these are extreme temperatures, humidity, shock, and vibration, which are sometimes more stressful than those characteristic of the operating environment. It may be noted that the impact of the latter categories of environmental interactions is addressed mainly in the engineering development stage.

3.5 INTERFACES AND INTERACTIONS

Interfaces: External and Internal

The previous section described the different ways in which a system interacts with its environment, including other systems. These interactions all occur at various boundaries of the system. Such boundaries are called the system's *external interfaces*. Their definition and control are a particular responsibility of the systems engineer because they require knowledge of both the system and its environment. Proper interface control is crucial for successful system operation.

A major theme of systems engineering is accordingly the management of interfaces. This involves

1. identification and description of interfaces as part of system concept definition and
2. coordination and control of interfaces to maintain system integrity during engineering development, production, and subsequent system enhancements.

Inside the system, the boundaries between individual components constitute the system's *internal interfaces*. Here, again, the definition of internal interfaces is the concern of the systems engineer because they fall between the responsibility boundaries of engineers concerned with the individual components. Accordingly, their definition and implementation must often include consideration of design trade-offs that impact on the design of both components.

Interactions

Interactions between two individual elements of the system are effected through the interface connecting the two. Thus, the interface between a car driver's hands and the steering wheel enables the driver to guide (interact with) the car by transmitting a force that turns the steering wheel and thereby the car's wheels. The interfaces between the tires of the car and the road both propel and steer the car by transmitting driving traction to the road, and also help cushion the car body from the roughness of the road surface.

The above examples illustrate how functional interactions (guiding or propelling the car) are effected by physical interactions (turning the steering wheel or the drive wheels) that flow across (physical) interfaces. Figure 3.5 illustrates the similar relations

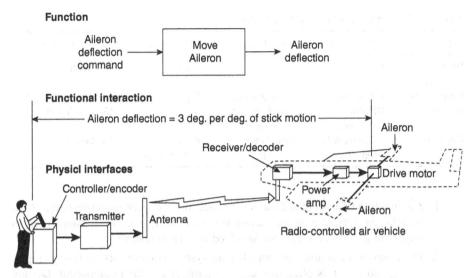

Figure 3.5. Functional interactions and physical interfaces.

between physical interfaces involved in steering an air vehicle and the resulting functional interactions.

An important and sometimes less than adequately addressed external system interaction occurs during system maintenance. This activity necessarily requires access to a number of vital system functions for testing purposes. Such access calls for the provision of special test points of the system, which can be sampled externally with a minimum of manipulation. In some complex systems, an extensive set of built-in tests (BITs) is incorporated, which may be exercised while the system is in its operational status. The definition of such interfaces is also the concern of the systems engineer.

Interface Elements

To systematize the identification of external and internal interfaces, it is convenient to distinguish three different types:

1. connectors, which facilitate the transmission of electricity, fluid, force, and so on, between components;
2. isolators, which inhibit such interactions; and
3. converters, which alter the form of the interaction medium. These interfaces are embodied in component parts or subcomponents, which can be thought of as interface elements.

Table 3.4 lists a number of common examples of interface elements of each of the three types, for each of four interaction media: electrical, mechanical, hydraulic, and human. The table brings out several points worthy of note:

TABLE 3.4. Examples of Interface Elements

Type	Electrical	Mechanical	Hydraulic	Human–machine
Interaction medium	Current	Force	Fluid	Information
Connectors	Cable switch	Joint coupling	Pipe valve	Display control panel
Isolator	RF shield insulator	Shock mount bearing	Seal	Cover window
Converter	Antenna A/D converter	Gear train piston	Reducing valve pump	Keyboard

1. The function of making or breaking a connection between two components (i.e., enabling or disabling an interaction between them) must be considered as an important design feature, often involved in system control.
2. The function of connecting nonadjacent system components by cables, pipes, levers, and so on, is often not part of a particular system component. Despite their inactive nature, such conducting elements must be given special attention at the system level to ensure that their interfaces are correctly configured.
3. The relative simplicity of interface elements belies their critical role in ensuring system performance and reliability. Experience has shown that a large fraction of system failures occurs at interfaces. Assuring interface compatibility and reliability is a particular responsibility of the systems engineer.

3.6 COMPLEXITY IN MODERN SYSTEMS

Earlier in the chapter, we described the system hierarchy—how systems are subdivided into subsystems, then components, subcomponents, and finally, parts (see Table 3.1). And as modern systems grow in complexity, the number, diversity, and complexity of these lower-level subsystems, components, and parts increase. Furthermore, the interactions between these entities also increase in complexity. Systems engineering principles, and their applied practices, are designed to deal with this complexity.

Increasingly, a single system may be, or become, a part of a larger entity. While there are many terms currently in use today to describe this supersystem concept, the term SoS seems to be accepted by a wide variety of organizations. Other terms are found in the literature—some meaning the same thing, some having different connotations.

This section provides a basic introduction to the engineering of entities that are considered "above," or more complex, than single systems: SoSs and enterprises.

SoS

For our purposes, we will use two definitions to describe what is meant by an SoS. Both come from the U.S. Department of Defense (DoD). The first is the simplest:

A set or arrangement of systems that results when independent and useful systems are integrated into a larger system that delivers unique capabilities

In essence, anytime a set of independently useful systems is integrated together to provide an enhanced capability beyond that of the sum of the individual systems' capabilities, we have an SoS. Of course, the level of integration could vary significantly. At one end of the spectrum, an SoS could be completely integrated from the earliest development phases, where the individual systems, while able to operate independently, are almost exclusively designed for the SoS. At the other end of the spectrum, multiple systems could be loosely joined for a limited purpose and time span to perform a needed mission, with no more than an agreement of the owners of each system. Thus, a method to capture this range of integration is necessary to fully describe the different nuances of SoSs.

The U.S. DoD produced a systems engineering guide in 2008 specifically for SoS environments and captured this spectrum using four categories. The categories are presented in the order of how tightly coupled the component systems are—from loosely to tightly.

- *Virtual.* Virtual SoSs lack a central management authority and a centrally agreed-upon purpose for the SoS. Large-scale behavior emerges—and may be desirable—but this type of SoS must rely upon relatively invisible mechanisms to maintain it.
- *Collaborative.* In collaborative SoSs, the component systems interact more or less voluntarily to fulfill agreed-upon central purposes. Standards are adopted, but there is no central authority to enforce them. The central players collectively decide how to provide or deny service, thereby providing some means of enforcing and maintaining standards.
- *Acknowledged.* Acknowledged SoSs have recognized objectives, a designated manager, and resources for an SoS; however, the constituent systems retain their independent ownership, objectives, funding, development and sustainment approaches. Changes in the systems are based on collaboration between the SoS and the system.
- *Directed.* Directed SoSs are those in which the integrated SoS is built and managed to fulfill specific purposes. It is centrally managed during long-term operation to continue to fulfill those purposes as well as any new ones the system owners might wish to address. The component systems maintain an ability to operate independently, but their normal operational mode is subordinated to the central managed purpose.

Although one could argue that the last category, the directed SoS, is closer to a single, complex system than an SoS, the definitions capture the range of situations that exist today when systems are integrated together to perform a function, or exhibit a capability, that is greater than any one system.

As the reader might surmise, engineering and architecting an SoS can be different than engineering and architecting a single system, especially for the two middle categories. System of systems engineering (SoSE) can be different because of the unique attributes of an SoS.

Maier first introduced a formal discussion of SoSs by identifying their character-istics in 1998. Since then, several publications have refined these characteristics; however, they have remained remarkably stable over time. Sage and Cuppan summa-rized these characteristics:

1. *Operational Independence of the Individual System.* An SoS is composed of systems that are independent and useful in their own right. If an SoS is disas-sembled into its associated component systems, these component systems are capable of independently performing useful operations independently of one another.

2. *Managerial Independence of the Individual System.* The component systems in an SoS not only can operate independently, but they also generally do operate independently to achieve an intended purpose. Often, they are individually acquired and integrated, and they maintain a continuing operational existence and serve purposes that may be independent of those served by the SoS.

3. *Geographic Distribution.* The geographic dispersion of component systems is often large. Often, these systems can readily exchange only information and knowledge with one another.

4. *Emergent Behavior.* The SoS performs functions and carries out purposes that are not necessarily associated with any component system. These behaviors are emergent properties of the entire SoS and not the behavior of any component system.

5. *Evolutionary Development.* The development of an SoS is generally evolution-ary over time. Components of structure, function, and purpose are added, removed, and modified as experience with the system grows and evolves over time. Thus, an SoS is usually never fully formed or complete.

These characteristics have since been refined to include additional characteristics. Although these refinements have not changed the basic characteristics, they did add two important features:

6. *Self-organization.* An SoS will have a dynamic organizational structure that is able to respond to changes in the environment and to changes in goals and objectives for the SoS.

7. *Adaptation.* Similar to a dynamic organization, the very structure of the SoS will be dynamic and respond to external changes and perceptions of the environment.

Engineering an SoS that falls into either the collaborative or acknowledged cate-gory must deal with the seven core attributes of SoS. Therefore, the basic tools that we have in systems engineering may not be sufficient. Additional methods, tools, and practices have been developed (and are continuing to be developed) to enable the engineer to develop these complex structures.

Some of these tools come from other branches of mathematics and engineering, such as complexity theory. Attributes such as emergent behavior, self-organization, and adaptation have been examined within this field, and various tools and methods have been developed to represent the inherent uncertainty these attributes bring. The challenge is to keep the mathematics simple enough for application to systems engineering.

Other areas that are being examined to support SoSE include social engineering, human behavior dynamics, and chaotic systems (chaos theory). These areas continue to be appropriate for further research.

Enterprise Systems Engineering

SoSE, by its nature, increases the complexity of developing single systems. However, it does not represent the highest level of complexity. In fact, just as Table 3.1 presented a hierarchy with the system at the apex, we can expand this hierarchy, and go beyond SoSs, to an enterprise. Figure 3.6 depicts this hierarchy.

Above an SoS lies the enterprise, which typically consists of multiple SoSs within its structure. Furthermore, an enterprise may consist of a varied collection of system types, not all of which are physical. For instance, an enterprise includes human or social systems that must be integrated with physical systems.

Formally, an enterprise is "anything that consists of people, processes, technology, systems, and other resources across organizations and locations interacting with each other and their environment to achieve a common mission or goal." The level of interaction between these entities varies, just as component systems within an SoS. And many entities fit into this definition. Almost all midsize to large organizations would satisfy this definition. In fact, suborganizations of some large corporations would themselves be defined as an enterprise.

Government agencies and departments would also fit into this definition. And finally, large social and physical structures, such as cities or nations, satisfy the definition.

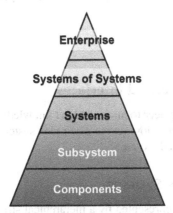

Figure 3.6. Pyramid of system hierarchy.

The source of complexity in enterprise systems engineering is primarily the integration of a diversity of systems and processes. The enterprise typically includes the following components that must be integrated together under the inherent uncertainty of today's enterprise:

- business strategy and strategic planning,
- business processes,
- enterprise services,
- governance,
- technical processes,
- people management and interactions,
- knowledge management,
- information technology infrastructure and investment,
- facility and equipment management,
- supplies management, and
- data and information management.

Enterprise systems engineering refers to the application of systems engineering principles and practices to engineering systems that are part of an enterprise. Developing the individual component systems of the enterprise is known by this term. Another broader term has also emerged: enterprise engineering. This term, with the "systems" omitted, typically refers to the architecting, development, implementation, and operation of the enterprise as a whole. Some have used the terms interchangeably; however, the two terms refer to different levels of abstraction.

The reason that enterprise systems engineering is deemed more complex than SoSE is that many of the components of an enterprise involve one or more SoSs. Therefore, the enterprise could be considered an integration of multiple SoSs.

Just as new tools and techniques are being developed for SoSE applications, so too are tools, methods, and techniques being developed for this relatively young field.

3.7 SUMMARY

System Building Blocks and Interfaces

The need for a systems engineer to attain a broad knowledge of the several interacting disciplines involved in the development of a complex system raises the question of how deep that understanding needs to be.

Hierarchy of Complex Systems

Complex systems may be represented by a hierarchical structure in that they are composed of subsystems, components, subcomponents, and parts.

The domain of the systems engineer extends down through the component level and extends across several categories. In contrast, the domain of the design specialist extends from the part level up through the component level, but typically within a single technology area or discipline.

System Building Blocks

System building blocks are at the level of components and are the basic building blocks of all engineered systems characterized by both functional and physical attributes. These building blocks are characterized by performing a distinct and significant function and are singular—they are within the scope of a single engineering discipline.

Functional elements are functional equivalents of components and are categorized into four classes by operating medium:

- signal elements, which sense and communicate information;
- data elements, which interpret, organize, and manipulate information;
- material elements, which provide structure and process material; and
- energy elements, which provide energy or power.

Components are the physical embodiment of functional elements, which are categorized into six classes by materials of construction:

- electronic,
- electro-optical,
- electromechanical,
- mechanical,
- thermomechanical, and
- software.

System building block models can be useful in identifying actions capable of achieving operational outcomes, facilitating functional partitioning and definition, identifying subsystem and component interfaces, and visualizing the physical architecture of the system.

The System Environment

The system environment, that is, everything outside the system that interacts with it, includes (1) system operators (part of system function but outside the delivered system); (2) maintenance, housing, and support systems; (3) shipping, storage, and handling; (4) weather and other physical environments; and (5) threats.

Interfaces and Interactions

Interfaces are a critical systems engineering concern, which effect interactions between components and can be classified into three categories: connect, isolate, or convert interactions. They require identification, specification, coordination, and control. Moreover, test interfaces typically are provided for integration and maintenance.

Complexity in Modern Systems

Each system is always part of a larger entity. At times, this larger entity can be classified as a separate system in itself (beyond simply an environment, or "nature"). These situations are referred to as "SoSs." They tend to exhibit seven distinct characteristics: operational independence of the individual system, managerial independence of the individual system, geographic distribution, emergent behavior, evolutionary development, self-organization, and adaptation.

Enterprise systems engineering is similar in complexity but focuses on an organizational entity. Since an enterprise involves social systems as well as technical systems, the complexity tends to become unpredictable.

PROBLEMS

3.1 Referring to Table 3.1, list a similar hierarchy consisting of a typical subsystem, component, subcomponent, and part for (1) a terminal air traffic control system, (2) a personal computer system, (3) an automobile, and (4) an electric power plant. For each system, you need only to name one example at each level.

3.2 Give three key activities of a systems engineer that require technical knowledge down to the component level. Under what circumstances should the systems engineer need to probe into the subcomponent level for a particular system component?

3.3 Referring to Figure 3.1, describe in terms of levels in the system hierarchy the knowledge domain of a design specialist. In designing or adapting a component for a new system, what typical characteristics of the overall system and of other components must the design specialist understand? Illustrate by an example.

3.4 The last column of Table 3.2 lists examples of the applications of the 23 functional elements. List one other example of an application than the one listed for three elements in each of the four classes of elements.

3.5 Referring to Figure 3.4, for each of the environments and interfaces illustrated, (1) list the principal interactions between the environment and the aircraft, (2) the nature of each interaction, and (3) describe how each affects the system design.

3.6 For a passenger automobile, partition the principal parts into four subsystems and their components. (Do not include auxiliary functions such as environmental or entertainment.) For the subsystems, group together components concerned with each primary function. For defining the components, use the principles of significance (performs an important function), singularity (largely falls within a simple discipline), and commonality (found in a variety of system types). Indicate where you may have doubts. Draw a block diagram relating the subsystems and components to the system and to each other.

3.7 In the cases selected in answering Problem 3.5, list the specific component interfaces that are involved in the above interactions.

3.8 Draw a context diagram for a standard coffeemaker. Make sure to identify all of the external entities and label all of the interactions.

3.9 Draw a context diagram for a standard washing machine. Make sure to identify all of the external entities and label all of the interactions.

3.10 In a context diagram, "maintainer" is typically an external entity, providing both activities (i.e., "maintenance") and materials (e.g., spare parts) to the system, and the system providing diagnostic data back to the maintainer. Describe the nature of the maintainer interfaces and what interactions could be done by the user.

3.11 List the test interfaces and BIT indicators in your automobile that are available to the user (do not include those only available to a mechanic).

FURTHER READING

D. Buede. *The Engineering Design of Systems: Models and Methods*, Second Edition, John Wiley & Sons, 2009.

Department of Defense. *Systems Engineering Guide for Systems of Systems*. DUSD (A&T) and OSD (AT&L), 2008.

M. Jamshidi, ed. *System of Systems Engineering: Innovations for the 21st Century*. John Wiley & Sons, 2008.

M. Jamshidi, ed. *Systems of Systems Engineering: Principles and Applications*. CRC Press, 2008.

M. Maier and E. Rechtin. *The Art of Systems Architecting*. CRC Press, 2009.

A. Sage and S. Biemer. Processes for system family architecting, design and integration. *IEEE Systems Journal*, 2007, 1(1), 5–16.

A. Sage and C. Cuppan. On the systems engineering and management of systems of systems and federations of systems. *Information Knowledge Systems Management*, 2001, 2(4), 325–345.

4

THE SYSTEM DEVELOPMENT PROCESS

4.1 SYSTEMS ENGINEERING THROUGH THE SYSTEM LIFE CYCLE

As was described in Chapter 1, modern engineered systems come into being in response to societal needs or because of new opportunities offered by advancing technology, or both. The evolution of a particular new system from the time when a need for it is recognized and a feasible technical approach is identified, through its development and introduction into operational use, is a complex effort, which will be referred to as the *system development process*. This chapter is devoted to describing the basic system development process and how systems engineering is applied at each step of this process.

A typical major system development exhibits the following characteristics:

- It is a complex effort.
- It meets an important user need.
- It usually requires several years to complete.
- It is made up of many interrelated tasks.

Systems Engineering Principles and Practice, Second Edition. Alexander Kossiakoff, William N. Sweet, Samuel J. Seymour, and Steven M. Biemer
© 2011 by John Wiley & Sons, Inc. Published 2011 by John Wiley & Sons, Inc.

- It involves several different disciplines.
- It is usually performed by several organizations.
- It has a specific schedule and budget.

The development and introduction into the use of a complex system inherently requires increasingly large commitments of resources as it progresses from concept through engineering, production, and operational use. Further, the introduction of new technology inevitably involves risks, which must be identified and resolved as early as possible. These factors require that the system development be conducted in a step-by-step manner, in which the success of each step is demonstrated, and the basis for the next one validated, before a decision is made to proceed to the next step.

4.2 SYSTEM LIFE CYCLE

The term "system life cycle" is commonly used to refer to the stepwise evolution of a new system from concept through development and on to production, operation, and ultimate disposal. As the type of work evolves from analysis in the early conceptual phases to engineering development and testing, to production and operational use, the role of systems engineering changes accordingly. As noted previously, the organization of this book is designed to follow the structure of the system life cycle, so as to more clearly relate systems engineering functions to their roles in specific periods during the life of the system. This chapter presents an overview of the system development process to create a context for the more detailed discussion of each step in the later chapters.

Development of a Systems Engineering Life Cycle Model for This Book

System life cycle models have evolved significantly over the past two decades. Furthermore, the number of models has grown as additional unique and custom applications were explored. Additionally, software engineering has spawned a significant number of development models that have been adopted by the systems community. The end result is that there is no single life cycle model that (1) is accepted worldwide and (2) fits every possible situation. Various standards organizations, government agencies, and engineering communities have published their particular models or frameworks that can be used to construct a model. Therefore, adopting one model to serve as an appropriate framework for this book was simply not prudent.

Fortunately, all life cycle models subdivide the system life into a set of basic steps that separate major decision milestones. Therefore, the derivation of a life cycle model to serve as an appropriate framework for this book had to meet two primary objectives. First, the steps in the life cycle had to correspond to the progressive transitions in the principal systems engineering activities. Second, these steps had to be capable of being mapped into the principal life cycle models in use by the systems engineering community. The derived model will be referred to as the "systems engineering life cycle,"

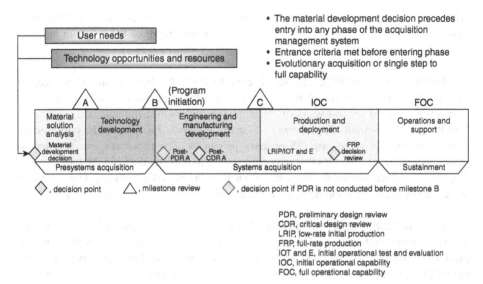

Figure 4.1. DoD system life cycle model.

and will be based on three different sources: the Department of Defense (DoD) Acquisition Management model (DoD 5000.2), the International model ISO/IEC 15288, and the National Society of Professional Engineers (NSPE) model.

DoD Acquisition Management Model. In the second half of the twentieth century, the United States was in the forefront of developing large-scale complex military systems such as warships, airplanes, tanks, and command and control systems. To manage the risks in the application of advanced technology, and to minimize costly technical or management failures, the DoD has evolved comprehensive system acquisition guidelines, which are contained in the DoD 5000 series of directives. The fall 2008 version of the DoD life cycle model, which reflects the acquisition guidelines, is displayed in Figure 4.1. It consists of five phases: material solution analysis, technology development, engineering and manufacturing development, production and deployment, and operations and support. The two activities of user need determination and technology opportunities and resources are considered to be part of the process but are not included in the formal portion of the acquisition cycle.

The DoD model is tailored toward managing large, complex system development efforts where reviews and decisions are needed at key events throughout the life cycle. The major reviews are referred to as milestones and are given letter designations: A, B, and C. Each of the three major milestones is defined with respect to entry and exit conditions. For example, at milestone A, a requirements document needs to be approved by a military oversight committee before a program will be allowed to transition to the next phase. In addition to milestones, the process contains four additional decision points: material development decision (MDD), preliminary design review (PDR), critical design review (CDR) and full-rate production (FRP) decision review. Therefore, DoD management is able to review and decide on the future of the program at up to seven major points within the life cycle.

International ISO/IEC 15288 Model. In 2002, the International Organization for Standardization (ISO) and the International Electrotechnical Commission (IEC) issued the result of several years of effort: a systems engineering standard designated ISO/IEC 15288, *Systems Engineering—System Life Cycle Processes*. The basic model is divided into six stages and 25 primary processes. The processes are intended to represent a menu of activities that may need to be accomplished within the basic stages. The ISO standard purposely does not align the stages and processes. The six basic stages are concept, development, production, utilization, support, and retirement.

Professional Engineering Model. The NSPE model is tailored to the development of commercial systems. This model is mainly directed to the development of new products, usually resulting from technological advances ("technology driven"). Thus, the NSPE model provides a useful alternative view to the DoD model of how a typical system life cycle may be divided into phases. The NSPE life cycle is partitioned into six stages: conceptual, technical feasibility, development, commercial validation and production preparation, full-scale production, and product support.

Systems Engineering Life Cycle Model. In structuring a life cycle model that corresponded to significant transitions in systems engineering activities throughout the system's active life, it was found most desirable to subdivide the life cycle into three broad stages and to partition these into eight distinct phases. This structure is shown in Figure 4.2 and will be discussed below. The names of these subdivisions were chosen

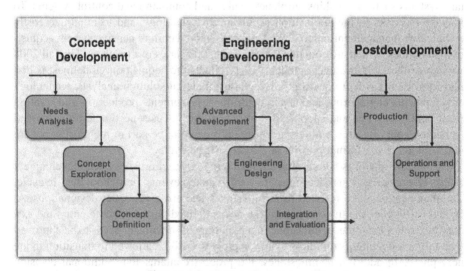

Figure 4.2. System life cycle model.

to reflect the primary activities occurring in each part of the process. Inevitably, some of these names are the same or similar to the names of corresponding parts of one or more of the existing life cycles.

Software Life Cycle Models. The system life cycle stages and their constituent phases represented by the above models apply to the majority of complex systems, including those containing significant software functionality at the component level. However, software-intensive systems, in which software performs virtually all the functionality, as in modern financial systems, airline reservation systems, the World Wide Web, and other information systems, generally follow life cycles similar in form but often involving iteration and prototyping. Chapter 11 describes the differences between software and hardware, discusses the activities involved in the principal stages of software system development, and contains a section dealing with examples of software system life cycles representing software-intensive systems. However, with that exception, the systems engineering life cycle model, as will be discussed in Chapters 5 through 15, provides a natural framework for describing the evolution of systems engineering activity throughout the active life of all engineered complex systems.

Systems Engineering Life Cycle Stages

As described above, and illustrated in Figure 4.2, the systems life cycle model consists of three stages, the first two encompassing the developmental part of the life cycle, and the third the postdevelopment period. These stages mark the more basic transitions in the system life cycle, as well as the changes in the type and scope of effort involved in systems engineering. In this book, these stages will be referred to as (1) The *concept development* stage, which is the initial stage of the formulation and definition of a system concept perceived to best satisfy a valid need; (2) the *engineering development* stage, which covers the translation of the system concept into a validated physical system design meeting the operational, cost, and schedule requirements; and (3) the *postdevelopment* stage, which includes the production, deployment, operation, and support of the system throughout its useful life. The names for the individual stages are intended to correspond generally to the principal type of activity characteristic of these stages.

The concept development stage, as the name implies, embodies the analysis and planning that is necessary to establish the need for a new system, the feasibility of its realization, and the specific system architecture perceived to best satisfy the user needs. Systems engineering plays the lead role in translating the operational needs into a technically and economically feasible system concept. Maier and Rechtin (2009) call this process "systems architecting," using the analogy of the building architect translating a client's needs into plans and specifications that a builder can bid on and build from. The level of effort during this stage is generally much smaller than in subsequent stages. This stage corresponds to the DoD activities of material solution analysis and technology development.

The principal objectives of the concept development stage are

1. to establish that there is a valid need (and market) for a new system that is technically and economically feasible;
2. to explore potential system concepts and formulate and validate a set of system performance requirements;
3. to select the most attractive system concept, define its functional characteristics, and develop a detailed plan for the subsequent stages of engineering, production, and operational deployment of the system; and
4. to develop any new technology called for by the selected system concept and to validate its capability to meet requirements.

The engineering development stage corresponds to the process of engineering the system to perform the functions specified in the system concept, in a physical embodiment that can be produced economically and maintained and operated successfully in its operational environment. Systems engineering is primarily concerned with guiding the engineering development and design, defining and managing interfaces, developing test plans, and determining how discrepancies in system performance uncovered during test and evaluation (T&E) should best be rectified. The main bulk of the engineering effort is carried out during this stage. The engineering development stage corresponds to the DoD activities of engineering and manufacturing development and is a part of production and deployment.

The principal objectives of the engineering development stage are

1. to perform the engineering development of a prototype system satisfying the requirements of performance, reliability, maintainability, and safety; and
2. to engineer the system for economical production and use and to demonstrate its operational suitability.

The postdevelopment stage consists of activities beyond the system development period but still requires significant support from systems engineering, especially when unanticipated problems requiring urgent resolution are encountered. Also, continuing advances in technology often require in-service system upgrading, which may be just as dependent on systems engineering as the concept and engineering development stages. This stage corresponds to a part of the DoD production and deployment phase and all of the operations and support phase.

The postdevelopment stage of a new system begins after the system successfully undergoes its operational T&E, sometimes referred to as *acceptance testing*, and is released for production and subsequent operational use. While the basic development has been completed, systems engineering continues to play an important supporting role in this effort.

The relations among the principal stages in the system life cycle are illustrated in the form of a flowchart in Figure 4.3. The figure shows the principal inputs and outputs of each of the stages. The legends above the blocks relate to the flow of information

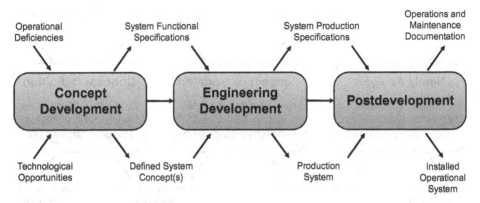

Figure 4.3. Principal stages in a system life cycle.

in the form of requirements, specifications, and documentation, beginning with operational needs. The inputs and outputs below the blocks represent the stepwise evolution of the design representations of an engineered system from the concept to the operational system. It is seen that both the documentation and design representations become increasingly complete and specific as the life cycle progresses. The later section entitled "System Materialization" is devoted to a discussion of the factors involved in this process.

Example: Development Stages of a New Commercial Aircraft. To illustrate the application of this life cycle model, consider the evolution of a new passenger aircraft. The concept development stage would include the recognition of a market for a new aircraft, the exploration of possible configurations, such as number, size, and location of engines, body dimensions, wing platform, and so on, leading to the selection of the optimum configuration from the standpoint of production cost, overall efficiency, passenger comfort, and other operational objectives. The above decisions would be based largely on analyses, simulations, and functional designs, which collectively would constitute justifications for selecting the chosen approach.

The engineering development stage of the aircraft life cycle begins with the acceptance of the proposed system concept and a decision by the aircraft company to proceed with its engineering. The engineering effort would be directed to validating the use of any unproven technology, implementing the system functional design into hardware and software components, and demonstrating that the engineered system meets the user needs. This would involve building prototype components, integrating them into an operating system and evaluating it in a realistic operational environment. The postdevelopment stage includes the acquisition of production tooling and test equipment, production of the new aircraft, customizing it to fit requirements of different customers, supporting regular operations, fixing any faults discovered during use, and periodically overhauling or replacing engines, landing gear, and other highly stressed components. Systems engineering plays a limited but vital supporting and problem-solving role during this stage.

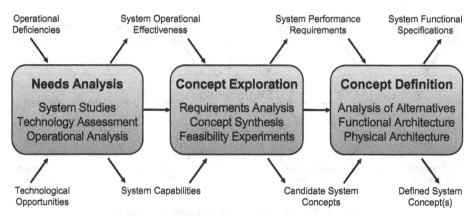

Figure 4.4. Concept development phases of a system life cycle.

Concept Development Phases

While the three stages described above constitute the dominant subdivisions of the system life cycle, each of these stages contains recognizable subdivisions with characteristically different objectives and activities. In the case of large programs, formal decision points also mark most of these subdivisions, similar to those marking the transition between stages. Furthermore, the roles of systems engineering tend to differ significantly among these intermediate subdivisions. Hence, to understand how the evolution of the system life cycle relates to the systems engineering process, it is useful to develop a model of its structure down to this second level of subdivision.

The concept development stage of the systems engineering life cycle encompasses three phases: *needs analysis*, *concept exploration*, and *concept definition*. Figure 4.4 shows these phases, their principal activities and inputs and outputs in a format analogous to Figure 4.3.

Needs Analysis Phase. The needs analysis phase defines the need for a new system. It addresses the questions "Is there a valid need for a new system?" and "Is there a practical approach to satisfying such a need?" These questions require a critical examination of the degree to which current and perceived future needs cannot be satisfied by a physical or operational modification of available means, as well as whether or not available technology is likely to support the increased capability desired. In many cases, the beginning of the life of a new system evolves from a continuing analysis of operational needs, or an innovative product development, without a sharply identified beginning.

The output of this phase is a description of the capabilities and operational effectiveness needed in the new system. In many ways, this description is the first iteration of the system itself, albeit a very basic conceptual model of the system. The reader should take note of how the "system" evolves from this very beginning phase throughout its life cycle. Although we would not yet call this description a set of requirements, they certainly are the forerunner of what will be defined as official requirements. Some communities refer to this early description as an initial capability description.

Several classes of tools and practices exist to support the development of the system capabilities and effectiveness description. Most fall into two categories of mathematics, known as operational analysis and operations research. However, technology assessments and experimentation are an integral part of this phase and will be used in conjunction with mathematical techniques.

Concept Exploration Phase. This phase examines potential system concepts in answering the questions "What performance is required of the new system to meet the perceived need?" and "Is there at least one feasible approach to achieving such performance at an affordable cost?" Positive answers to these questions set a valid and achievable goal for a new system project prior to expending a major effort on its development.

The output of this phase includes our first "official" set of requirements, typically known as system performance requirements. What we mean by official is that a contractor or agency can be measured against this set of required capabilities and performance. In addition to an initial set of requirements, this phase produces a set of candidate system concepts. Note the plural—more than one alternative is important to explore and understand the range of possibilities in satisfying the need.

A variety of tools and techniques are available in this phase and range from process methods (e.g., requirements analysis) to mathematically based (e.g., decision support methods) to expert judgment (e.g., brainstorming). Initially, the number of concepts can be quite large from some of these techniques; however, the set quickly reduces to a manageable set of alternatives. It is important to understand and "prove" the feasibility of the final set of concepts that will become the input of the next phase.

Concept Definition Phase. The concept definition phase selects the preferred concept. It answers the question "What are the key characteristics of a system concept that would achieve the most beneficial balance between capability, operational life, and cost?" To answer this question, a number of alternative concepts must be considered, and their relative performance, operational utility, development risk, and cost must be compared. Given a satisfactory answer to this question, a decision to commit major resources to the development of the new system can be made.

The output is really two perspectives on the same system: a set of functional specifications that describe what the system must do, and how well, and a selected system concept. The latter can be in two forms. If the complexity of the system is rather low, a simple concept description is sufficient to communicate the overall design strategy for the development effort to come. However, if the complexity is high, a simple concept description is insufficient and a more comprehensive system architecture is needed to communicate the various perspectives of the system. Regardless of the depth of description, the concept needs to be described in several ways, primarily from a functional perspective and from a physical perspective. Further perspectives may very well be needed if complexity is particularly high.

The tools and techniques available fall into two categories: analysis of alternatives (a particular method pioneered by the DoD, but fully part of operations research), and systems architecting (pioneered by Ebbert Rechtin in the early 1990s).

As noted previously, in commercial projects (NSPE model), the first two phases are often considered as a single preproject effort. This is sometimes referred to as a "feasibility study" and its results constitute a basis for making a decision as to whether or not to invest in a concept definition effort. In the defense acquisition life cycle, the second and third phases are combined, but the part corresponding to the second phase is performed by the government, resulting in a set of system performance requirements, while that corresponding to the third can be conducted by a government–contractor team or performed by several contractors competing to meet the above requirements.

In any case, before reaching the engineering development stage, only a fractional investment has usually been made in the development of a particular system, although some years and considerable effort may have been spent in developing a firm understanding of the operational environment and in exploring relevant technology at the subsystem level. The ensuing stages are where the bulk of the investment will be required.

Engineering Development Phases

Figure 4.5 shows the activities, inputs, and outputs of the constituent phases of the engineering development stage of the system life cycle in the same format as used in Figure 4.3. These are referred to as *advanced development, engineering design,* and *integration and evaluation.*

Advanced Development Phase. The success of the engineering development stage of a system project is critically dependent on the soundness of the foundation laid

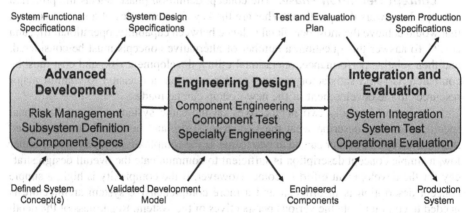

Figure 4.5. Engineering development phases in a system life cycle.

during the concept development stage. However, since the conceptual effort is largely analytical in nature and carried out with limited resources, significant unknowns invariably remain that are yet to be fully defined and resolved. It is essential that these "unknown unknowns" be exposed and addressed early in the engineering stage. In particular, every effort must be made to minimize the number of as yet undisclosed problems prior to translating the functional design and associated system requirements into engineering specifications for the individual system hardware and software elements.

The advanced development phase has two primary purposes: (1) the identification and reduction of development risks and (2) the development of system design specifications. The advanced development phase is especially important when the system concept involves advanced technology not previously used in a similar application, or where the required performance stresses the system components beyond proven limits. It is devoted to designing and demonstrating the undeveloped parts of the system, to proving the practicality of meeting their requirements, and to laying the basis for converting the functional system requirements into system specifications and component design requirements. Systems engineering is central to the decisions of what needs to be validated and how, and to the interpretation of the results.

This phase corresponds to the defense acquisition phase called "engineering and manufacturing development," once referred to as "demonstration and validation." When the risks of using unproven technology are large, this phase is often contracted separately, with contracts for the remaining engineering phase contingent on its success.

Matching the purpose of this phase, the two primary outputs are the design specifications and a validated development model. The specifications are a refinement and evolution of the earlier function specifications. The development model is the final outcome of a very comprehensive risk management task—where those unknowns mentioned above have been identified and resolved. This is what we mean when we use the adjective "validated." The systems engineer needs to be convinced that this system can be designed and manufactured before transitioning from this phase. Therefore, all risks at this phase must be rated as manageable before proceeding.

Modern risk management tools and techniques are essential to reduce and ultimately to mitigate risks inherent in the program. As these risks are managed, the level of definition continues to migrate down, from the system to the subsystem. Furthermore, a set of specifications for the next level of decomposition, at the component level, occurs. In all of these cases, both experimental models and simulations are often employed at this stage to validate component and subsystem design concepts at minimum cost.

Engineering Design Phase. The detailed engineering design of the system is performed during this phase. Because of the scale of this effort, it is usually punctuated by formal design reviews. An important function of these reviews is to provide an opportunity for the customer or user to obtain an early view of the product, to monitor its cost and schedule, and to provide valuable feedback to the system developer.

While issues of reliability, producibility, maintainability, and other "ilities" have been considered in previous phases, they are of paramount importance in the engineering design phase. These types of issues are typically known as "specialty engineering." Since the product consists of a set of components capable of being integrated and tested as a system, the systems engineer is responsible for ensuring that the engineering design of the individual components faithfully implements the functional and compatibility requirements, and for managing the engineering change process to maintain interface and configuration control.

The tasks of this phase deals with converting the component specifications into a set of component designs. Of course, testing these components is essential to occur immediately after design, or in some cases, concurrently with design. One additional task is performed during this phase: the refinement of the system T&E plan. We use the term *refinement* to distinguish between the initiation and continuation. The T&E plan is initially developed much earlier in the life cycle. At this phase, the T&E plan is largely finished, using the knowledge gained from the previous phases.

The two primary outputs are the T&E plan and an engineered prototype. The prototype can take many forms and should not be thought of in the same way as we think of a software prototype. This phase may produce a prototype that is virtual, physical, or a hybrid, depending on the program. For example, if the system is an ocean-going cargo vessel, the prototype at this stage may be a hybrid of virtual and physical mockups. A full-scale prototype of a cargo ship may not be possible or prudent at this phase. On the other hand, if the system is a washing machine, a full-scale prototype may be totally appropriate.

Modern computer-aided design tools are available as design engineers perform their trade. System models and simulations are also updated as designs are finalized and tested.

Integration and Evaluation Phase. The process of integrating the engineered components of a complex system into a functioning whole, and evaluating the system's operation in a realistic environment, is nominally part of the engineering design process because there is no formal break in the development program at this point. However, there is a basic difference between the role and responsibility of systems engineering during the engineering design of the system elements and that during the integration and evaluation process. Since this book is focused on the functions of systems engineering, the system integration and evaluation process is treated as a separate phase in the system life cycle.

It is important to realize that the first time a new system can be assembled and evaluated as an operating unit is after all its components are fully engineered and built. It is at this stage that all the component interfaces must fit and component interactions must be compatible with the functional requirements. While there may have been prior tests at the subsystem level or at the level of a development prototype, the integrity of the total design cannot be validated prior to this point.

It should also be noted that the system integration and evaluation process often requires the design and construction of complex facilities to closely simulate operational stimuli and constraints and to measure the system's responses. Some of these facilities may be adapted from developmental equipment, but the magnitude of the task should not be underestimated.

The outputs of this phase are twofold: (1) the specifications to guide the manufacturing of the system, typically called the system production specifications (sometimes referred to as the production baseline), and (2) the production system itself. The latter includes everything necessary to manufacture and assemble the system and may include a prototype system.

Modern integration techniques and T&E tools, methods, facilities, and principles are available to assist and enable the engineers in these tasks. Of course, before full-scale production can occur, the final production system needs to be verified and validated through an evaluation within the operational environment or a sufficient surrogate for the operational environment.

Postdevelopment Phases

Production Phase. The production phase is the first of the two phases comprising the postdevelopment stage, which are exactly parallel to the defense acquisition phases of "production and deployment" and "operations and support."

No matter how effectively the system design has been engineered for production, problems inevitably arise during the production process. There are always unexpected disruptions beyond the control of project management, for example, a strike at a vendor's plant, unanticipated tooling difficulties, bugs in critical software programs, or an unexpected failure in a factory integration test. Such situations threaten costly disruptions in the production schedule that require prompt and decisive remedial action. Systems engineers are often the only persons qualified to diagnose the source of the problem and to find an effective solution. Often a systems engineer can devise a "work-around" that solves the problem for a minimal cost. This means that an experienced cadre of systems engineers intimately familiar with the system design and operation needs to be available to support the production effort. Where specialty engineering assistance may be required, the systems engineers are often best qualified to decide who should be called in and when.

Operations and Support Phase. In the operations and support phase, there is an even more critical need for systems engineering support. The system operators and maintenance personnel are likely to be only partially trained in the finer details of system operation and upkeep. While specially trained field engineers generally provide support, they must be able to call on experienced systems engineers in case they encounter problems beyond their own experience.

Proper planning for the operational phase includes provision of a logistic support system and training programs for operators and maintenance personnel. This planning should have major participation from systems engineering. There are always unanticipated problems that arise after the system becomes operational that must be recognized and included in the logistic and training systems. Very often, the instrumentation required for training and maintenance is itself a major component of the system to be delivered.

Most complex systems have lifetimes of many years, during which they undergo a number of minor and major upgrades. These upgrades are driven by evolution in the system mission, as well as by advances in technology that offer opportunities to improve operation, reliability, or economy. Computer-based systems are especially subject to periodic upgrades, whose cumulative magnitude may well exceed the initial system development. While the magnitude of an individual system upgrade is a fraction of that required to develop a new system, it usually entails a great many complex decisions requiring the application of systems engineering. Such an enterprise can be extremely complex, especially in the conceptual stage of the upgrade effort. Anyone that has undergone a significant home alteration, such as the addition of one bedroom and bath, will appreciate the unexpected difficulty of deciding just how this can be accomplished in such a way as to retain the character of the original structure and yet realize the full benefits of the added portion, as well as be performed for an affordable price.

4.3 EVOLUTIONARY CHARACTERISTICS OF THE DEVELOPMENT PROCESS

The nature of the system development process can be better understood by considering certain characteristics that evolve during the life cycle. Four of these are described in the paragraphs below. The section The Predecessor System discusses the contributions of an existing system on the development of a new system that is to replace it. The section System Materialization describes a model of how a system evolves from concept to an engineered product. The section The Participants describes the composition of the system development team and how it changes during the life cycle. The section System Requirements and Specifications describes how the definition of the system evolves in terms of system requirements and specifications as the development progresses.

The Predecessor System

The process of engineering a new system may be described without regard to its resemblance to current systems meeting the same or similar needs. The entire concept and all of its elements are often represented as starting with a blank slate, a situation that is virtually never encountered in practice.

In the majority of cases, when new technology is used to achieve radical changes in such operations as transportation, banking, or armed combat, there exist predecessor systems. In a new system, the changes are typically confined to a few subsystems, while the existing overall system architecture and other subsystems remain substantially unchanged. Even the introduction of automation usually changes the mechanics but not the substance of the process. Thus, with the exception of such breakthroughs as the first generation of nuclear systems or of spacecraft, a new system development can expect to have a predecessor system that can serve as a point of departure.

A predecessor system will impact the development of a new system in three ways:

1. The deficiencies of the predecessor system are usually recognized, often being the driving force for the new development. This focuses attention on the most

important performance capabilities and features that must be provided by the new system.

2. If the deficiencies are not so serious as to make the current system worthless, its overall concept and functional architecture may constitute the best starting point for exploring alternatives.

3. To the extent that substantial portions of the current system perform their function satisfactorily and are not rendered obsolete by recent technology, great cost savings (and risk reduction) may be achieved by utilizing them with minimum change.

Given the above, the average system development will almost always be a hybrid, in that it will combine new and undemonstrated components and subsystems with previously engineered and fully proven ones. It is a particular responsibility of systems engineering to ensure that the decisions as to which predecessor elements to use, which to reengineer, which to replace by new ones, and how these are to be interfaced are made through careful weighing of performance, cost, schedule, and other essential criteria.

System Materialization

The steps in the development of a new system can be thought of as an orderly progressive "materialization" of the system from an abstract need to an assemblage of actual components cooperating to perform a set of complex functions to fulfill that need. To illustrate this process, Table 4.1 traces the growth of materialization throughout the phases of the project life cycle. The rows of the table represent the levels of system subdivision, from the system itself at the top to the part level at the bottom. The columns are successive phases of the system life cycle. The entries are the primary activities at each system level and phase, and their degree of materialization. The shaded areas indicate the focus of the principal effort in each phase.

It is seen that each successive phase defines (materializes) the next lower level of system subdivision until every part has been fully defined. Examining each row from left to right, say, at the component level, it is also seen that the process of definition starts with visualization (selecting the general type of system element), then proceeds to defining its functions (functional design, what it must do), and then to its implementation (detailed design, how it will do it).

The above progression holds true through the engineering design phase, where the components of the system are fully "materialized" as finished system building blocks. In the integration and evaluation phase, the materialization process takes place in a distinctly different way, namely, in terms of the materialization of an integrated and validated operational system from its individual building blocks. These differences are discussed further in Chapter 13.

It is important to note from Table 4.1 that while the detailed design of the system is not completed until near the end of its development, its general characteristics must be visualized very early in the process. This can be understood from the fact that the selection of the specific system concept requires a realistic estimate of the cost to

TABLE 4.1. Evolution of System Materialization through the System Life Cycle

	Phase					
	Concept development			Engineering development		
Level	Needs analysis	Concept exploration	Concept definition	Advanced development	Engineering design	Integration and evaluation
System	Define system capabilities and effectiveness	Identify, explore, and synthesize concepts	Define selected concept with specifications	Validate concept		Test and evaluate
Subsystem		Define requirements and ensure feasibility	Define functional and physical architecture	Validate subsystems		Integrate and test
Component			Allocate functions to components	Define specifications	Design and test	Integrate and test
Subcomponent	Visualize			Allocate functions to subcomponents	Design	
Part					Make or buy	

develop and produce it, which in turn requires a visualization of its general physical implementation as well as its functionality. In fact, it is essential to have at least a general vision of the physical embodiment of the system functions during even the earliest investigations of technical feasibility. It is of course true that these early visualizations of the system will differ in many respects from its final materialization, but not so far as to invalidate conclusions about its practicality.

The role of systems architecting fulfills this visualization requirement by providing visual perspectives into the system concept early in the life cycle. As a system project progresses through its life cycle, the products of the architecture are decomposed to ever-lower levels.

At any point in the cycle, the current state of system definition can be thought of as the current system model. Thus, during the concept development stage, the system model includes only the system functional model that is made up entirely of descriptive material, diagrams, tables of parameters, and so on, in combination with any simulations that are used to examine the relationships between system-level performance and specific features and capabilities of individual system elements. Then, during the engineering development stage, this model is augmented by the gradual addition of hardware and software designs for the individual subsystems and components, leading finally to a completed engineering model. The model is then further extended to a production model as the engineering design is transformed into producible hardware designs, detailed software definition, production tooling, and so on. At every stage of the process, the current system model necessarily includes models of all externally imposed interfaces as well as the internal system interfaces.

The Participants

A large project involves not only dozens or hundreds of people but also several different organizational entities. The ultimate user may or may not be an active participant in the project but plays a vital part in the system's origin and in its operational life. The two most common situations are when (1) the government serves as the system acquisition agent and user, with a commercial prime contractor supported by subcontractors as the system developer and producer, and (2) a commercial company serves as the acquisition manager, system developer, and producer. Other commercial companies or the general public may be the users. The principal participants in each phase of the project are also different. Therefore, one of the main functions of systems engineering is to provide the continuity between successive participating levels in the hierarchy and successive development phases and their participants through both formal documentation and informal communications.

A typical distribution of participants in an aerospace system development is shown in Figure 4.6. The height of the columns represents the relative number of engineering personnel involved. The entries are the predominant types of personnel in each phase. It is seen that, in general, participation varies from phase to phase, with systems engineering providing the main continuity.

The principal participants in the early phases are analysts and architects (system and operations/market). The concept definition phase is usually carried out by an

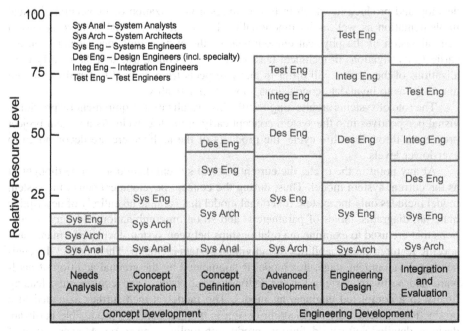

Sys Anal – System Analysts
Sys Arch – System Architects
Sys Eng – Systems Engineers
Des Eng – Design Engineers (incl. specialty)
Integ Eng – Integration Engineers
Test Eng – Test Engineers

Figure 4.6. Principal participants in a typical aerospace system development.

expedited team effort, representing all elements necessary to select and document the most cost-effective system concept for meeting the stated requirements.

The advanced development phase usually marks the initial involvement of the system design team that will carry the project through the engineering stage and on into production. It is led by systems engineering, with support from the design and test engineers engaged in the development of components and subsystems requiring development.

The engineering design phase further augments the effort with a major contribution from specialty engineering (reliability, maintainability, etc.), as well as test and production engineering. For software, this phase involves designers, as well as coders, to the extent that prototyping is employed.

The integration and evaluation phase relies heavily on test engineering with guidance from systems engineering and support from design engineers and engineering specialists.

System Requirements and Specifications

Just as the system design gradually materializes during the successive steps of system development, so the successive forms of system requirements and specifications become more and more specific and detailed. These start with a set of operational requirements and end with a complete set of production specifications, operation, maintenance, and training manuals and all other information needed to replicate, operate, maintain, and repair the system. Thus, each phase can be thought of as producing a more detailed description of the system: what it does, how it works, and how it is built.

Since the above documents collectively determine both the course of the development effort and the form and capabilities of the system as finally delivered, oversight of their definition and preparation is a primary responsibility of systems engineering. This effort must, however, be closely coordinated with the associated design specialists and other involved organizations.

The evolution of system requirements and specifications is shown in the first row of Table 4.2 as a function of the phases in the system life cycle. It should be emphasized that each successive set of documents does not replace the versions defined during the previous phases but rather supplements them. This produces an accumulation rather than a succession of system requirements and other documents. These are "living documents," which are periodically revised and updated.

The necessity for an aggregation of formal requirements and specifications developed during successive phases of the system development can be better understood by recalling the discussion of "Participants" and Figure 4.6. In particular, not only are there many different groups engaged in the development process, but many, if not most, of the key participants change from one phase to the next. This makes it essential that a complete and up-to-date description exists that defines what the system must do and also, to the extent previously defined, how it must do it.

The system description documents not only lay the basis for the next phase of system design but they also specify how the results of the effort are to be tested in order to validate compliance with the requirements. They provide the information base needed for devising both the production tools and the tools to be used for inspecting and testing the product of the forthcoming phase.

The representations of system characteristics also evolve during the development process, as indicated in the second row of Table 4.2. Most of these will be recognized as architecture views and conventional engineering design and software diagrams and models. Their purpose is to supplement textual descriptions of successive stages of system materialization by more readily understandable visual forms. This is especially important in defining interfaces and interactions among system elements designed by different organizations.

4.4 THE SYSTEMS ENGINEERING METHOD

In the preceding sections, the engineering of a complex system was seen to be divisible into a series of steps or phases. Beginning with the identification of an opportunity to achieve a major extension of an important operational capability by a feasible technological approach, each succeeding phase adds a further level of detailed definition (materialization) of the system, until a fully engineered model is achieved that proves to meet all essential operational requirements reliably and at an affordable cost. While many of the problems addressed in a given phase are peculiar to that state of system definition, the systems engineering principles that are employed, and the relations

TABLE 4.2. Evolution of System Representation

	Concept development			Engineering development		
	Needs analysis	Concept exploration	Concept definition	Advanced development	Engineering design	Integration and evaluation
Documents	System capabilities and effectiveness	System performance requirements	System functional requirements	System design specifications	Design documents	Test plans and evaluation reports
System models	Operational diagrams, mission simulations	System diagrams, high-level system simulations	Architecture products and views, simulations, mock-ups	Architecture products and views, detailed simulations, breadboards	Architecture drawings and schematics, engineered components, computer-aided design (CAD) products	Test setups, simulators, facilities, and test articles

among them, are fundamentally similar from one phase to the next. This fact, and its importance in understanding the system development process, has been generally recognized; the set of activities that tends to repeat from one phase to the next has been referred to in various publications on systems engineering as the "systems engineering process," or the "systems engineering approach," and is the subject of the sections below. In this book, this iterative set of activities will be referred to as the "systems engineering method."

The reason for selecting the word "method" in place of the more widely used "process" or "approach" is that it is more definitive and less ambiguous. The word method is more specific than process, having the connotation of an orderly and logical process. Furthermore, the term systems engineering process is sometimes used to apply to the total system development. Method is also more appropriate than approach, which connotes an attitude rather than a process. With all this said, the use of a more common terminology is perfectly acceptable.

Survey of Existing Systems Engineering Methods and Processes

The first organization to codify a formal systems engineering process was the U.S. DoD, captured in the military standard, MIL-STD-498. Although the process evolved through several iterations, the last formal standard to exist (before being discontinued) was MIL-STD-499B. This process is depicted in Figure 4.7 and contains four major activities: requirements analysis, functional analysis and allocation, synthesis, and systems analysis and control. The component tasks are presented within each activity.

While this military standard is no longer in force, it is still used as a guide by many organizations and is the foundation for understanding the basics of today's systems engineering processes.

Three relevant commercial standards describe a systems engineering process: IEEE-1220, the EIA-STD-632, and the ISO-IEC-IEEE-STD-15288. As these three processes are presented, notice that each commercial standard blends aspects of a systems engineering process with the life cycle model describe above. The order that we present these three methods is important—they are presented in order of the level of convergence with the life cycle model of system development. And in fact, the military standard discussed above could be placed first in the sequence. In other words, MIL-STD-499B is the most divergent from the life cycle model. In contrast, ISO-15288 could easily be thought of as a life cycle model for system development.

Figure 4.8 presents the IEEE-1220 process. The main control activity is located in the middle of the graph. The general flow of activities is then clockwise, starting from the bottom left, beginning with "process inputs" and ending with "process outputs." This process could also be thought of as an expansion of the military standard—the four basic activities are present, with a verification or validation step in between.

Figure 4.9 presents the EIA-632 process. Actually, the EIA-632 standard presents a collection of 13 processes that are linked together. One can easily recognize the iterative and circular nature of these linkages. Although the general flow is top–down, the processes are repeated multiple times throughout the system life cycle.

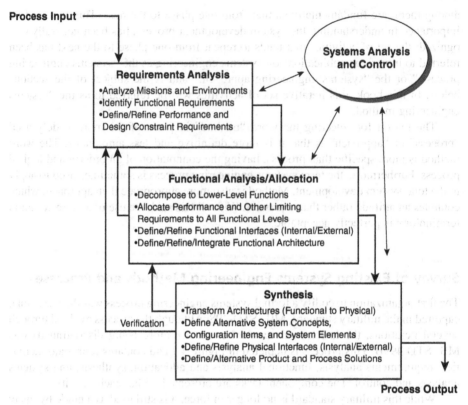

Figure 4.7. DoD MIL-STD499B.

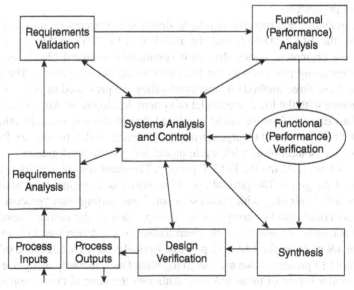

Figure 4.8. IEEE-1220 systems engineering process.

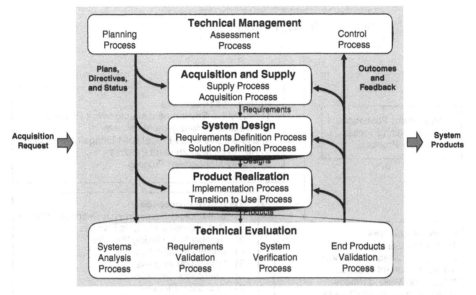

Figure 4.9. EIA-632 systems engineering process.

The 13 processes are further categorized into five sets: technical management, acquisition and supply, system design, product realization, and technical evaluation. The first and last process sets occur almost continuously throughout the system development life cycle. Planning, assessment, and control do not stop after the initial development phases, and systems analysis, requirements validation, system verification, and end-product validation commence well before a physical product is available. The three middle sets occur linearly, but with feedback and iterations.

Figure 4.10 presents the ISO-15288 process. This standard presents processes for both the system life cycle and systems engineering activities. In addition, the philosophy behind this standard is based on the systems engineer's and the program manager's ability to tailor the processes presented into a sequence of activities that is applicable to the program. Thus, no specific method is presented that sequences a subset of processes.

Our Systems Engineering Method

The *systems engineering method* can be thought of as the systematic application of the scientific method to the engineering of a complex system. It can be considered as consisting of four basic activities applied successively, as illustrated in Figure 4.11:

1. requirements analysis,
2. functional definition,
3. physical definition, and
4. design validation.

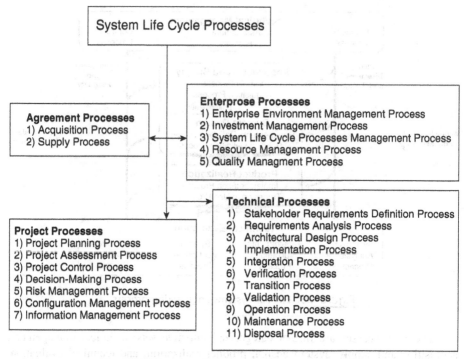

Figure 4.10. ISO-15288 Systems engineering process.

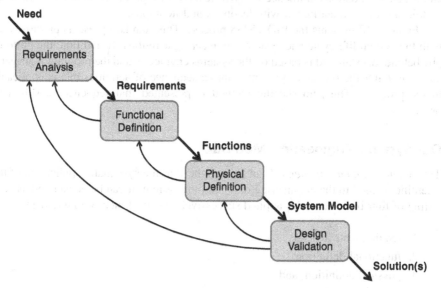

Figure 4.11. Systems engineering method top-level flow diagram.

These steps will vary in their specifics depending on the type of system and the phase of its development. However, there is enough similarity in their operating principles that it is useful to describe the typical activities of each step in the method. Such brief descriptions of the activities in the four steps are listed below.

Requirements Analysis (Problem Definition). Typical activities include

- assembling and organizing all input conditions, including requirements, plans, milestones, and models from the previous phase;
- identifying the "whys" of all requirements in terms of operational needs, constraints, environment, or other higher-level objectives;
- clarifying the requirements of what the system must do, how well it must do it, and what constraints it must fit; and
- correcting inadequacies and quantifying the requirements wherever possible.

Functional Definition (Functional Analysis and Allocation). Typical activities include

- translating requirements (why) into functions (actions and tasks) that the system must accomplish (what),
- partitioning (allocating) requirements into functional building blocks, and
- defining interactions among functional elements to lay a basis for their organization into a modular configuration.

Physical Definition (Synthesis, Physical Analysis, and Allocation). Typical activities include

- synthesizing a number of alternative system components representing a variety of design approaches to implementing the required functions, and having the most simple practicable interactions and interfaces among structural subdivisions;
- selecting a preferred approach by trading off a set of predefined and prioritized criteria (measures of effectiveness [MOE]) to obtain the best "balance" among performance, risk, cost, and schedule; and
- elaborating the design to the necessary level of detail.

Design Validation (Verification and Evaluation). Typical activities include

- designing models of the system environment (logical, mathematical, simulated, and physical) reflecting all significant aspects of the requirements and constraints;
- simulating or testing and analyzing system solution(s) against environmental models; and
- iterating as necessary to revise the system model or environmental models, or to revise system requirements if too stringent for a viable solution until the design and requirements are fully compatible.

The elements of the systems engineering method as described above are displayed in the form of a flow diagram in Figure 4.12, which is an expanded view of Figure

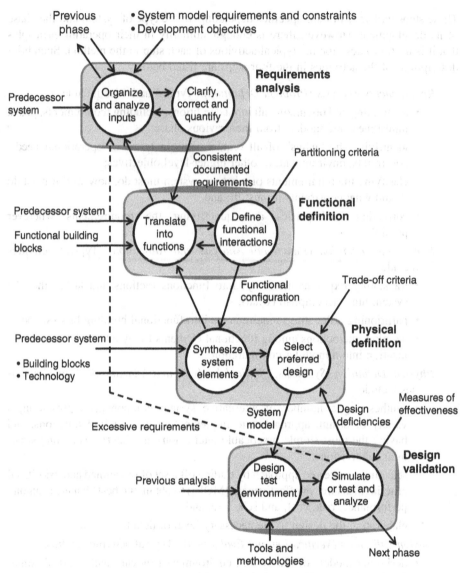

Figure 4.12. Systems engineering method flow diagram.

4.11. The rectangular blocks are seen to represent the above four basic steps in the method: requirements analysis, functional definition, physical definition, and design validation. At the top are shown inputs from the previous phase, which include requirements, constraints, and objectives. At the left of each block are shown external inputs, such as the predecessor system, system building blocks, and previous analyses. At the upper right of the top blocks and at the very bottom are inputs from systems engineering methodology.

The circles inside each block are simplified representations of key processes in that step of the method. The interfacing arrows represent information flow. It is seen that there are feedbacks throughout the process, iteration within the elements as well as to prior elements, and indeed all the way back to the requirements.

Each of the elements of the method is described more fully in the remainder of this section.

Requirements Analysis (Problem Definition)

In attempting to solve any problem, it is first necessary to understand exactly what is given, and to the extent that it appears to be incomplete, inconsistent, or unrealistic, to make appropriate amplifications or corrections. This is particularly essential in the system development process, where a basic characteristic of systems engineering is that everything is not necessarily what it seems and that important assumptions must be verified before they are accepted as being valid.

Thus, in a system development project, it is the responsibility of systems engineering to thoroughly analyze all requirements and specifications, first in order to understand them vis-à-vis the basic needs that the system is intended to satisfy, and then to identify and correct any ambiguities or inconsistencies in the definition of capabilities for the system or system element being addressed.

The specific activities of requirements analysis vary as the system development progresses, as the inputs from the previous phase evolve from operational needs and technological opportunities (see Fig. 4.3) to increasingly specific representations of requirements and system design. The role of systems engineering is essential throughout, but perhaps more so in the early phases, where an understanding of the operational environment and the availability and maturity of applicable technology are most critical. In later phases, environmental, interface, and other interelement requirements are the special province of systems engineering.

Organization and Interpretation. In a well-structured acquisition process, a new phase of the system life cycle begins with three main inputs, which are defined during or upon completion of the previous phase:

1. the system model, which identifies and describes all design choices made and validated in the preceding phases;
2. requirements (or specifications) that define the design, performance, and interface compatibility features of the system or system elements to be developed during the next phase; these requirements are derived from previously developed higher-level requirements, including any refinements and/or revisions introduced during the latest phase; and
3. specific progress to be achieved by each component of the engineering organization during the next phase, including the identification of all technical design data, hardware/software products, and associated test data to be provided; this information is usually presented in the form of a series of interdependent task statements.

Clarification, Correction, and Quantification. It is always difficult to express objectives in unambiguous and quantitative terms, so it is therefore common that stated requirements are often incomplete, inconsistent, and vague. This is especially true if the requirements are prepared by those who are unfamiliar with the process of converting them to system capabilities, or with the origins of the requirements in terms of operational needs. In practice, the completeness and accuracy of these inputs can be expected to vary with the nature of the system, its degree of departure from predecessor systems, the type of acquisition process employed, and the phase itself.

The above analysis must include interaction with the prospective users of the system to gain a first-hand understanding of their needs and constraints and to obtain their inputs where appropriate. The result of the analysis may be modifications and amplifications of the requirements documents so as to better represent the objectives of the program or the availability of proposed technological improvements. The end objective is to create a firm basis from which the nature and location of the design changes needed to meet the requirements may be defined.

Functional Definition (Functional Analysis and Allocation)

In the systems engineering method, functional design precedes physical or product design to ensure a disciplined approach to an effective organization (configuration) of the functions and to the selection of the implementation that best balances the desired characteristics of the system (e.g., performance and cost).

Translation into Functions. The system elements that may serve as functional building blocks are briefly discussed in Chapter 3. The basic building blocks are at the component level representing elements that perform a single significant function and deal with a single medium, that is, either signals, data, material, or energy. They, in turn, consist of subelements performing lower-level functions and aggregate into functional subsystems. Thus, functional design can be thought of as selecting, subdividing, or aggregating functional elements appropriate to the required tasks and level of system materialization (see Table 4.1).

Decomposition and allocation of each iterative set of requirements and functions for implementation at the next lower level of system definition is a prime responsibility of systems engineering. This first takes place during the concept development stage as follow-on to the definition of the system architecture. It includes identification and description of all functions to be provided, along with the associated quantitative requirements to be met by each subsystem, in order that the prescribed system-level capabilities can in fact be achieved. This information is then reflected in *system functional specifications*, which serve as the basis for the follow-on engineering development stage.

As part of the advanced development phase, these top-level subsystem functions and requirements are further allocated to individual system components within each subsystem. This, as noted earlier, is the lowest level in the design hierarchy that is of direct concern to systems engineering, except in special cases where lower-level elements turn out to be critical to the operation of the system.

Trade-Off Analysis. The selection of appropriate functional elements, as all aspects of design, is an inductive process, in which a set of postulated alternatives is examined, and the one judged to be best for the intended purpose is selected. The systems engineering method relies on making design decisions by the use of trade-off analysis. Trade-off analysis is widely used in all types of decision making, but in systems engineering, it is applied in a particularly disciplined form, especially in the step of physical definition. As the name implies, trade-offs involve the comparison of alternatives, which are superior in one or more required characteristics, with those that are superior in others. To ensure that an especially desirable approach is not overlooked, it is necessary to explore a sufficient number of alternative implementations, all defined to a level adequate to enable their characteristics to be evaluated relative to one another. It is also necessary that the evaluation be made relative to a carefully formulated set of criteria or "MOE." Chapters 8 and 9 contain more detailed discussions of trade-off analysis.

Functional Interactions. One of the single most important steps in system design is the definition of the functional and physical interconnection and interfacing of its building blocks. A necessary ingredient in this activity is the early identification of all significant functional interactions and the ways in which the functional elements may be aggregated so as to group strongly interacting elements together and to make the interactions among the groups as simple as possible. Such organizations (architectures) are referred to as "modular" and are the key to system designs that are readily maintainable and capable of being upgraded to extend their useful life. Another essential ingredient is the identification of all external interactions and the interfaces through which they affect the system.

Physical Definition (Synthesis or Physical Analysis and Allocation)

Physical definition is the translation of the functional design into hardware and software components, and the integration of these components into the total system. In the concept development stage, where all design is still at the functional level, it is nevertheless necessary to visualize or imagine what the physical embodiment of the concept would be like in order to help ensure that the solution will be practically realizable. The process of selecting the embodiment to be visualized is also governed by the general principles discussed below, applied more qualitatively than in the engineering development stage.

Synthesis of Alternative System Elements. The implementation of functional design elements requires decisions regarding the specific physical form that the implementation should take. Such decisions include choice of implementation media, element form, arrangement, and interface design. In many instances, they also offer a choice of approaches, ranging from exploiting the latest technology to relying on proven techniques. As in the case of functional design, such decisions are made by the use of trade-off analysis. There usually being more choices of different physical implementations than functional configurations, it is even more important that good systems engineering practice be used in the physical definition process.

Selection of Preferred Approach. At various milestones in the system life cycle, the selection of a preferred approach, or approaches, will need to be made. It is important to understand that this selection process changes depending on the phase within the life cycle. Early phases may require selecting a several approaches to explore, while later phases may require a down-select to a single approach. Additionally, the level of decisions evolves. Early decisions relate to the system as a whole; later decisions focus on subsystems and components.

As stated previously, to make a meaningful choice among design alternatives, it is necessary to define a set of evaluation criteria and to establish their relative priority. Among the most important variables to be considered in the physical definition step is the relative affordability or cost of the alternatives and their relative risk of successful accomplishment. In particular, early focus on one particular implementation concept should be avoided.

Risk as a component of trade-off analysis is basically an estimate of the probability that a given design approach will fail to produce a successful result whether because of deficient performance, low reliability, excessive cost, or unacceptable schedule. If the component risk appears substantial, the risk to the overall project must be reduced (risk abatement) by either initiating an intensive component development effort, by providing a backup using a proven but somewhat less capable component, by modifying the overall technical approach to eliminate the need for the particular component that is in doubt, or, if these fail, by relaxing the related system performance specification. Identifying significantly high-risk system elements and determining how to deal with them are an essential systems engineering responsibility. Chapter 5 discusses the risk management process and its constituent parts.

Proper use of the systems engineering method thus ensures that

1. all viable alternatives are considered;
2. a set of evaluation criteria is established; and
3. the criteria are prioritized and quantified where practicable.

Whether or not it is possible to make quantitative comparisons, the final decision should be tempered by judgment based on experience.

Interface Definition. Implicit in the physical definition step is the definition and control of *interfaces*, both internal and external. Each element added or elaborated in the design process must be properly connected to its neighboring elements and to any external inputs or outputs. Further, as the next lower design level is defined, adjustments to the parent elements will inevitably be required, which must in turn be reflected in adjustments to their previously defined interfaces. All such definitions and readjustments must be incorporated into the model design and interface specifications to form a sound basis for the next level of design.

Design Validation (Verification and Evaluation)

In the development of a complex system, even though the preceding steps of the design definition may have been carried out apparently in full compliance with requirements, there still needs to be an explicit validation of the design before the next phase is undertaken. Experience has shown that there are just too many opportunities for undetected errors to creep in. The form of such validation varies with the phase and degree of system materialization, but the general approach is similar from phase to phase.

Modeling the System Environment. To validate a model of the system, it is necessary to create a model of the environment with which the system can interact to see if it produces the required performance. This task of modeling the system environment extends throughout the system development cycle. In the concept development stage, the model is largely functional, although some parts of it may be physical, as when an experimental version of a critical system component is tested over a range of ambient conditions.

In later stages of development, various aspects of the environment may be reproduced in the laboratory or in a test facility, such as an aerodynamic wind tunnel or inertial test platform. In cases where the model is dynamic, it is more properly called a simulation, in which the system design is subjected to a time-varying input to stimulate its dynamic response modes.

As the development progresses into the engineering development stage, modeling the environment becomes increasingly realistic, and environmental conditions are embodied in system and component test equipment, such as environmental chambers, or shock and vibration facilities. During operational evaluation testing, the environment is, insofar as is practicable, made identical to that in which the system will eventually operate. Here, the model has transitioned into greater reality.

Some environments that are of great significance to system performance and reliability can only be imperfectly understood and are very difficult to simulate, for example, the deep ocean and exoatmospheric space. In such cases, defining and simulating the environment may become a major effort in itself. Even environments that were thought to be relatively well understood can yield surprises, for example, unusual radar signal refraction over the Arabian Desert.

At each step, the system development process requires a successively more detailed definition of the requirements that the system must meet. It is against these environmental requirements that the successive models of the system are evaluated and refined. A lesson to be learned is that the effort required to model the environment of a system for the purpose of system T&E needs to be considered at the same level of priority as the design of the system itself and may even require a separate design effort comparable to the associated system design activity.

Tests and Test Data Analysis. The definitive steps in the validation of the system design are the conduct of tests in which the system model (or a significant portion of it) is made to interact with a model of its environment in such a way that the effects can be measured and analyzed in terms of the system requirements. The scope of such tests evolves with the degree of materialization of the system, beginning with paper calculations and ending with operational tests in the final stages. In each case, the objective is to determine whether or not the results conform to those prescribed by the requirements, and if not, what changes are required to rectify the situation.

In carrying out the above process, it is most important to observe the following key principles:

1. All critical system characteristics need to be stressed beyond their specified limits to uncover incipient weak spots.
2. All key elements need to be instrumented to permit location of the exact sources of deviations in behavior. The instruments must significantly exceed the test articles in precision and reliability.
3. A test plan and an associated test data analysis plan must be prepared to assure that the requisite data are properly collected and are then analyzed as necessary to assure a realistic assessment of system compliance.
4. All limitations in the tests due to unavoidable artificialities need to be explicitly recognized and their effect on the results compensated or corrected for, as far as possible.
5. A formal test report must be prepared to document the degree of compliance by the system and the source of any deficiencies.

The test plan should detail each step in the test procedure and identify exactly *what information* will be recorded prior to, during, and at the conclusion of each test step, as well as *how* and *by whom* it will be recorded. The test data analysis plan should then define how the data would be reduced, analyzed, and reported along with specific criteria that will be employed to demonstrate system compliance.

To the extent that the validation tests reveal deviations from required performance, the following alternatives need to be considered:

1. Can the deviation be due to a deficiency in the environmental simulation (i.e., test equipment)? This can happen because of the difficulty of constructing a realistic model of the environment.
2. Is the deviation due to a deficiency in the design? If so, can it be remedied without extensive modifications to other system elements?
3. Is the requirement at issue overly stringent? If so, a request for a deviation may be considered. This would constitute a type of feedback that is characteristic of the system development process.

Preparation for the Next Phase

Each phase in the system development process produces a further level of requirements or specifications to serve as a basis for the next phase. This adds to, rather than replaces, previous levels of requirements and serves two purposes:

1. It documents the design decisions made in the course of the current phase.
2. It establishes the goals for the succeeding phase.

Concurrent with the requirements analysis and allocation activity, systems engineering, acting in concert with project management, is also responsible for the definition of specific technical objectives to be met, and for the products (e.g., hardware/software components, technical documentation, and supporting test data) that will be provided in response to the stated requirements for inputs to the next phase. These identified end products of each phase are also often accompanied by a set of intermediate technical milestones that can be used to judge technical progress during each particular design activity.

The task of defining these requirements or specifications and the efforts to be undertaken in implementing the related design activities is an essential part of system development. Together, these constitute the official guide for the execution of each phase of the development.

It must be noted, however, that in practice, the realism and effectiveness of this effort, which is so critical to the ultimate success of the project, depends in large part on good communication and cooperation between systems engineering and project management on the one hand, and on the other, the design specialists who are ultimately the best judges of what can and cannot be reasonably accomplished given the stated requirements, available resources, and allotted time scale.

Since the nature of the preparation for the next phase varies widely from phase to phase, it is not usually accorded the status of a separate step in the systems engineering method; most often, it is combined with the validation process. However, this does not diminish its importance because the thoroughness with which it is done directly affects the requirements analysis process at the initiation of the next phase. In any event, the definition of the requirements and tasks to be performed in the next phase serves an important interface function between phases.

Systems Engineering Method over the System Life Cycle

To illustrate how the systems engineering method is applied in successive phases of the system life cycle, Table 4.3 lists the primary focus of each of the four steps of the method for each of the phases of the system life cycle. As indicated earlier in Table 4.1, it is seen that as the phases progress, the focus shifts to more specific and detailed (lower-level) elements of the system until the integration and evaluation phase.

The table also highlights the difference in character of the physical definition and design validation steps in going from the concept development to the engineering development stage. In the concept development stage (left three columns), the defined concepts are still in functional form (except where elements of the previous or other systems are applied without basic change). Accordingly, physical implementation has not yet begun, and design validation is performed by analysis and simulation of the functional elements. In the engineering stage, implementation into hardware and software proceeds to lower and lower levels, and design validation includes tests of experimental, prototype, and finally production system elements and the system itself.

TABLE 4.3. Systems Engineering Method over Life Cycle

	Phase					
	Concept development			Engineering development		
Step	Needs analysis	Concept exploration	Concept definition	Advanced development	Engineering design	Integration and evaluation
Requirements analysis	Analyze needs	Analyze operational requirements	Analyze performance requirements	Analyze functional requirements	Analyze design requirements	Analyze tests and evaluation requirements
Functional definition	Define system objectives	Define subsystem functions	Develop functional architecture component functions	Refine functional architecture subcomponent functions140	Define part functions	Define functional tests
Physical definition	Define system capabilities; visualize subsystems, ID technology	Define system concepts, visualize components	Develop physical architecture components	Refine physical architecture; specify component construction	Specify subcomponent construction	Define physical tests; specify test equipment and facilities
Design validation	Validate needs and feasibility	Validate operational requirements	Evaluate system capabilities	Test and evaluate critical subsystems	Validate component construction	Test and evaluate system

In interpreting both Tables 4.3 and 4.1, it should be borne in mind that in a given phase of system development, some parts of the system might be prototyped to a more advanced phase to validate critical features of the design. This is particularly true in the advanced development phase, where new potentially risky approaches are prototyped and tested under realistic conditions. Normally, new software elements are also prototyped in this phase to validate their basic design.

While these tables present a somewhat idealized picture, the overall pattern of the iterative application of the systems engineering method to successively lower levels of the system is an instructive and valid general view of the process of system development.

Spiral Life Cycle Model

The iterative nature of the system development process, with the successive applications of the systems engineering method to a stepwise materialization of the system has been captured in the so-called spiral model of the system life cycle. A version of this model as applied to life cycle phases is shown in Figure 4.13. The sectors representing the four steps in the systems engineering method defined in the above section are shown separated by heavy radial lines. This model emphasizes that each phase of the development of a complex system necessarily involves an iterative application of the systems engineering method and the continuing review and updating of the work performed and conclusions reached in the prior phases of the effort.

4.5 TESTING THROUGHOUT SYSTEM DEVELOPMENT

Testing and evaluation are not separate functions from design but rather are inherent parts of design. In basic types of design, for example, as of a picture, the function of T&E is performed by the artist as part of the process of transferring a design concept to canvas. To the extent that the painting does not conform to the artist's intent, he or she alters the picture by adding a few brushstrokes, which tailor the visual effect (performance) to match the original objective. Thus, design is a closed-loop process in which T&E constitutes the feedback that adjusts the result to the requirements that it is intended to meet.

Unknowns

In any new system development project, there are a great many unknowns that need to be resolved in the course of producing a successful product. For each significant departure from established practice, the result cannot be predicted with assurance. The project cost depends on a host of factors, none of them known precisely. The resolution of interface incompatibilities often involves design adjustment on both sides of the interface, which frequently leads to unexpected and sometimes major technical difficulties.

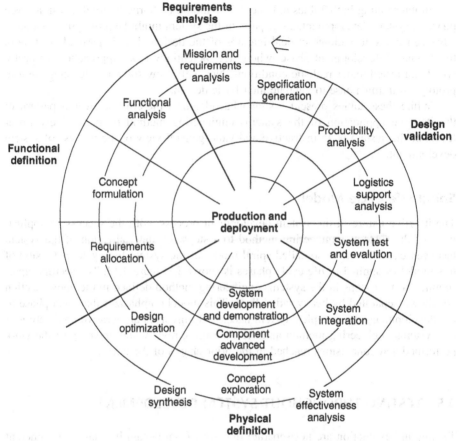

Figure 4.13. Spiral model of the system life cycle.

An essential task of systems engineering is to guide the development of the system so that the unknowns are turned into knowns as early in the process as possible. Any surprises occurring late in the program can prove to be many times more costly than those encountered in its early phases.

Many unknowns are evident at the beginning, and may be called "known unknowns." These are identified early as potential problem areas and are therefore singled out for examination and resolution. Usually, this can be accomplished through a series of critical experiments involving simulations and/or experimental hardware and software. However, many other problem areas are only identified later when they are discovered during system development. These unanticipated problems are often identified as unknown unknowns or "unk-unks" to distinguish them from the group of known unknowns that were recognized at the outset and dealt with before they could seriously impact the overall development process.

Transforming the Unknown into the Known

The existence of unk-unks makes the task of attempting to remove all the unknowns far more difficult. It forces an active search for hidden traps in the favored places of technical problems. It is the task of the systems engineer to lead this search based on experience gathered during previous system developments and supported by a high degree of technical insight and a "What if …?" attitude.

Since every unknown poses an uncertainty in the accomplishment of the final objective, it represents a potential risk. In fact, unknowns present the principal risks in any development program. Hence, the task of risk assessment and integration is one and the same as that of identifying unknowns and resolving them.

The tools for resolving unknowns are analysis, simulation, and test, these being the means for discovering and quantifying critical system characteristics. This effort begins during the earliest conceptual stages and continues throughout the entire development, only changing in substance and character and not in objective and approach.

In designing a new system or a new element of a system that requires an approach never attempted before under the same circumstances (as, e.g., the use of new materials for making a highly stressed design element), the designer faces a number of unknowns regarding the exact manner in which the new design when implemented will perform (e.g., the element made of a new material may not be capable of being formed into the required shape by conventional tools). In such cases, the process of testing serves to reveal whether or not the unknown factors create unanticipated difficulties requiring significant design changes or even abandonment of the approach.

When a new design approach is undertaken, it is unwise to wait until the design is fully implemented before determining whether or not the approach is sound. Instead, testing should first be done on a theoretical or experimental model of the design element, which can be created quickly and at a minimum cost. In doing so, a judgment must be made as to the balance between the potential benefit of a greater degree of realism of the model and the time and cost of achieving it. This is very often a system-level rather than a component-level decision, especially if the performance of the element can have a system impact. If the unknowns are largely in the functional behavior of the element, then a computational model or a simulation is indicated. If, on the other hand, the unknowns are concerned with the material aspects, an experimental model is required.

Systems Engineering Approach to Testing

The systems engineering approach to testing can be illustrated by comparing the respective views of testing by the design engineer, the test engineer, and the systems engineer. The design engineer wants to be sure that a component passes the test, wanting to know, "Is it OK?" The test engineer wants to know that the test is thorough so as to be sure the component is stressed enough. The systems engineer wants to be sure to find and identify all deficiencies present in the component. If the component fails a test, the systems engineer wants to know why, so that there will be a basis for devising changes that will eliminate the deficiency.

It is evident from the above that the emphasis of systems engineering is not only on the test conditions but also on the acquisition of data showing exactly how the various parts of the system did or did not perform. Furthermore, the acquisition of data itself is not enough; it is necessary to have in hand procedures for analyzing the data. These are often complicated and require sophisticated analytical techniques, which must be planned in advance.

It also follows that a systems engineer must be an active participant in the formulation of the test procedures and choice of instrumentation. In fact, the prime initiative for developing the test plan should lie with systems engineering, working in close cooperation with test engineering. To the systems engineer, a test is like an experiment is to a scientist, namely, a means of acquiring critical data on the behavior of the system under controlled circumstances.

System T&E

The most intensive use of testing in the system life cycle takes place in the last phase of system development, integration and evaluation, which is the subject of Chapter 13. Chapter 10 also contains a section on T&E during the advanced development phase.

4.6 SUMMARY

Systems Engineering through the System Life Cycle

A major system development program is an extended complex effort to satisfy an important user need. It involves multiple disciplines and applies new technology, requires progressively increasing commitment of resources, and is conducted in a stepwise manner to a specified schedule and budget.

System Life Cycle

The system life cycle may be divided into three major stages.

Concept Development. Systems engineering establishes the system need, explores feasible concepts, and selects a preferred system concept. The concept development stage may be further subdivided into three phases:

1. *Needs Analysis:* defines and validates the need for a new system, demonstrates its feasibility, and defines system operational requirements;
2. *Concept Exploration:* explores feasible concepts and defines functional performance requirements; and
3. *Concept Definition:* examines alternative concepts, selects the preferred concept on the basis of performance, cost, schedule, and risk, and defines system functional specifications (A-Spec).

Engineering Development. Systems engineering validates new technology, transforms the selected concept into hardware and software designs, and builds and tests production models. The engineering development stage may be further subdivided into three phases:

1. *Advanced Development:* identifies areas of risk, reduces these risks through analysis, development, and test, and defines system development specifications (B-Spec);
2. *Engineering Design:* performs preliminary and final design and builds and tests hardware and software components, for example, configuration items (CIs); and
3. *Integration and Evaluation:* integrates components into a production prototype, evaluates the prototype system, and rectifies deviations.

Postdevelopment. Systems engineering produces and deploys the system and supports system operation and maintenance. The postdevelopment stage is further subdivided into two phases:

1. *Production:* develops tooling and manufactures system products, provides the system to the users, and facilitates initial operations; and
2. *Operations and Support:* supports system operation and maintenance, and develops and supports in-service updates.

Evolutionary Characteristics of the Development Process

Most new systems evolve from predecessor systems—their functional architecture and even some components may be reusable.

A new system progressively "materializes" during its development. System descriptions and designs evolve from concepts to reality. Documents, diagrams, models, and products all change correspondingly. Moreover, key participants in system development change during development; however, systems engineering plays a key role throughout all phases.

The Systems Engineering Method

The systems engineering method involves four basic steps:

1. *Requirements Analysis*—identifies why requirements are needed,
2. *Functional Definition*—translates requirements into functions,
3. *Physical Definition*—synthesizes alternative physical implementations, and
4. *Design Validation*—models the system environment.

These four steps are applied repetitively in each phase during development. Application of the systems engineering method evolves over the life cycle—as the system progressively materializes, the focus shifts from system level during needs analysis down to component and part levels during engineering design.

Testing throughout System Development

Testing is a process to identify unknown design defects in that it verifies resolution of known unknowns and uncovers unknown unknowns (unk-unks) and their causes. Late resolution of unknowns may be extremely costly; therefore, test planning and analysis is a prime systems engineering responsibility.

PROBLEMS

4.1 Identify a recent development (since 2000) of a complex system (commercial or military) of which you have some knowledge. Describe the need it was developed to fill and the principal ways in which it is superior to its predecessor(s). Briefly describe the new conceptual approach and/or technological advances that were employed.

4.2 Advances in technology often lead to the development of a new or improved system by exploiting an advantage not possessed by its predecessor. Name three different types of advantages that an advanced technology may offer and cite an example of each.

4.3 If there is a feasible and attractive concept for satisfying the requirements for a new system, state why it is important to consider other alternatives before deciding which to select for development. Describe some of the possible consequences of failing to do so.

4.4 The space shuttle was an example of an extremely complicated system using leading edge technology. Give three examples of shuttle components that you think represented unproven technology at the time of its development, and which much have required extensive prototyping and testing to reduce operational risks to an acceptable level.

4.5 What steps can the systems engineer take to help ensure that system components designed by different technical groups or contractors will fit together and interact effectively when assembled to make up the total system? Discuss in terms of mechanical, electrical, and software system elements.

4.6 For six of the systems listed in Tables 1.1 and 1.2, list their "predecessor systems." For each, indicate the main characteristics in which the current systems are superior to their predecessors.

4.7 Table 4.2 illustrates the evolution of system models during the system development process. Describe how the evolution of requirements documents illustrates the materialization process described in Table 4.1.

4.8 Look up a definition of the "scientific method" and relate its steps to those postulated for the systems engineering method. Draw a functional flow diagram of the scientific method parallel to that of Figure 4.11.

4.9 Select one of the household appliances listed below:
  - automatic dishwasher
  - washing machine

- television set
  - (a) State the *functions* that it performs during its operating cycle. Indicate the primary medium (signals, data, material, or energy) involved in each step and the basic function that is performed on this medium.
  - (b) For the selected appliance, describe the physical elements involved in the implementation of each of the above functions.

FURTHER READING

S. Biemer and A. Sage. Chapter 4: Systems engineering: Basic concepts and life cycle. In *Agent-Directed Simulation and Systems Engineering*, L. Yilmaz and T. Oren, eds. John Wiley & Sons, 2009.

B. Blanchard. *System Engineering Management*, Third Edition. John Wiley & Sons, 2004.

B. Blanchard and W. Fabrycky. *System Engineering and Analysis*, Fourth Edition. Prentice Hall, 2006.

D. Buede. *The Engineering Design of Systems: Models and Methods*, Second Edition. John Wiley & Sons, 2009.

H. Chesnut. *System Engineering Methods*. John Wiley, 1967.

P. DeGrace and L. H. Stahl. *Wicked Problems, Righteous Solutions*. Yourdon Press, Prentice Hall, 1990.

H. Eisner. *Computer-Aided Systems Engineering*. Prentice Hall, 1988, Chapter 10.

H. Eisner. *Essentials of Project and Systems Engineering Management*, Second Edition. John Wiley & Sons, 2002.

A. D. Hall. *A Methodology for Systems Engineering*. Van Nostrand, 1962, Chapter 4.

M. Maier and E. Rechtin. *The Art of Systems Architecting*, Third Edition. CRC Press, 2009.

J. N. Martin. *Systems Engineering Guidebook: A Process for Developing Systems and Products*. CRC Press, 1997, Chapters 2–5.

E. Rechtin. *Systems Architecting: Creating and Building Complex Systems*. Prentice Hall, 1991, Chapters 2 and 4.

N. B. Reilly. *Successful Systems for Engineers and Managers*. Van Nostrand Reinhold, 1993, Chapter 3.

A. P. Sage. *Systems Engineering*. McGraw Hill, 1992, Chapter 2.

A. P. Sage and J. E. Armstrong, Jr. *Introduction to Systems Engineering*. Wiley, 2000, Chapter 2.

A. Sage and S. Biemer. Processes for system family architecting, design and integration. *IEEE Systems Journal*, 2007, 1, 5–16.

S. M. Shinners. *A Guide for System Engineering and Management*. Lexington Books, 1989, Chapter 1.

R. Stevens, P. Brook, K. Jackson, and S. Arnold. *Systems Engineering, Coping with Complexity*. Prentice Hall, 1998, Chapters 7 and 8.

5

SYSTEMS ENGINEERING MANAGEMENT

5.1 MANAGING SYSTEM DEVELOPMENT AND RISKS

As noted in the first chapter, systems engineering is an integral part of the management of a system development project. The part that systems engineering plays in the project management function is pictured in the Venn diagram of Figure 5.1. The ovals in the diagram represent the domain of *project management* and those of its principal constituents: *systems engineering* and *project planning and control*. It is seen that both constituents are wholly contained within the project management domain, with technical guidance being the province of systems engineering, while program, financial, and contract guidance are the province of project planning and control. The allocation of resources and the definition of tasks are necessarily shared functions.

To better understand the many different functions of systems engineering, this chapter describes some of the main features of the project management framework, such as the work breakdown structure (WBS), project organization, and the systems engineering management plan (SEMP). It also discusses the subject of risk management, the organization of systems engineering effort, and the capability maturity model integrated as it applies to systems engineering.

Systems Engineering Principles and Practice, Second Edition. Alexander Kossiakoff, William N. Sweet, Samuel J. Seymour, and Steven M. Biemer
© 2011 by John Wiley & Sons, Inc. Published 2011 by John Wiley & Sons, Inc.

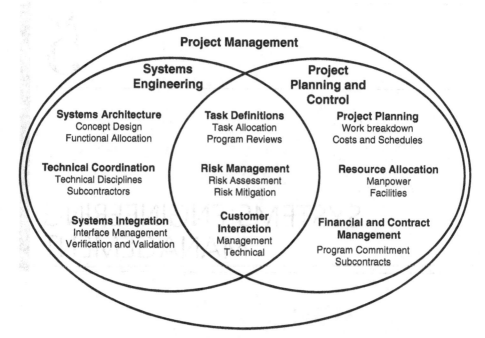

Figure 5.1. Systems engineering as a part of project management.

The engineering of a complex system requires the performance of a multitude of interrelated tasks by dozens or hundreds of people and a number of contractors or other organizational entities. These tasks include not only the entire development process but also usually everything needed to support system operation, such as maintenance, documentation, training, and so on, which must be provided for. Test equipment, facilities, and transportation have to be developed and acquired. The tasks involved in project management and systems engineering, including planning, scheduling, costing, and configuration control, need to be explicitly dealt with.

The sections in this chapter are intended to apply to the management of all systems engineering activities for all types of complex systems. However, in the management of software-intensive systems, in which essentially all of the functionality is performed by software, there are a number of special characteristics that need to be considered. These are noted in Chapter 11, in particular, in the section Software Engineering Management.

Proposal Development and Statement of Work (SOW)

System development often starts with someone who has a need, a customer, who requests support often in the form of a request for proposal (RFP) when in a competitive environment. Following a corporate decision to respond to the RFP, a program manager or a professional proposal team is assigned to generate the proposal. While a systems engineer may not be officially assigned to the team, it is essential that the technical concepts and implied design and interfaces are feasible. Hence, even in the early phases of a project, the integration of systems engineering with project management is evident.

A critical element of the proposal is the SOW. This is a narrative description of the work that is needed to develop the system to meet the customer needs. The systems engineer concerns will focus on the product to be developed; ensuring the scope of work in the SOW includes all the products and services needed to complete the effort. Specifically, the systems engineer focuses on being responsive to the customer needs, ensures the SOW is based on a credible concept of operations, reviews the implied design for the use of legacy components and their availability, and examines to see if the proposed system integrates commercial off-the-shelf (COTS) components and determines the technology readiness levels for the important subsystems envisioned in the preliminary system design. This early planning sets the stage for the work the technical contributors will have "to live with" throughout the life of the project.

5.2 WBS

The successful management of the system development effort requires special techniques to ensure that all essential tasks are properly defined, assigned, scheduled, and controlled. One of the most important techniques is the systematic organization of project tasks into a form called the *WBS* or, less commonly, the project or system breakdown structure. It defines all of the tasks in terms of goods and services to be accomplished during the project in terms of a hierarchical structure. Its formulation begins early in the concept definition phase to serve as a point of reference for concept trade-off studies. It is then more fully articulated in the latter stages to serve as a basis for system life cycle costing. The WBS is often a contractual requirement in competitive system developments.

The WBS typically defines the whole system to be developed, produced, tested, deployed, and supported, including hardware, software, services, and data. It defines a skeleton or framework on which the project is to be implemented.

Elements of a Typical WBS

The WBS format is generally tailored to the specific project in hand, but always follows a hierarchical tree structure designed to ensure a specific place for every significant portion of work under the project. For purposes of illustration, the following paragraphs describe the main elements of a typical system WBS.

With the system project at level 1 in the hierarchy (some WBS structures begin at Level 0), the level 2 categories may be broken down as follows:

1.1. system product,
1.2. system support,
1.3. system testing,

    1.4. project management, and

    1.5. systems engineering.

Note that these categories are not parallel in content or scope, but collectively, they are designed to encompass all the work under the system project.

    1.1. *System Product* is the total effort required to develop, produce, and integrate the system itself, together with any auxiliary equipment required for its operation. Table 5.1 shows an example of the WBS breakdown of the system product. The level 3 entries are seen to be the several subsystems, as well as the equipment required for their integration (assembly equipment), and other auxiliary equipment used by more than one subsystem. The figure also shows an example of the level 4 and 5 breakdown of one of the subsystems into its

TABLE 5.1. System Product WBS Partial Breakdown Structure

Level 1	Level 2	Level 3	Level 4	Level 5
1. System product				
	1.1 System product			
		1.1.1 Subsystem A		
			1.1.1.1 Component A$_1$	
				1.1.1.1.1 Functional design
				1.1.1.1.2 Engineering design
				1.1.1.1.3 Fabrication
				1.1.1.1.4 Unit text
				1.1.1.1.5 Documentation
			1.1.1.2 Component A$_2$	
				1.1.1.2.1 Functional design ... (etc.)
		1.1.1 Subsystem B		
			1.1.2.1 Component B$_1$	
				1.1.2.1.1 Functional design ... (etc.)
		1.1.3 Subsystem C		
		1.1.4 Assembly equipment		
		1.1.5 Assembly equipment		

constituent components, which represent definable products of development, engineering, and production effort. It is preferred that integration and test of hardware and software component is done separately for each subsystem, and then the tested subsystems are integrated in the final system for testing (1.3 below). Finally, for cost allocation and control purposes, each component is further broken down at level 5 into work packages that define the several steps of the component's design, development, and test. From this level and below the WBS, elements are often expressed with action words, for example, purchase, design, integrate, and test.

1.2. *System Support* (or integrated logistic support) provides equipment, facilities, and services necessary for the development and operation of the system product. These items can be categorized (level 3 categories) under six headings:

1.2.1. Supply support

1.2.2. Test equipment

1.2.3. Transport and handling

1.2.4. Documentation

1.2.5. Facilities

1.2.6. Personnel and training

Each of the system support categories applies to both the development process and system operation, which may involve quite different activities.

1.3. *System Testing* begins after the design of the individual components has been validated via component tests. A very significant fraction of the total test effort is usually allocated to system-level testing, which involves four categories of tests as follows:

1.3.1. *Integration Testing.* This category supports the stepwise integration of components and subsystems to achieve a total system.

1.3.2. *System Testing.* This category provides for overall system tests and the evaluation of test results.

1.3.3. *Acceptance Testing.* This category provides for factory and installation tests of delivered systems.

1.3.4. *Operational Testing and Evaluation.* This category tests the effectiveness of the entire system in a realistic operational environment.

Individual tests to be performed at each level are prescribed in a series of separate test plans and procedures. However, an overall description of test objectives and content and a listing of the individual tests to be performed should also be set forth in an integrated test planning and management document, the "test and evaluation management plan" (TEMP) in defense acquisition terminology. Chapter 13 is devoted to the subject of system integration and evaluation.

1.4 *Project Management* tasks include all activities associated with project planning and control, including the management of the WBS, costing, scheduling, performance measurement, project reviews and reports, and associated activities.

1.5 *Systems Engineering* tasks include the activities of the systems engineering staff in guiding the engineering of the system through all its conceptual and engineering phases. This specifically includes activities such as requirements analysis, trade-off studies (analysis of alternatives), technical reviews, test requirements and evaluations, system design requirements, configuration management, and so on, which are identified in the SEMP. Another important activity is the integration of specialty engineering into the early phases of the engineering effort, in other words, concurrent engineering.

The WBS is structured so that every task is identified at the appropriate place within the WBS hierarchy. Systems engineering plays an important role in helping the project manager to structure the WBS so as to achieve this objective. The use of the WBS as a project-organizing framework generally begins in the concept exploration phase. In the concept definition phase, the WBS is defined in detail as the basis for organizing, costing, and scheduling. At this point, the subsystems have been defined and their constituent components identified. Also, decisions have been made, at least tentatively, regarding outside procurement of elements of the system. Accordingly, the level down to which the WBS needs to be defined in detail should have been established.

It is, of course, to be expected that the details of the WBS evolve and change as the system is further engineered. However, its main outline should remain stable.

Cost Control and Estimating

The WBS is the heart of the project cost control and estimating system. Its organization is arranged so that the lowest indenture work packages correspond to cost allocation items. Thus, at the beginning of the project, the target cost is distributed among the identified work packages and is partitioned downward as lower-level packages are defined. Project cost control is then exercised by comparing actual reported costs against estimated costs, identifying and focusing attention on those work packages that deviate seriously from initial estimates.

The collection of project costs down to the component level and their distribution among the principal phases of project development, engineering, and fabrication is essential also for contributing to a database, which is used by the organization for estimating the costs of future projects. For new components, cost estimates must be developed by adapting the previously experienced costs of items directly comparable to those in the projected system, at the lowest level of aggregation for which cost figures are available. At higher levels, departures from one system to the next become too large to reliably use data derived from previous experience without major correction.

It should not be expected that the lowest indenture level would be uniform throughout the various subsystems and their components. For example, if a subsystem is being obtained on a fixed price subcontract, it may well be appropriate to terminate the lowest indenture in the WBS at that subsystem. In general, program control, including costing, is exercised at the level at which detailed specifications, interface definitions, and work assignments are available, representing in effect a contract between the project and the organization charged with the responsibility for developing, engineering, or fabricating given elements of the system.

Critical Path Method (CPM)

Network scheduling techniques are often used in project management to aid in the planning and control of the project. Networks are composed of events and activities needed to carry out the project. Events are equivalent to a milestone indicating when an activity starts and finishes. Activities represent the element of work or task, usually derived from the WBS that needs to be accomplished. Critical path analysis is an essential project management tool that traces each major element of the system back through the engineering of its constituent parts. Estimates are made of not only the size but also the duration of effort required for each step. The particular path that is estimated to require the longest time to complete its constituent activities is called the "critical path." The differences between this time and the times required for other paths are called "slack" for those paths. The resulting critical path network is a direct application of the WBS. The systems engineer uses the CPM to understand the dependences of task activities, to help prioritize the work of the technical teams, and to communicate graphically the work of the entire program.

5.3 SEMP

In the development of a complex system, it is essential that all of the key participants in the system development process not only know their own responsibilities but also know how they interface with one another. Just as special documentation is required to control system interfaces, so the interfacing of responsibilities and authority within the project must also be defined and controlled. This is usually accomplished through the preparation and dissemination of a SEMP or its equivalent. The primary responsibility of creating such a plan for guiding the engineering effort is that of the systems engineering component of project management.

The importance of having formalized plans for managing the engineering effort has been recognized in defense acquisition programs by requiring the contractor to prepare a SEMP as part of the concept definition effort. The most important function of the SEMP is to ensure that all of the many active participants (subsystem managers, component design engineers, test engineers, systems analysts, specialty engineers, subcontractors, etc.) know their responsibilities to one another. This is an exact analogue of the component interface function of systems engineering defining the interactions among all parts of the system so that they fit together and operate smoothly. It also serves as a reference for the procedures that are to be followed in carrying out the numerous systems engineering tasks. The place of the SEMP in the program management planning is shown in Figure 5.2.

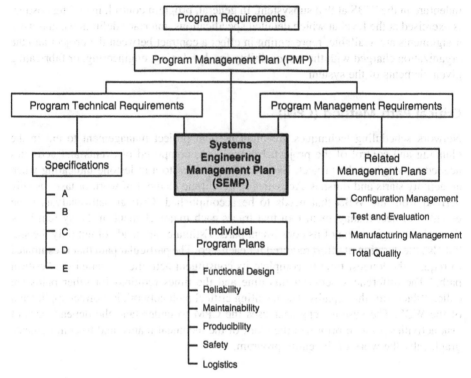

Figure 5.2. Place of SEMP in program management plans.

The SEMP is intended to be a living document, starting as an outline, and being elaborated and otherwise updated as the system development process goes on. Having a formal SEMP also provides a control instrument for comparing the planned tasks with those accomplished.

Elements of a Typical SEMP

The SEMP contains a detailed statement of how the systems engineering functions are to be carried out in the course of system development. It can be considered to consist of three types of activity:

1. *Development Program Planning and Control:* describes the systems engineering tasks that must be implemented in managing the development program, including
   - statements of work;
   - organization;
   - scheduling;

- program, design, and test readiness reviews;
- technical performance measurement; and
- risk management.

2. *Systems Engineering Process:* describes the systems engineering process as it applies to the development of the system, including
   - operational requirements,
   - functional analysis,
   - systems analysis and trade-off strategy, and
   - system test and evaluation strategy.

3. *Engineering Specialty Integration:* describes how the areas of specialty engineering are to be integrated into the primary system design and development, including
   - reliability, maintainability, availability (RMA) engineering;
   - producibility engineering;
   - safety engineering; and
   - human factors engineering.

A typical SEMP outline is tailored to the development system but could include the following:

Introduction
    Scope, Purpose, Overview, Applicable Documents
Program Planning and Control
    Organizational Structure
    Responsibilities, Procedures, Authorities
    WBS, Milestones, Schedules
    Program Events
    Program, Technical, Test Readiness Reviews
    Technical and Schedule Performance Metrics
    Engineering Program Integration, Interface Plans
Systems Engineering Process
    Mission, System Overview Graphic
    Requirements and Functional Analysis
    Trade Studies (Analysis of Alternatives)
    Technical Interface Analysis/Planning
    Specification Tree/Specifications
    Modeling and Simulation
    Test Planning
    Logistic Support Analysis
    Systems Engineering Tools

Engineering Integration
    Integration Design/Plans
    Specialty Engineering
    Compatibility/Interference Analysis
    Producibility Studies

5.4   RISK MANAGEMENT

The development of a new complex system by its nature requires acquiring knowledge about advanced but not fully developed devices and processes so as to wisely guide the system design to a product that performs its intended mission reliably and at an affordable cost. At every step, however, unpredictable outcomes can be encountered that pose risks of performance shortfalls, environmental susceptibility, unsuitability for production, or a host of other unacceptable consequences that may require a change in course with impacts on program cost and schedule. One of the greatest challenges to systems engineering is to steer a course that poses minimum risks while still achieving maximum results.

At the outset of the development, there are uncertainties and hence risks in every aspect. Are the perceived operational requirements realistic? Will they remain valid throughout the new system's operational life? Will the resources required to develop and produce the system be available when needed? Will the advanced technology necessary to achieve the required operational goals perform as expected? Will the anticipated advances in production automation materialize? Will the development organization be free from work stoppages?

It is the special task of systems engineering to be aware of such possibilities and to guide the development so as to minimize (mitigate) their impact if and when they may occur. The methodology that is employed to identify and minimize risk in system development is called *risk management*. It has to begin at the outset of the system development and progress throughout its duration.

Risk Reduction through the System Life Cycle

Reducing program risks is a continual process throughout the life cycle. For example, the needs analysis phase reduces the risk of embarking on the development of a system that does not address vital operational needs. The concept exploration phase reduces the risk of deriving irrelevant or unrealistic system performance requirements. And the system definition phase selects a system concept that utilizes technical approaches that are neither excessively immature nor unaffordable, but rather one that has the best chance of meeting all system goals.

Figure 5.3 is a schematic representation of how the program risk of a hypothetical system development (in arbitrary units) decreases as the development progresses through the phases of the life cycle. The abscissa is time, sectioned into the phases of system development. In the same figure is plotted a curve of the typical relative effort expended during each phase.

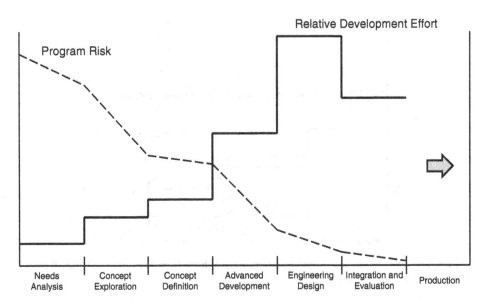

Figure 5.3. Variation of program risk and effort throughout system development.

The descending risk curve conveys the fact that as development progresses, uncertainties (unknowns), which constitute risks of unforeseen adverse events, are systematically eliminated or reduced by analysis, experiment, test, or change in course. A variant of this curve is referred to as the "risk mitigation waterfall" (Figure 5.4). The ascending effort curve represents the stepwise increases in the costs of succeeding phases of system development, showing the progression of activity from conceptual to engineering to integration and evaluation.

Figure 5.3 is intended to illustrate several key principles:

1. As the development progresses, the investment in program effort typically rises steeply. To maintain program support, the risk of failure must be correspondingly reduced so as to maintain the financial risk at reasonable levels.

2. The initial stages in the program produce major reductions in risk, when the basic decisions are made regarding the system requirements and the system concept. This demonstrates the importance of investing adequate effort in the formative phases.

3. The two phases that typically produce the greatest risk reduction are *concept exploration* and *advanced development*. Concept exploration provides a solid conceptual basis for the system approach and architecture. Advanced development matures new advanced technologies to insure their meeting performance goals.

4. By the time the development is complete and the system is ready for production and distribution, the residual level of risk must be extremely low if the system is to be successful.

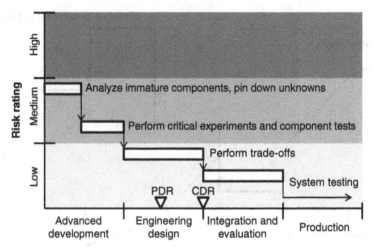

<u>Figure 5.4.</u> Example of a risk mitigation waterfall chart. PDR, Preliminary Design Review; CDR, Critical Design Review.

Components of Risk Management

Risk management is formally recognized in systems engineering standards, and especially in government acquisition programs. Each program is expected to prepare a risk management plan. Risk management for a major system is expected to have its own organization, staffing, database, reporting, and independent review, and to extend to all phase of program development, production, operation, and support. A detailed description of risk management as defined by the DoD is contained in the *Risk Management Guide for DoD Acquisition* published by the Defense Acquisition University.

The *Risk Management Guide* divides the subject of risk management into risk planning, risk assessment, risk prioritization, risk handling, and risk monitoring. The discussion to follow will combine these into two categories: *risk assessment*, which will include risk planning and prioritization, and *risk mitigation*, which will include risk handling and monitoring. The subject of risk planning is addressed by the risk management plan, which is part of the SEMP.

Risk Assessment

The general process of risk assessment is inherent in all decisions involving prospective uncertainty. As will be described in Chapter 10, risk assessment is used to eliminate alternative concepts that are overly dependent on immature technologies, unproven technical approaches, or other ambitious advances that do not appear to be warranted by their projected benefits compared to the uncertainty of their realization. Some of the more common sources of program risk are listed in Chapter 12.

In the advanced development phase, risk assessment will be seen to be a useful approach to the identification and characterization of proposed design features that represent a sufficient development risk (i.e., likelihood of failing to meet requirements) and a significant program impact to warrant analysis and, if necessary, development and test. Thus, risk assessment identifies the weakest and most uncertain features of the design and focuses attention on means for eliminating the possibility that these features will present complications and will require design changes during the subsequent phases of development.

Once the system components possessing questionable design features have been identified, the task of systems engineering is to define a program of analysis, development, and test to eliminate these weaknesses or to take other actions to reduce their potential danger to the program to an acceptable level. In this process, the method of risk assessment can be of further value by providing a means for determining how to best allocate available time and effort among the identified areas of risk. For this purpose, risk assessment can be applied to judge the relative risks posed by the design features in question.

To compare the potential importance of different sources of program risk, it is necessary to consider two risk components: the *likelihood* that a given component will fail to meet its goals and the *impact* or *criticality* of such a failure to the success of the program. Thus, if the impact of a given failure would be catastrophic, even a low likelihood of its occurring cannot be tolerated. Alternatively, if the likelihood of failure of a given approach is high, it is usually prudent to take a different approach even if its impact may be low but significant.

These risk components are often displayed in the form of a "risk cube" typically of three or five dimensions. The five-dimension cube is shown in Figure 5.5, and the three-dimension cube is discussed below. Since the probabilities are usually qualitative in nature, experienced judgment is needed to develop an informed assignment of risk. The relative nature is also important to understand since work in foundational research areas is naturally more risky than work that is developing a system to well-defined specifications. The risk tolerance of customers will also vary by domain and experience.

Risk Likelihood: Probability of Failure. There are too many uncertainties to be able to compute a numerical value for the likelihood that a specific program goal will be achieved, and hence it is not useful to attempt to quantify risks beyond a relatively rough measure to assist in their relative prioritization.

In the case of unproven technology, it is possible to estimate very roughly the relative degree of maturity from the engineering status of the technology. This may be carried out by identifying one or more cases where the technology is used in connection with a similar functional application and by determining its level of development (e.g., in the range from a laboratory design to an experimental prototype to a qualified production component). High, medium, and low risk is about as fine a scale as is normally useful. Beyond that, it is good practice to rank order the parts of the system that appear to be risky and to concentrate on the few that are judged to be most immature and complex. If the candidates are numerous, it may be a sign that the entire system design approach is too ambitious and should be reconsidered.

Risks associated with highly complex components and interfaces are even more difficult to quantify than those using advanced technology. Interfaces always require

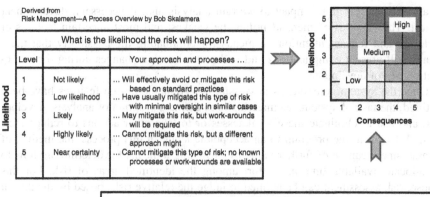

Figure 5.5. An example of a risk cube display.

special attention, especially in human–machine interactions. The latter always warrant early prototyping and testing. Here again, rank ordering of the relative complexity is an effective way of prioritizing the effort required for risk management.

The prioritization of software risks is again a matter of judgment. Real-time programs with numerous external interrupts always require special attention, as do concurrent processes. New or significantly altered operating systems can be particularly complicated. Programs with high logic content tend to be more likely to malfunction as a consequence of undetected faults than those that are largely computational.

Table 5.2 lists some of the considerations discussed above in arriving at a general prioritization of risk probabilities.

Risk Criticality: Impact of Failure. It was stated earlier that the seriousness of the risk of a particular failure might be considered in terms of two factors—the likelihood that a failure will occur and the criticality of its impact on the success of the program. In a semiquantitative sense, the seriousness of the risk can be thought of as a combination of those two factors.

As in the case of risk likelihood, there is no accepted numerical scale for risk criticality, and one may consider the same relative levels as those for likelihood: high, medium, or low. Some agreed-upon definitions need to be assigned to these levels, such as those listed in Table 5.3.

TABLE 5.2. Risk Likelihood

Risk likelihood	Design status
High	• Significant extension from past designs
	• Multiple new and untried components
	• Complex components and/or interfaces
	• Marginal analytical tools and data
Medium	• Moderate extension from past designs
	• Components complex but not highly stressed
	• Analytical tools available
Low	• Application of qualified components
	• Components of medium complexity
	• Mature technologies and tools

TABLE 5.3. Risk Criticality

Criticality	System impact	Program impact
High	• Major degradation in performance (50–90%)	• Major increase in cost and/or schedule (30–70%)
	• Serious safety problem	• Production cutbacks
Medium	• Significant degradation in performance (10–50%)	• Significant increases in cost and/or schedule (10–30%)
	• Short losses of operability	• Intense reviews and oversight
	• Costly operational support	• Production delays
Low	• Minor degradation in performance (>10%)	• Minor increase in cost and/or schedule (<10%)
	• Occasional brief delays	• Vigorous reviews and oversight
	• Increased maintenance	

The middle column of the table lists expected impacts on system operation if the system component at risk failed to perform its function. The right column lists the types of impacts on the overall program that could be expected if the system component was discovered to be faulty late in development and indicates the likely effects on the program.

While some systems engineering textbooks advocate the derivation of an overall risk factor by assigning numerical values to the estimates of risk likelihood and risk criticality and taking their product, the disadvantages of this practice are believed to outweigh the presumed advantage of a seemingly simple single risk factor. In the first place, assignment of numerical estimates creates the illusion of quantitative knowledge, which has no real basis. In the second place, combining the two indices into one has the effect of diminishing the net information content, as was noted in connection with combining figures of merit of individual parts of the system into a single score. Accordingly, it is recommended that the individual ratings be retained as abstractions, such as high, medium, and low, and the two components retain their identity, such as medium–low, and so on.

In connection with the criticality scale, the highest level of criticality listed in Table 5.3 stops short of including the case of near total loss in system performance resulting in mission failure. Such an eventuality would likely risk program cancellation, and as such would be considered unacceptable. This implies that design risks of this degree of criticality would not be considered as feasible options.

Role of Systems Engineering. The task of risk assessment (and the subsequent task of risk management) is clearly the responsibility of systems engineering. This is because the judgments that are involved require a breadth of knowledge of system characteristics and the constituent technologies beyond that possessed by design specialists, and also because judgments of risk criticality are at the system and program levels. The process of risk assessment thus helps the systems engineer to identify the system features that need to be most thoroughly understood and raised to a level of design maturity suitable for full-scale engineering.

Risk Mitigation

The most common methods of dealing with identified program risks are the following, listed in order of increasing seriousness of the perceived risk:

1. intensified technical and management reviews of the engineering process,
2. special oversight of designated component engineering,
3. special analysis and testing of critical design items,
4. rapid prototyping and test feedback,
5. consideration of relieving critical design requirements, and
6. initiation of fallback parallel developments.

Each of the above methods is briefly described below.

Technical and Management Reviews. Formal design reviews may address entire subsystems, but the depth of coverage is mainly on design aspects considered of greatest importance. It is the responsibility of systems engineering to ensure that the significant risk items are fully presented and discussed so that special management attention and resources may be directed to issues warranting additional effort. The aim should be to resolve problems at the earliest possible time, so full disclosure of experienced or anticipated difficulties is essential. The process of design reviews is further described in the Component Design section of Chapter 12 (Section 12.4).

Oversight of Designated Component Engineering. Regularly scheduled design reviews are neither frequent enough nor detailed enough to provide adequate oversight of known design problem areas. Each designated problem area should be assigned a special status, subjected to appropriately frequent reviews, and overseen by designated senior design and systems engineers. Where appropriate, outside consultants should be engaged in the process. A risk mitigation plan should be prepared and tracked until the problem areas are resolved.

Special Analysis and Testing. For components whose design involves issues not resolved in the advanced technology phase, additional analysis and, if necessary, fabrication and test should be carried out to obtain sufficient design data to validate the technical approach. This will require assigning additional resources and modifying the engineering schedule to accommodate the results of such analysis and testing.

Rapid Prototyping. For unproven components for which analysis and limited testing cannot adequately validate the design, it may be necessary to construct and test prototypes to ensure their validity. Such action would normally be taken in the advanced technology phase, but sometimes, the problem is not recognized at that time, and in other cases, the action fails to resolve the problem.

Relief of Excessive Requirements. Experience has shown that attempting to meet all initially posed requirements often fails to achieve a practical overall solution and requires an adjustment of some performance or compatibility requirement. This possibility should be explored whenever efforts to meet fully a requirement result in a solution that is inordinately complex, costly, unreliable, or otherwise undesirable from a practical standpoint. This problem is uniquely a task for systems engineering since all factors of performance, cost, and schedule need to be considered together. It is an option that should be invoked only in exceptional cases, but neither should it be put off until excessive resources and time have been committed to vain efforts to fulfill the requirement.

Fallback Alternatives. The development of alternative design approaches is most appropriate for components using new technology whose successful development cannot be fully assured. In such cases, adequate alternative approaches should be established during the advanced development phase to serve as fallbacks in the event that the new designs do not fulfill expectations. Such fallback alternatives almost always result in reduced performance, greater cost, or some other perceived deficiency compared to the selected approach, but are more conservative in their design and hence are more certain to succeed.

It happens not infrequently that the engineering design phase begins before a clear resolution is reached as to the ultimate success of a given technical approach, and hence before a final decision as to whether or not to fall back to a more conservative alternative. In such cases, an expedited program to reach such a decision by further development, analysis, and test must be invoked. Again the decision is one for systems engineering. Often the choice also involves reexamination of the initial requirements, as discussed in the previous paragraphs.

The above methods may be applied singly, but most often work best in combination. Their oversight is a program manager's responsibility, and their planning and direction are a systems engineering function.

TABLE 5.4. Sample Risk Plan Worksheet

Risk title:	Project name:	
Risk owner:	Last updated:	
Team:		
Date submitted:		

Description of risk:	Risk type:	Place X, 1, 2, … in the appropriate cells.
	☐ Technical	
Statement of basic cause:	☐ Schedule	
	☐ Cost	
Consequence if risk is realized:	☐ Other	

Likelihood (5, 4, 3, 2, 1) vs. Consequences (1 2 3 4 5)

Risk reduction plan

Action/milestone event.	Date	Success criteria	Risk level if successful L	C	Comments
1.					
2.					
3.					
4.					

Risk Management Plan

The importance of the actions described above to the overall success of a system development requires that it be part of the overall program management process. To this end, a formal risk management plan should be developed and progressively updated, in which mitigation is a major part.

For every significant risk, there should be a plan that minimizes its potential impact through specific actions to be taken, either concurrently with the engineering or to be invoked should the anticipated risk materialize. The formulation of such a plan must be predicated on the objective of minimizing the total expected program cost, which means that the planned activities to contain program risks must not be more costly than the expected impact of the risks, should they eventuate. For items for which a fallback approach is to be developed, the plan should define the conditions under which the backup will be activated, or if activated at the outset, how far it is to be carried in the absence of evidence that the main approach will prove unsatisfactory. A diagram of a risk mitigation plan known as a "risk mitigation waterfall chart" is shown in Figure 5.4. An example of a risk plan worksheet is pictured in Table 5.4.

5.5 ORGANIZATION OF SYSTEMS ENGINEERING

Despite decades of study, there are many opinions, but no general agreement, on which the organizational form is most effective for a given type of enterprise. For this reason, the organizations participating in a system development project are likely to employ a variety of different organizational styles. Each individual style has evolved as a result of history, experience, and the personal preferences of upper management. Accordingly, despite its central importance to the success of a given system development project, the systems engineering function will usually need to adapt to preexisting organizational structures.

Virtually all system development projects are managed by a single industrial company. Hence, it is the organizational form of this company that drives the organization of systems engineering. In most cases, this company will develop some subsystems in-house, and contract for other subsystems with subcontractors. We will refer to the first company as the prime contractor or system contractor, and to the collection of participating contractors as the "contractor team." This means that the systems engineering activity must span not only a number of different disciplines but also several independent companies.

The organizational structure of the prime contractor is usually some form of a "matrix" organization. In a matrix organization, most of the engineering staff is organized in discipline- or technology-oriented groups. Major projects are managed by project management teams reporting to a "vice president for project management" or an equivalent. At times, these teams are called integrated product teams (IPTs) (see Chapter 7). A technical staff is assigned to individual projects as required, but employees retain affiliation with their engineering groups.

The main variations in matrix-type organizations relates to whether the bulk of the technical staff assigned to a project are physically relocated to an area dedicated to the project and remain as full-time participants throughout much of the development or whether they remain in their home group areas. A related difference is the degree to which authority for the direction of the technical work assignments is retained by their home group supervisors.

As stated earlier, the organization of the systems engineering function is necessarily dependent on the system contractor's organizational structure. There should be some common practices, however. Referring to Figure 5.1, a major system project should have a single focus of responsibility for the systems engineering function (a project systems engineer), as apart from the project planning and control function. As an integral part of project management, an appropriate title might be "associate (or deputy) project manager for systems engineering" or, more simply, "systems engineering manager." Since the systems engineering function is that of guidance, authority is exercised by establishing goals (requirements and specifications), formulating task assignments, conducting evaluations (design reviews, analyses, and tests), and controlling the configuration.

Effective technical communications are difficult to maintain in any organization for a variety of reasons, many of them inherent in human behavior. They are, nevertheless, absolutely vital to the ultimate success of the development project. Perhaps the single most important task of the project systems engineer is to establish and maintain effective communication among the many individuals and groups, inside and outside the company, whose work needs to interact with others. This is a human interface function corresponding to the system physical interface functions that make the system elements fit together and operate as one. Since the systems engineer usually works in parallel with rather than through established lines of authority, he or she must exercise extraordinary leadership to bring together those individuals who need to interact.

There are several different means of communication, all of which need to be exercised as appropriate:

1. All key participants need to know what they are expected to do, when, and why: the "what" is expressed in task assignments and WBS; the "when" is contained in schedules, milestones, and critical path networks; and the "why" should be answered in the requirements and specifications. A clear and complete statement of the "why" is essential to ensure that the designers, analysts, and testers understand the objectives and constraints of the task assignments.

2. Participants must be aware of how their portions of the system interact with other key elements and of the nature of their mutual interdependence. Such interactions, and particularly their underlying causes, can never be sufficiently covered in specification documents. This awareness can only be provided by periodic personal communication among the responsible participants and the documentation of any resulting agreements, interface definitions, and so on, however small or tentative. Systems engineering must provide the glue that binds these items of system design through the formation of *interface working groups* and the development of *interface control documents*, and such less formal communications as may be needed in special cases.

3. Subcontractors and other key participants at remote sites must be integrated into the project communication framework. At the management level, this is the task of the system project manager, but at the engineering level, it is the responsibility of the project systems engineering staff. It is essential that the same two coordinating functions described above be provided for the entire contractor team. For this purpose, conventional formal contractual mechanisms never suffice and sometimes hinder. Accordingly, special efforts should be made to integrate the team members effectively into the total system development effort. This needs to be carried out at two levels: (1) periodic program management reviews attended by top-level representatives of the contractor team and (2) frequent technical coordination meetings concerned with specific ongoing aspects of the program.

4. The principal leaders of the system design effort must have a regular and frequent means of communication with one another to keep the program closely coordinated and to react quickly to problems. This is discussed in the following paragraphs.

Systems Analysis Staff

An essential part of any systems engineering organization is a highly competent and experienced analytical staff. Such a staff need not be a single entity, nor does it need to be organizationally colocated with the project staff itself, but it must be part of the systems engineering organization, at least during the conceptual and early engineering phases of the project. The systems analysis staff must have a deep understanding of the system environment, with respect to both its operational and physical characteristics. In both instances, it must be able to model the system environment, by use of mathematical and computer models, to provide a basis for analyzing the effectiveness of system models. In the concept exploration phase, the systems analysis staff is the source of much of the quantitative data involved in defining the system performance required to meet its operational requirements. In the concept definition phase, the analysis staff is responsible for constructing the system simulations used in the trade-off studies and in the selection of the best system concept. Throughout the engineering development stage of the program, the analysis staff is involved in numerous component trade-off studies. It conducts test analyses to derive quantitative measures of the performance of system prototypes and contributes to defining the quantitative aspects of system design specifications.

While the systems analysis staff must be skilled in mathematical modeling, software design, and other specialized techniques, its members are also required to have a system perspective and a thorough knowledge of the operational requirements of the system under development.

System Design Team

The exercise of leadership and coordination in any large program requires one or more teams of key individuals working closely together, maintaining a general consensus on the conduct of the engineering program. A system design team for a complex system development project may have the following membership:

- systems engineer,
- lead subsystems engineers,
- software systems engineers,
- support engineers,
- test engineers,
- customer representative, and
- specialty and concurrent engineers.

The customer representative is an advocate for the system requirements. The advantage of the team approach is that it generally increases the esprit de corps and motivation of the participants and broadens their understanding of the status and problems of other related aspects of the system development. This develops a sense of ownership of the team members in the overall system rather than the limitation of responsibility that is the rule in many organizations. It makes the response to unexpected problems and other program changes more effective.

In a particular application, the leadership of the system development needs to be tailored to the prime contractor's organization and to the customer's level of involvement in the process. The most important common denominators are

1. quality of leadership of the team leader,
2. representation of those with key responsibilities, and
3. participation of key technical contributors.

Without energetic leadership, the members of the system design team will flounder or go their separate ways. If for some reason the person designated as the project systems engineer does not have the required personal leadership qualities, either the project engineer or a deputy systems engineer should assume the team leadership role.

The presence of the leaders of the major portion of the development effort is necessary to bring them into the design decision process, as well as to have them available to use their resources to resolve problems. There are usually several senior systems engineers whose experience and knowledge are of great value to the project. Their presence adds a necessary ingredient of wisdom to the design process.

Involvement of the customer in the design process is essential but, in many cases, may be an inhibiting influence on free discussion in team meetings. Frequent but more formal meetings with the customer may be preferred to team membership.

5.6 SUMMARY

Managing System Development and Risks

Systems engineering is a part of project management that provides technical guidelines, system integration, and technical coordination.

WBS

The systems engineer's role also involves contributing to resource allocation, task definition, and customer interaction, with the initial focus on the development of the WBS, a hierarchical task organization that subdivides total effort into successively smaller work elements. This provides the basis for scheduling, costing, and monitoring, and enables cost control and estimating.

One key tool used for program scheduling is the CPM. CPM is based on WBS work elements and creates a network of sequential activities. Analyzing this network enables the systems engineer and program manager to identify paths that take the longest to complete.

SEMP

The SEMP plans the implementation of all systems engineering tasks. In the process, it defines the roles and responsibilities of all participants.

Risk Management

Risk management is a major challenge to systems engineering since all new system developments present uncertainties and risks. Reducing program risks is a continual process throughout the life cycle; moreover, risk must be reduced as program investment rises.

A risk management plan is important to support risk management. Risk assessment identifies the importance of risk in terms of risk likelihood (probability of occurrence) and risk criticality (impact and consequences of risk realization).

Risk mitigation of a critical area may include one or more of the following: management reviews, special engineering oversight, special analysis and tests, rapid prototyping, retry of rigorous requirements, and/or fallback developments.

Organization of Systems Engineering

The systems engineering organization spans disciplines and participating organizations, but also adapts to the company organizational structure. Therefore, systems engineering must communicate effectively "what, when, and why" to the proper stakeholders and must provide technical reviews for all participants. In large programs, systems engineering is supported by a systems analysis staff.

Large programs will require formal system design teams, which integrate major subsystems and subcontractors, and the products of software systems engineering. These teams contain members from support engineering and the test organization, and typically contain specialty (concurrent engineering) members as appropriate. They may also include user representation when appropriate. A key role for systems engineering involvement in these design teams is to keep their focus on the success of the entire enterprise.

PROBLEMS

5.1 Developing a detailed WBS for a system development project is a basic function of project management. What part should be played by systems engineering in the definition of the WBS in addition to detailing the section named "Systems Engineering"?

5.2 The preparation of a formal SEMP is usually a required portion of a contractor's proposal for a competitive system development program. Since at this time the system design is still in a conceptual state, explain where you would get the information to address the elements of a typical SEMP as listed in this chapter.

5.3 Define the two main components of risk management discussed in this chapter and give two examples of each. Show by an example how you would apply risk management processes to a system development project that proposes to use one or more components that utilize unproven technology.

5.4 One of the methods for estimating the risk likelihood (probability of failure) of a system development is to compare the current design status with comparable situations in existing systems. Table 5.2 shows some basic characteristics that are useful in making these estimates. For each of the first three conditions associated with high-risk projects, briefly describe an example of such a condition in a real system, or describe a hypothetical example having such characteristics.

5.5 Suppose you are a new systems engineer for a major new system development effort that involves new technology. Obviously, this represents a major technical (if not programmatic) risk area. What activities would you recommend early in the system development effort to mitigate these technical risks? For each mitigation activity, describe whether the activity will lower the likelihood of the risk, or the consequences of the risk, or both.

5.6 There are a number of risk mitigation methods for dealing with program risks. Referring to the description of high and low program impact in Table 5.3, discuss how you would best use risk mitigation methods to reduce their risk criticality.

5.7 Suppose you are a systems engineer on a new system development project in which your design engineers have never developed the subsystems and components required for this new system. Obviously, this represents a major risk area. What activities would you recommend early in the system development effort to mitigate these technical risks? For each mitigation activity, describe whether the activity will lower the likelihood of the risk or the consequences of the risk, or both.

5.8 This chapter presents a method for quantifying risk in two elements, likelihood and criticality, and for plotting these two metrics on a risk matrix. Suppose you wanted to combine these two metrics into a single, combined metric for risk. Suggest three methods for combining likelihood and criticality into a single metric. List the advantages and disadvantages for each.

5.9 Research the building of the tunnel under the English Channel in the late twentieth century.
    (a) What risks were present with this project?
    (b) What successful activities were undertaken to mitigate these risks that led to the tunnel's completion?

5.10 Describe the general type of the organizational structure in which you work. Discuss instances where this structure has been beneficial and those where it has not been so beneficial to programs you have been involved in or have some knowledge of.

5.11 Discuss the advantages of using the system design team approach for a large development project. List and discuss six requirements that are needed to make this approach successful.

FURTHER READING

B. Blanchard. *System Engineering Management*, Third Edition. John Wiley & Sons, 2004.

B. Blanchard and W. Fabrycky. *System Engineering and Analysis*, Fourth Edition. Prentice Hall, 2006, Chapters 18 and 19.

W. P. Chase. *Management of Systems Engineering*. John Wiley, 1974, Chapters 2 and 8.

D. Cooper, S. Grey, G. Raymon, and P. Walker. *Managing Risk in Large Projects and Complex Procurements*. John Wiley & Sons, 2005.

H. Eisner. *Essentials of Project and Systems Engineering Management*, Second Edition. John Wiley & Sons, 2002, Chapters 1–6.

A. D. Hall. *A Methodology for Systems Engineering*. Van Nostrand, 1962.

International Council on Systems Engineering. *Systems Engineering Handbook. A Guide for System Life Cycle Processes and Activities*. Version 3.2, July 2010.

T. Kendrick. *Indentifying and Managing Project Risk: Essential Tools for Failure-Proofing Your Project*. American Management Association, 2003.

R. S. Pressman. *Software Engineering: A Practitioner's Approach*, Sixth Edition. McGraw–Hill, 2005, Chapters 3, 5, and 6.

E. Rechtin. *Systems Architecting: Creating and Building Complex Systems*. Prentice Hall, 1991, Chapter 4.

A. P. Sage. *Systems Engineering*. McGraw–Hill, 1992, Chapter 3.

A. P. Sage and J. E. Armstrong, Jr. *Introduction to Systems Engineering*. Wiley, 2000, Chapter 6.

P. Smith and G. Merritt. *Proactive Risk Management: Controlling Uncertainty in Product Development*. Productivity Press, 2002.

R. Stevens, P. Brook, K. Jackson, and S. Arnold. *Systems Engineering, Coping with Complexity*. Prentice Hall, 1998, Chapter 6.

PART II

CONCEPT DEVELOPMENT STAGE

Part II begins the systematic account of the key roles played by systems engineering throughout the three stages of the systems engineering life cycle. This initial stage of the life cycle is where systems engineering makes its greatest contribution to the success of the system development project by performing the function of "systems architecting." The system decisions made during this stage in most cases determine the success or failure of the project.

Chapter 6 introduces the origins of a new system, whether driven by new needs or by technological opportunities. The chapter focuses on the role of systems engineering in the validation of an operational need for a new system and the development of a definitive set of operational requirements.

Chapter 7 presents the concept exploration phase, which explains how system concepts are developed from the requirements, and how several alternative concepts are examined for the purpose of deriving a set of necessary and sufficient performance requirements suitable for defining a system meeting the operational needs.

The final phase in the concept development stage is selecting a preferred system architecture that meets the performance requirements established previously. Chapter 8 describes how systems engineering uses modeling, visualization, and analysis to

Systems Engineering Principles and Practice, Second Edition. Alexander Kossiakoff, William N. Sweet, Samuel J. Seymour, and Steven M. Biemer
© 2011 by John Wiley & Sons, Inc. Published 2011 by John Wiley & Sons, Inc.

accomplish this result. In the acquisition of major systems, the satisfactory completion of this process leads to a commitment to proceed with engineering development and a possible ultimate production of the new system.

The final chapter in this part describes the process and activities involved in engineering-level decision making. A detailed description of the trade-off analysis is provided to provide formality to a systems engineer's thinking about decisions.

6

NEEDS ANALYSIS

6.1 ORIGINATING A NEW SYSTEM

The primary objective of the needs analysis phase of the system life cycle is to show clearly and convincingly that a valid operational need (or potential market) exists for a new system or a major upgrade to an existing system, and that there is a feasible approach to fulfilling the need at an affordable cost and within an acceptable level of risk. It answers the question of why a new system is needed and shows that such a system offers a sufficient improvement in capability to warrant the effort to bring it into being. This is achieved, in part, by devising at least one conceptual system that can be shown to be functionally capable of fulfilling the perceived need, and by describing it in sufficient detail to persuade decision makers that it is technically feasible and can be developed and produced at an acceptable cost. In short, this whole process must produce persuasive and defensible arguments that support the stated needs and create a "vision of success" in the minds of those responsible for authorizing the start of a new system development.

Systems Engineering Principles and Practice, Second Edition. Alexander Kossiakoff, William N. Sweet, Samuel J. Seymour, and Steven M. Biemer
© 2011 by John Wiley & Sons, Inc. Published 2011 by John Wiley & Sons, Inc.

Place of the Needs Analysis Phase in the System Life Cycle

The exact beginning of the active development of a new system is often difficult to identify. This is because the earliest activities in the origin of a new system are usually exploratory and informal in nature, without a designated organizational structure, specified objectives, or established timetable. Rather, the activities seek to determine whether or not a dedicated effort would be warranted, based on an assessment of a valid need for a new system and a feasible technological approach to its implementation.

The existence of a discrete phase corresponding to that defined as *needs analysis* in Chapter 4 is more characteristic of need-driven system developments than of those that are technology driven. In defense systems, for example, "material solution analysis" (see Department of Defense [DoD] life cycle of Fig. 4.1) is a required prerequisite activity for the official creation of a specific item in the budget for the forthcoming fiscal year, thereby allocating funds for the initiation of a new system project. Within this activity, a need determination task produces an initial capability description (ICD), which attests to the validity of the system objective or need, and gives evidence that meeting the stated objective will yield significant operational gains and is feasible of realization. Its completion culminates in the first official milestone of the defense acquisition life cycle.

In a technology-driven system development, typical of new commercial systems, the needs analysis phase is considered to be part of the conceptual development stage (Fig. 4.2). However, in this case too, there must be similar activities, such as market analysis, assessment of competitive products, and assessment of deficiencies of the current system relative to the proposed new system, that establish a bona fide need (potential market) for a product that will be the object of the development. Accordingly, the discussion to follow will not distinguish between needs-driven and technology-driven system developments except where specifically noted.

The place of the needs analysis phase in the system life cycle is illustrated in Figure 6.1. Its inputs are seen to be *operational deficiencies* and/or *technological opportunities*.

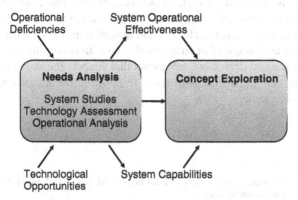

Figure 6.1. Needs analysis phase in the system life cycle.

Its outputs to the following phase, *concept exploration*, are an estimate of *system operational effectiveness* that specify what results a new system should achieve to meet the identified need, together with *system capabilities*, the output of various operational analyses and system studies, which provide evidence that an affordable system capable of meeting the effectiveness target is feasible.

As discussed above and depicted in the figure, the impetus for the initiation of a new system development generally comes from one of two sources: (1) the perception of a serious deficiency in a current system designed to meet an important operational need (need driven) or (2) an idea triggered by a technological development whose application promises a major advance over available systems in satisfying a need (technology driven). Either of these may then lead to investigations and analyses that eventually culminate in a program to develop a new system. Quite often, both factors contribute to the final decision.

Examples of New System Needs

The automobile industry is a prime example where changing conditions have forced the need for system improvements. Government laws require manufacturers to make substantial improvements in fuel economy, safety, and pollution control. Almost overnight, existing automobile designs were rendered obsolete. These regulations posed a major challenge to the automobile industry because they required technically difficult trade-offs and the development of many new components and materials. While the government gave manufacturers a number of years to phase in these improvements, the need for innovative design approaches and new components was urgent. In this case, the need for change was triggered by legislative action based on the needs of society as a whole.

Examples of technology-driven new systems are applications of space technology to meet important public and military needs. Here, the development of a range of advanced devices, such as powerful propulsion systems, lightweight materials, and compact electronics, made the engineering of reliable and affordable spacecraft a practical reality. In recent years, satellites have become competitive and often superior platforms for communication relays, navigation (GPS), weather surveillance, and a host of surveying and scientific instruments.

A more pervasive example of technology-driven system developments is the application of computer technology to the automation of a wide range of commercial and military systems. Information systems, in particular (e.g., banking, ticketing, routing, and inventory), have been drastically altered by computerization. System obsolescence in these cases has come not from recognized deficiencies but rather from opportunities to apply rapidly advancing technology to enhance system capabilities, to reduce cost, and to improve competitive position.

External Events. As will be seen later in this section, analysis of needs goes on more or less continuously in most major mission or product areas. However, external events often precipitate intensification and focusing of the process; this results in the formulation of a new operational requirement. In the defense area, this may be an intelligence finding of a new potential enemy threat, a local conflict that exposes the deficiency in a system, a major technological opportunity uncovered in a continuing program of concept exploration, or a major deficiency uncovered in periodic operational testing. In the civil products area, a triggering event might be a sudden shift in customer demand or a major technological change, such as the discovery of a radically new product, or an opportunity to automate a labor-intensive process. The drastic increase in the price of petroleum has triggered an intensive and successful effort to develop more fuel-efficient commercial aircraft: the wide-bodied jets.

Competitive Issues

Going from a perceived need to the initiation of a development program requires more than a statement of that need. Regardless of the source of funding (government or private), there is likely to be competition for the resources necessary to demonstrate a bona fide need. In the case of the military, it is not unusual for competition to come from another department or service. For example, should maritime superiority be primarily a domain of the surface or air navy, or a combination of the two? Should cleaner air be achieved by more restrictions on the automobile engine combustion process or on the chemical composition of the fuel? The answers to these types of questions can have a major impact on the direction of any resulting development. For these reasons, strong competition can be expected from many sectors when it is publicly known that a new system development is under consideration. The task of sorting out these possibilities for further consideration is a major systems engineering responsibility.

Design Materialization Status

As described in Chapter 4, the phases of the system development process can be considered as steps in which the system gradually materializes, that is, progresses from a general concept to a complex assembly of hardware and software that performs an operational function. In this initial phase of the system life cycle, this process of materialization has only just started. Its status is depicted in Table 6.1, an overlay of Table 4.1 in Chapter 4.

The focus of attention in this phase is on the system operational objectives and goes no deeper than the subsystem level. Even at that level, the activity is listed as "visualize" rather than definition or design. The term *visualize* is used here and elsewhere in the book in its normal sense of "forming a mental image or vision," implying a conceptual rather than a material view of the subject. It is at this level of generality that most designs first originate, drawing on analogies from existing system elements.

Table 6.1 (and Table 4.1) oversimplifies the representation of the evolving state of a system by implying that all of its elements begin as wholly conceptual and evolve at a uniform rate throughout the development. This is very seldom, if ever, the case in practice. To take an extreme example, a new system based on rectifying a major deficiency in one of the subsystems of its predecessor may well retain the majority of the other subsystems with little change, except perhaps in the selection of production parts.

TABLE 6.1. Status of System Materialization at the Needs Analysis Phase

| | Phase | | | | | |
| | Concept development | | | Engineering development | | |
Level	Needs analysis	Concept exploration	Concept definition	Advanced development	Engineering design	Integration and evaluation
System	Define system capabilities and effectiveness	Identify, explore, and synthesize concepts	Define selected concept with specifications	Validate concept		Test and evaluate
Subsystem		Define requirements and ensure feasibility	Define functional and physical architecture	Validate subsystems		Integrate and test
Component			Allocate functions to components	Define specifications	Design and test	Integrate and test
Subcomponent		Visualize		Allocate functions to subcomponents	Design	
Part					Make or buy	

Such a new system would start out with many of its subsystems well advanced in materialization status, and with very few, if any, in a conceptual status.

Similarly, if a new system is technology driven, as when an innovative technical approach promises a major operational advance, it is likely that parts of the system not directly involved in the new technology will be based on existing system components. Thus, the materialization status of the system in both examples will not be uniform across its parts but will differ for each part as a function of its derivation. However, the general principle illustrated in the table is nevertheless valuable for the insight it provides into the system development process.

Applying the Systems Engineering Method in Needs and Requirements Analysis

Being the initial phase in the system development cycle, the needs analysis phase is inherently different from most of the succeeding phases. There being no preceding phase, the inputs come from different sources, especially depending on whether the development is needs driven or technology driven, and on whether the auspices are the government or a commercial company.

Nevertheless, the activities during the needs analysis phase can be usefully discussed in terms of the four basic steps of the systems engineering method described in Chapter 4, with appropriate adaptations. These activities are summarized below: the generic names of the individual steps as used in Figure 4.12 are listed in parentheses.

Operations Analysis (Requirements Analysis). Typical activities include
- analyzing projected needs for a new system, either in terms of serious deficiencies of current systems or the potential of greatly superior performance or lower cost by the application of new technology;
- understanding the value of fulfilling projected needs by extrapolating over the useful life of a new system; and
- defining quantitative operational objectives and the concept of operation.

The general products of this activity are a list of *operational objectives* and *system capabilities*.

Functional Analysis (Functional Definition). Typical activities include
- translating operational objectives into functions that must be performed and
- allocating functions to subsystems by defining functional interactions and organizing them into a modular configuration.

The general product of this activity is a list of initial *functional requirements*.

Feasibility Definition (Physical Definition). Typical activities include
- visualizing the physical nature of subsystems conceived to perform the needed system functions and

- defining a feasible concept in terms of capability and estimated cost by varying (trading off) the implementation approach as necessary.

The general product of this activity is a list of initial *physical requirements*.

Needs Validation (Design Validation). Typical activities include
- designing or adapting an effectiveness model (analytical or simulation) with operational scenarios, including economic (cost, market, etc.) factors;
- defining validation criteria;
- demonstrating the cost-effectiveness of the postulated system concept, after suitable adjustment and iteration; and
- formulating the case for investing in the development of a new system to meet the projected need.

The general product of this activity are a list of operational *validation criteria*.

Given a successful outcome of the needs analysis process, it is necessary to translate the operational objectives into a formal and quantitative set of *operational requirements*. Thus, this phase produces four primary products. And since three of these have the name "requirements" as part of their description, it can be confusing to separate the three. The primary output of the needs analysis phase is the set of operational requirements. But let us introduce four types of requirements so as not to confuse the reader.

Operational Requirements. These refer largely to the mission and purpose of the system. The set of operational requirements will describe and communicate the end state of the world after the system is deployed and operated. Thus, these types of requirements are broad and describe the overall objectives of the system. All references relate to the system as a whole. Some organizations refer to these requirements as capability requirements, or simply required capabilities.

Functional Requirements. These refer largely to what the system should do. These requirements should be action oriented and should describe the tasks or activities that the system performs during its operation. Within this phase, they refer to the system as a whole, but they should be largely quantitative. These will be significantly refined in the next two phases.

Performance Requirements. These refer largely to how well the system should perform its requirements and affect its environment. In many cases, these requirements correspond to the two types above and provide minimal numerical thresholds. These requirements are almost always objective and quantitative, though exceptions occur. These will be significantly refined in the next two phases.

Physical Requirements. These refer to the characteristics and attributes of the physical system and the physical constraints placed upon the system design. These may include appearance, general characteristics, as well as volume, weight, power, and material and external interface constraints to which the system must adhere. Many organizations do not have a special name for these and refer to them simply as *constraints*, or even *system requirements*. These will be significantly refined in the next two phases.

For new start systems, the first iteration through the needs and requirements analysis phase results in a set of operational requirements that are rather broad and are not completely defined. In the military, for example, the requirements-like document that emerges from the needs analysis is formally known as the ICD. This term is also used in the non-DoD community as a generic description of capabilities desired. In either case, the ICD document contains a broad description of the system concept needed and focuses on operational, or capability, requirements. Only top-level functional, performance, and physical requirements are included. Later documents will provide detail to this initial list.

The elements of the systems engineering method as applied to the needs analysis phase described above are displayed in the flow diagram of Figure 6.2. It is a direct adaptation of Figure 4.12, with appropriate modifications for the activities in this phase. Rectangular blocks represent the four basic steps, and the principal activities are shown as circles, with the arrows denoting information flow.

The inputs at the top of the diagram are operational deficiencies and technological opportunities. Deficiencies in current systems due to obsolescence or other causes are need drivers. Technological opportunities resulting from an advance in technology that offers a potential major increase in performance or a decrease in cost of a marketable system are technology drivers. In the latter case, there must also be a projected concept of operation for the application of the new technology.

The two middle steps are concerned with determining if there is at least one possible concept that is likely to be feasible at an affordable cost and at an acceptable risk. The validation step completes the above analysis and also seeks to validate the significance of the need being addressed in terms of whether or not it is likely to be worth the investment in developing a new system. Each of these four steps is further detailed in succeeding sections of this chapter.

6.2 OPERATIONS ANALYSIS

Whether the projected system development is need driven or technology driven, the first issue that must be addressed is the existence of a valid need (potential market) for a new system. The development of a new system or a major upgrade is likely to be very costly and will usually extend over several years. Accordingly, a decision to initiate such a development requires careful and deliberate study.

Analysis of Projected Needs

In the commercial sector, market studies are continuously carried out to assess the performance of existing products and the potential demand for new products. Customer reactions to product characteristics are solicited. The reason for lagging sales is sys-

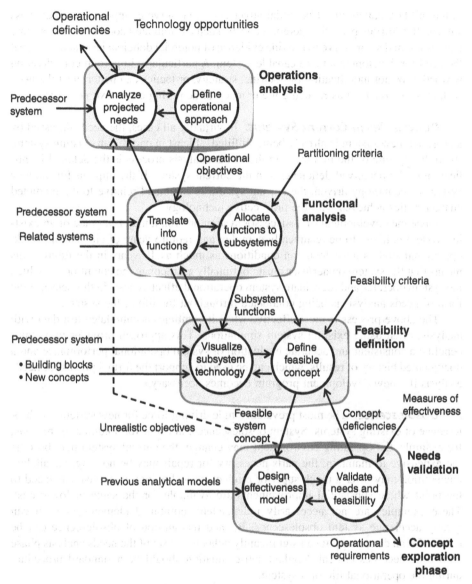

Figure 6.2. Needs analysis phase flow diagram.

tematically probed. The strengths and weaknesses of competing systems and their likely future growth are analyzed.

For military systems, each service has one or more systems analysis organizations whose responsibility is to maintain a current assessment of their operational capability and readiness. These organizations have access to intelligence assessments of changes in the military capability of potential adversaries that serve as inputs to effectiveness studies. In addition, periodic operational tests, such as simulated combat at sea, landing operations, and so on, serve to provide evidence of potential deficiencies that may signal the need for developing a more capable system. A particularly important consideration is whether or not modification of doctrine, strategy, or tactics can better meet the need with existing assets, thus reducing the urgency of acquiring expensive new assets.

Deficiencies in Current Systems. In virtually all cases, the need addressed by a projected new system is already being fulfilled, at least in part, by an existing system. Accordingly, one of the first steps in the needs analysis process is the detailed identification of the perceived deficiencies in the current system. If the impetus for the new system is technology driven, the current system is examined relative to the predicted characteristics achieved with the prospective technology.

Since the development of a successor system or even a major upgrade of an existing system is likely to be technically complex and require years of challenging work, operational studies must focus on conditions as much as 10 years in the future. This means that the system owner/user must continually extrapolate the conditions in which the system operates and reevaluate system operational effectiveness. In this sense, some form of needs analysis is being conducted throughout the life of the system.

The above process is most effective when it combines accumulated test data with analysis, often using existing system simulations. This approach provides two major benefits: a consistent and accurate evaluation of system operational performance and a documented history of results, which can be used to support the formal process of needs analysis if a new development program becomes necessary.

Obsolescence. The most prevalent single driving force for new systems is obsolescence of existing systems. System obsolescence can occur for a number of reasons; for example, the operating environment may change; the current system may become too expensive to maintain; the parts necessary for repair may be no longer available; competition may offer a much superior product; or technology may have advanced to the point where substantial improvements are available for the same or lower cost. These examples are not necessarily independent; combined elements of each can greatly accelerate system obsolescence. Belated recognition of obsolescence can be painful for all concerned. It can significantly delay the onset of the needs analysis phase until time becomes critical. Vigilant self-evaluation should be a standard procedure during the operational life of a system.

An essential factor in maintaining a viable system is keeping aware of advances in technology. Varied research and development (R & D) activities are carried out by many agencies and industry. They receive support from government or private funding or combinations of both. In the defense sector, contractors are authorized to use a percentage of their revenues on relevant research as allowable overhead. Such activity is called independent research and development (IRAD). There are also a number of wholly or partially government-funded exploratory development efforts. Most large producers of commercial products support extensive applied R & D organizations. In any case, the wise system sponsor, owner, or operator should continually keep abreast of these activities and should be ready to capitalize on them when the opportunity presents itself. Competition at all levels is a potent driver of these activities.

Operational Objectives

The principal outcome of operational studies is the definition of the objectives, in operational terms, that a new system must meet in order to justify its development. In a needs-driven development, these objectives must overcome such changes in the environment or deficiencies in the current system as have generated the pressure for an improved system. In a technology-driven development, the objectives must embody a concept of operations that can be related to an important need.

The term "objectives" is used in place of "requirements" because at this early stage of system definition, the latter term is inappropriate; it should be anticipated that many iterations (see Fig. 6.2) would take place before the balance between operational performance and technical risk, cost, and other developmental factors will be finally established.

To those inexperienced in needs analysis, the development of objectives can be a strange process. After all, engineers typically think in terms of requirements and specifications, not high-level objectives. Although objectives should be quantifiable and objective, the reality is that most are qualitative and subjective at this early stage. Some rules of thumb can be helpful:

- Objectives should address the end state of the operational environment or scenario—it focuses on what the system will accomplish in the large sense.
- Objectives should address the purpose of the system and what constitutes the satisfaction of the need.
- Taken together, objectives answer the "why" question—why is the system needed?
- Most objectives start with the infinitive word "provide," but this is not mandatory.

Objectives Analysis. The term objectives analysis is the process of developing and refining a set of objectives for a system. Typically, the product of this effort is an objectives tree, where a single or small set of top-level objectives are decomposed into a set of primary and secondary objectives. Figure 6.3 illustrates this tree. Decomposition is appropriate until an objective becomes verifiable, or you begin to define functions of the system. When that occurs, stop at the objective. The figure illustrates functions by graying the boxes—they would not be part of your objectives tree. In our experience, most objectives trees span one or two levels deep; there is no need to identify extensive depth.

As an example of an objectives analysis, think about a new automobile. Suppose an auto company wants to design a new passenger vehicle, which it can market as "green" or environmentally friendly. Understanding the objectives of this new car establishes priorities for the eventual design. Thus, company management begins an objectives exercise. Objectives analysis forces the company, both management and the

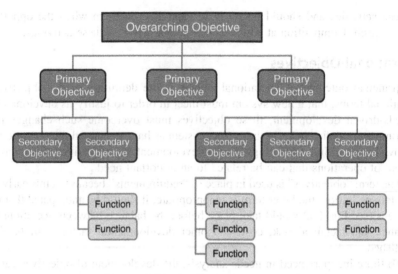

Figure 6.3. Objectives tree structure.

technical staff, to evaluate and decide what is important when developing a new system. Therefore, it is worth investing some time, energy, and capital in determining what the overall objectives of the system are. Moreover, agreeing to a concise single statement helps focus the development team to the job at hand.

In the automobile example, the company might soon realize that the overall objective of this new vehicle is to provide users with clean transportation. The top-level objective does not include performance, cargo capacity, off-road capability, and so on. In the overall objective are two key words: clean and transportation. Both imply various aspects or attributes of this new car. Since both words are not yet well defined, we need to decompose them further. But the overall goal is clear: this vehicle is going to be environmentally "clean" and will provide sufficient transportation.

The first decomposition focuses the thinking of the development team. Clearly, the two key words need to be "fleshed out." In this case, "clean" may mean "good gas mileage" as well as "comfortable." Transportation also implies a safe and enjoyable experience in the vehicle as it travels from one point to the next. There may also be another objective that is loosely tied to *clean* and *transportation*—cost.

Thus, in our example, the development team focuses on four primary objectives that flow from our overarching objective: comfort, mileage, safety, and cost. These four words need to be worded as an objective of course. Figure 6.4 presents one possibility of an objectives tree.

In determining whether an objective needs further decomposition, one should ask a couple of questions:

- Does the objective stand on its own in terms of clarity of understanding?
- Is the objective verifiable?

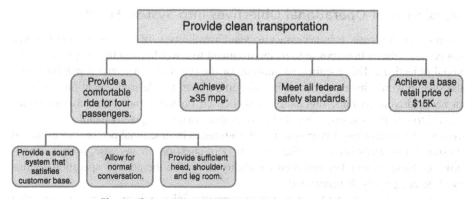

Figure 6.4. Example objectives tree for an automobile.

- Would decomposition lead to better understanding?
- Are requirements and functions readily implied by the objective?

In our example, one could argue that three of the primary objectives are sufficient as stated, and all three are verifiable. Only the subjective objective relating to comfort needs further decomposition. In this case, comfort can be divided into three components: a sound system, noise levels that allow conversation, and physical space. As worded in the figure, these three could all be verified by various methods (a satisfaction survey in the first, a definition of noise levels for normal conversation, and volume requirements). Having an objectives tree focuses the development effort on the priorities. In our example, the four primary objectives communicate what is important with this new automobile.

In many cases where objectives trees are used, an initial tree will be similar to our example, listing only those objectives that are the highest priorities. These trees would then be expanded to include other areas that will need to be addressed. For our automobile, these "other" areas would include maintenance considerations, human–system interaction expectations, and cargo space, to name a few. An objective of having an objectives tree is ultimately to identify the functions and their performance requirements. Therefore, the logical next step after objectives analysis is functional analysis.

6.3 FUNCTIONAL ANALYSIS

At this initial phase of the system development process, functional analysis is an extension of operational studies, directed to establishing whether there is a feasible technical approach to a system that could meet the operational objectives. At this stage, the term "feasible" is synonymous with "possible" and implies making a case that there is a good likelihood that such a system could be developed within the existing state of the art, without having to prove it beyond reasonable doubt.

Translation of Operational Objectives into System Functions

To make such a case, it is necessary to visualize the type of system that could carry out certain actions in response to its environment that would meet the projected operational objectives. This requires an analysis of the types of functional capabilities that the system would have to possess in order to perform the desired operational actions. In needs-driven systems, this analysis is focused on those functional characteristics needed to satisfy those operational objectives that are not adequately handled by current systems. In technology-driven systems, the advances in functional performances would presumably be associated with the technology in question. In any case, both the feasibility of these approaches and their capability to realize the desired operational gains must be adequately demonstrated.

The visualization of a feasible system concept is inherently an abstract process that relies on reasoning on the basis of analogy. This means that all the elements of the concept should be functionally related to elements of real systems. A helpful approach to the translation of operational objectives to functions is to consider the type of primary media (signals, data, material, or energy) that are most likely to be involved in accomplishing the various operational objectives. This association usually points to the class of subsystems that operate on the medium, as, for example, sensor or communication subsystems in the case of signals, computing subsystems for data, and so on. In the above process, it must be shown that all of the principal system functions, especially those that represent advances over previous systems, are similar to those already demonstrated in some practical context. An exception to this process of reasoning by analogy is when an entirely new type of technology or application is a principal part of a proposed system; in this case, it may be necessary to go beyond analysis and to demonstrate its feasibility by modeling and, ultimately, experimentation.

In identifying the top-level functions that the system needs to perform, it is important even at this early stage to visualize the entire system life cycle, including its nonoperational phases.

The above discussion is not meant to imply that all considerations at this stage are qualitative. On the contrary, when primarily quantitative issues are involved, as in the example of automobile pollution, it is necessary to perform as much quantitative analysis as available resources and existing knowledge permit.

Allocation of Functions to Subsystems

In cases where all operational objectives can be directly associated with system-level functions that are analogous to those presently exhibited by various real systems, it is still essential to visualize just how these might be allocated, combined, and implemented in the new system. For this purpose, it is not necessary to visualize some best system configuration. Rather, it need only be shown that the development and production of an appropriate system is, in fact, feasible. Toward this end, a top-level system concept that implements all the prescribed functions should be visualized in order to demonstrate that the desired capabilities can be obtained by a plausible combination of the prescribed functions and technical features. Here it is particularly important that all interactions and interfaces, both external and internal to the system, be identified and associated with the system functions, and that a trade-off process be employed to ensure that the consideration of the various system attributes is thorough and properly balanced. This is typically done in terms of an initial concept of operation.

6.4 FEASIBILITY DEFINITION

The feasibility of a system concept (and therefore of meeting the projected need) cannot be established solely on the basis of its functional design. The issue of feasibility must also address the physical implementation. In particular, system cost is always a dominant consideration, especially as it may compare to that of other alternatives, and this cannot be judged at the functional level. Accordingly, even at this initial phase of system development, it is necessary to visualize the physical makeup of the system as it is intended to be produced. It is also necessary to visualize all external constraints and interactions, including compatibility with other systems.

While it is necessary to consider the physical implementation of the projected system in the needs analysis phase, this does not imply that any design decisions are made at this time. In particular, no attempt should be made to seek optimum designs; those issues are dealt with much later in the development process. The focus at this point is to establish feasibility to meet a given set of operational objectives. It is the validation of these objectives that is the primary purpose of the needs analysis phase. The paragraphs that follow discuss some of the issues that need to be considered, but only in an exploratory way.

Visualization of Subsystem Implementation

Given the allocation of functions to subsystems, it is necessary to envision how these might be implemented. At this stage, it is only necessary to find examples of similar functional units in existing systems so that the feasibility of applying the same type of technology to the new system may be assessed. The identification of the principal media involved in each major function (signal, data, material, and energy), as discussed in the previous section, is also helpful in finding systems with similar functional elements and, hence, with physical implementations representative of those required in the new system.

Relation to the Current System. Where there exists a system that has been meeting the same general need for which the new system is intended, there are usually a number of subsystems that may be candidates for incorporation in modified form in the new system. Whether or not they will be utilized as such, they are useful in building a case for system feasibility and for estimating part of the development and production cost of the new system.

Existing models and simulations of the current system are especially useful tools in this type of analysis since they will usually have been verified against data gathered over the life of the system. They may be used to answer "What if?" questions and to find the driving parameters, which helps to focus the analysis process. Another important tool, used in conjunction with the system simulation, is an effectiveness model and the analytic techniques of effectiveness analysis, as described in the next section.

Other less tangible factors can also come into play, such as the existence of a support infrastructure. In the case of the automobile engine, many years of successful operation have established a very wide base of support for conventional reciprocating engines in terms of repair sites, parts suppliers, and public familiarity. Because of the prospective cost for changing this base, innovative changes, such as the Wankel rotary engine and designs based on the Stirling cycle, have been resisted. The point here is that beneficial technological innovations are often overridden by economic or psychological resistance to change.

Application of Advanced Technology. In technology-driven systems, it is more difficult to establish feasibility by reference to existing applications. Instead, it may be necessary to build the case on the basis of theoretical and experimental data available from such research and development work as has been done on the candidate technology. In case this proves to be insufficient, limited prototyping may be required to demonstrate the basic feasibility of the application. Consultation with outside experts may be helpful in adding credibility to the feasibility investigation.

Unfortunately, highly touted technical advances may also come with unproven claims and from unreliable sources. Sometimes, a particular technology may offer a very substantial gain but lacks maturity and an established knowledge base. In such situations, the case for incorporating the technology should be coupled with a comparably capable backup alternative. Systems engineers must be intimately involved in the above process to keep the overall system priorities foremost.

Cost. The assessment of cost is always an important concern in needs analysis. This task is particularly complicated when there is a mix of old, new, and modified subsystems, components, and parts. Here again, cost models and maintenance records of the current system, combined with inflation factors, can be helpful. By comparing similar components and development activities, cost estimation will at least have a credible base from which to work. In the case of new technology, cost estimates should contain provisions for substantial development and testing prior to commitment for its use.

Definition of a Feasible Concept

To satisfy the objectives of the needs analysis phase, the above considerations should culminate in the definition and description of a plausible system concept, and a well-documented substantiation of its technical feasibility and affordability. The system description should include a discussion of the development process, anticipated risks, general development strategy, design approach, evaluation methods, production issues, and concept of operations. It should also describe how the cost of system development and production had been assessed. It need not be highly detailed but should show that all major aspects of system feasibility have been addressed.

6.5   NEEDS VALIDATION

The final and most critical step in the application of the systems engineering method is the systematic examination of the validity of the results of the previous steps. In the case of the needs analysis phase, the validation step consists of determining the basic soundness of the case that has been made regarding the existence of a need for a new system and for the feasibility of meeting this need at an affordable cost and at an acceptable risk.

Operational Effectiveness Model

In the concept development stage the analyses that are designed to estimate the degree to which a given system concept may be expected to meet a postulated set of operational requirements is called *operational effectiveness analysis*. It is based on a mathematical model of the operational environment and of the candidate system concept being analyzed.

In effectiveness analysis, the operational environment is modeled in terms of a set of scenarios—postulated actions that represent a range of possible encounters to which the system must react. Usually, initial scenarios are selected to present the more likely situations, followed by more advanced cases for testing the limits of the operational requirements. For each scenario, the acceptable responses of the system in terms of operational outcomes are used as evaluation criteria. To animate the engagements between the system model and the scenarios, an effectiveness model is designed with the capability of accepting variable system performance parameters from the system model. A more extensive treatment of operational scenarios is contained in the next section.

Effectiveness analysis must include not only the operational modes of the system but also must represent its nonoperating modes, such as transport, storage, installation, maintenance, and logistics support. Collectively, all the significant operational requirements and constraints need to be embodied in operational scenarios and in the accompanying documentation of the system environment.

System Performance Parameters. The inputs from the system model to the effectiveness analysis are values of performance characteristics that define the system's response to its environment. For example, if a radar device needs to sense the presence of an object (e.g., an aircraft), its predicted sensing parameters are entered to determine the distance at which the object will be detected. If it needs to react to the presence of the object, its response processing time will be entered. The effectiveness model ensures that all of the significant operational functions are addressed in constructing the system model.

Measures of Effectiveness (MOE). To evaluate the results of effectiveness simulations, a set of criteria is established that identifies those characteristics of the system response to its environment that are critical to its operational utility. These are called "MOE." They should be directly associated with specific objectives and prioritized according to their relative operational importance. MOE and measures of performance (MOP) are described in more detail below.

While the effort required to develop an adequate effectiveness model for a major system is extensive, once developed, it will be valuable throughout the life of the system, including potential future updates. In the majority of cases where there is a current system, much of the new effectiveness model may be derived from its predecessor.

The Analysis Pyramid. When estimating or measuring the effectiveness of a system, the analyst needs to determine the perspective within which the system's effectiveness will be described. For example, the system effectiveness may be described within a larger context, or mission, where the system is one of many working loosely or tightly together to accomplish a result. On the other hand, effectiveness can be described in terms of an individual system's performance in a given situation in response to selected stimuli, where interaction with other systems is minimal.

Figure 6.5 depicts a common representation of what is known as the analysis pyramid. At the base of the pyramid is the foundational physics and physical phenomenology knowledge. Analysis at this end of the spectrum involves a detailed evaluation of environmental interactions, sometimes down to the molecular level.

As the analyst travels up the pyramid, details are abstracted and the perspective of the analyst broadens, until he reaches the apex. At this level, technical details have been completely abstracted and the analysis focuses on strategy and policy alternatives and implications.

The systems engineer will find that typically, analysis perspectives during the needs analysis phase tend to be near the top of the pyramid. Although strategy may not be in the domain of the system development effort, certainly the system's effectiveness within a multiple-mission or a single-mission context would need to be explored. The lower

Figure 6.5. Analysis pyramid.

part of the pyramid is usually not analyzed due to lack of system definition. As the system becomes more defined, the analysis performed will tend to migrate down the pyramid. We will explore the analysis pyramid more as we continue our look at systems engineering within the development phases.

MOE and MOP

With the introduction of operational effectiveness analysis, we need to explore the concept and meaning of certain metrics. Metrics are key to ultimately defining the system, establishing meaningful and verifiable requirements, and testing the system. Therefore, defining these metrics appropriately and consistently through the development life cycle is essential.

Many terms exist to describe these effectiveness and performance metrics. Two commonly used terms (and ones we will use throughout this book) are MOEs and MOPs. Unfortunately, no universal definitions exist for these terms. But the basic concept behind them is crucial to understanding and communicating a system concept.

We propose the following definitions for this book:

MOE: a qualitative or quantitative metric of a system's overall performance that indicates the degree to which it achieves it objectives under specified conditions. An MOE always refers to the system as a whole.

MOP: a quantitative metric of a system's characteristics or performance of a particular attribute or subsystem. An MOP typically measures a level of physical performance below that of the system as a whole.

Regardless of the definition you use, it is a universal axiom that an MOE is superior to MOP. In other words, if the two are placed in a hierarchy, MOEs will always be above MOPs.

Typically, an MOE or MOP will have three parts: the metric, its units, and the conditions or context under which the metric applies. For example, an MOE of a new recreational aircraft (such as a new version of Piper Cub) would be maximum range, in nautical miles at sea level on a standard atmospheric day. The metric is "maximum range"; the units are "nautical miles"; and the conditions are "a standard atmospheric day (which is well defined) at sea level." This MOE relates to the aircraft as a whole and describes one aspect of its performance in achieving the objective of aerial flight.

MOEs can be of many forms, but we can define three general categories: measurement, likelihood, or binary. Measurement is an MOE that can be directly measured (either from an actual system, subsystem, or mathematical or physical model). It may be deterministic or random. Likelihood MOEs correspond to a probability of an event occurring and may include other MOEs. For example, a likelihood MOE could be the probability of an aircraft achieving a maximum altitude of 20,000 ft. In this case, the likelihood is defined in terms of another measurement MOE. Finally, a binary MOE is a logical variable of the occurrence of an event. Either the event occurs or not.

When an MOE is measured or determined, we call the resultant measurement the *value* of the MOE. Thus, in our aircraft example, if we measure the maximum range of a new aircraft as 1675 nm, then "1675" is the value. Of course, MOEs, as any metrics, can have multiple values under different conditions or they could be random values.

Finally, engineers use binary MOEs to determine whether a particular characteristic of a system exceeds a threshold. For example, we could define a threshold for the maximum range of an aircraft as 1500 nm at sea level on a standard day. A binary MOE could then be defined to determine whether a measured value of the MOE exceeds our threshold. For example, the binary MOE would be "yes"; our measured value of 1675 nm exceeds our threshold of 1500 nm.

MOEs and MOPs are difficult concepts to grasp! Unless one has worked with metrics before, they tend to be confusing. Many students of systems engineering will provide a requirement when asked for an MOE. Others provide values. Still others simply cannot identify MOEs for a new system. However, the concept of measures is utilized throughout the systems engineering discipline. We will revisit these concepts in the subsequent chapters.

Validation of Feasibility and Need

Finally, the effectiveness analysis described above is mainly directed to determining whether or not a system concept, derived in the functional and physical definition process, is (1) feasible and (2) satisfies the operational objectives required to meet a projected need. It assumes that the legitimacy of the need has been established previously. This assumption is not always a reliable one, especially in the case of technology-driven system developments, where the potential application is new and its acceptance depends on many intangible factors. A case in point, of which there are hundreds of examples, is the application of automation to a system previously operated mainly by people. (The airplane reservation and ticketing system is one of the larger successful ones.) The validation of the need for such a system requires technical, operational, and market analyses that seek to take into account the many complex factors likely to affect the acceptability of an automated system and its probable profitability.

In complex cases such as the above example, only a very preliminary validation can be expected before considerable exploratory development and experimentation should take place. However, even a preliminary validation analysis will bring out most of the critical issues and may occasionally reveal that the likelihood of meeting some postulated needs may be too problematical to warrant a major investment at the current state of the technology.

6.6 SYSTEM OPERATIONAL REQUIREMENTS

The primary product of needs analysis is a set of operational objectives, which are then translated into a set of operational requirements. The system operational requirements that result from the needs analysis phase will establish the reference against which the subsequent development of a system to meet the projected needs will be judged. Accordingly, it is essential that these requirements be clear, complete, consistent, and feasible of accomplishment. The feasibility has presumably been established by the identification of at least one system approach that is judged to be both feasible and capable of meeting the need. It remains to make certain that the operational requirements are adequate and consistent.

Operational Scenarios

A logical method of developing operational requirements is to postulate a range of scenarios that together are representative of the full gamut of expected operational situations. These scenarios must be based on an extensive study of the operational environment, discussions with experienced users of the predecessor and similar systems, and a detailed understanding of past experience and demonstrated deficiencies of current systems. It is especially important to establish the user priorities for the required improvements, in particular, those that appear most difficult to achieve.

While scenarios range widely in their content depending on their application, we are able to define five basic components of almost all scenarios.

1. *Mission Objectives.* The scenario should identify the overall objectives of the mission represented, and the purpose and role of the system(s) in focus in accomplishing those objectives. In some cases, this component is system independent, meaning that the role of any one system is not presented—only a general description of the mission at stake and the objectives sought. In a commercial example, the mission could be to capture market share. In a government example, the mission might be to provide a set of services to constituents. In a military example, the mission might be to take control of a particular physical installation.

2. *Architecture.* The scenario should identify the basic system architecture involved. This includes a list of systems, organizations, and basic structural information. If governance information is available, this would be included. This component could also include basic information on system interfaces or a description of the information technology infrastructure. In essence, a description of the resources available is provided. In a commercial scenario, the resources of the organization are described. If this is a government scenario, the organizations and agencies involved in the mission are described. If this is a military scenario, these resources could include the units involved, with their equipment.

3. *Physical Environment.* The scenario should identify the environment in which the scenario takes place. This would include the physical environment (e.g., terrain, weather, transportation grid, and energy grid) as well as the business environment (e.g., recession and growth period). "Neutral" entities are described in this section. For example, customers and their attributes would be defined, or neutral nations and their resources.

4. *Competition.* The scenario should identify competition to your efforts. This may be elements that are directly opposed to your mission success, such as a

software hacker or other type of "enemy." This may be your competition in the market or outside forces that influence your customers. This could also include natural disasters, such as a tsunami or hurricane.

5. *General Sequence of Events.* The scenario should describe a general sequence of events within the mission context. We are careful to use the term *general* though. The scenario should allow for freedom of action on the part of the players. Since we use scenarios to generate operational requirements and to estimate system effectiveness, we need the ability to alter various parameters and events within the overall scenario description. Scenarios should not "script" the system; they are analysis tools, not shackles to restrain the system development. Thus, scenarios typically provide a general sequence of events and leave the details to an analyst using the scenario. At times, a scenario may provide a detailed sequence of events leading up to a point in time, whereby the analysis starts and actions may be altered from that time forward.

A scenario could include much more, depending on its application and intended purpose. They come in all sizes, from a short, graphic description of a few pictures to hundreds of pages of text and data.

Even though the operational scenarios developed during this phase are frequently not considered a part of the formal operational requirements document, in complex systems, they should be an essential input to the concept exploration phase. Experience has shown that it is seldom possible to encompass all of the operational parameters into a requirements document. Further, the effectiveness analysis process requires operational inputs in scenario form. Accordingly, a set of operational scenarios should be appended to the requirements document, clearly stating that they are representative and not a comprehensive statement of requirements.

As noted above, the scenarios should include not only the active operational interactions of the system with its environment but also the requirements involved in its transport, storage, installation, maintenance, and logistics support. These phases often impose physical and environmental constraints and conditions that are more severe than normal operations. The only means for judging whether or not requirements are complete is to be sure that all situations are considered. For example, the range of temperature or humidity of the storage site may drastically affect system life.

Operational Requirements Statements

Operational requirements must initially be described in terms of operational outcomes rather than system performance. They must not be stated in terms of implementation nor biased toward a particular conceptual approach. All requirements should be expressed in measurable (testable) terms. In cases where the new system is required to use substantial portions of an existing system, this should be specifically stated.

The rationale for all requirements must be stated or referenced. It is essential for the systems engineers leading the system development to understand the requirements in terms of user needs so that inadvertent ambiguities do not result in undue risks or costs.

The time at which a new system needs to be available is not readily derived from purely operational factors but may be critical in certain instances due to financial factors, obsolescence of current systems, schedules of system platforms (e.g., airplanes and airports), and other considerations. This may place constraints on system development time and hence on the degree of departure from the existing system.

Since the initially stated operational requirements for a new system are seldom based on an exhaustive analysis, it should be understood by both the customer and potential developer that these requirements will be refined during the development process, as further knowledge is gained concerning the system needs and operating environment.

From the above considerations, it is seen that work carried out during the needs analysis phase must be regarded as preliminary. Subsequent phases will treat all system aspects in more detail. However, experience has shown that the basic conceptual approach identified during needs analysis often survives into subsequent phases. This is to be expected because considerable time and effort is usually devoted to this process, which may last for 2 or 3 years. Even though only limited funds are expended, many organizations are often involved.

Feasibility Validation

Effectiveness analysis is intrinsically concerned with the functional performance of a system and therefore cannot in itself validate the feasibility of its physical implementation. This is especially true in the case where unproven technology is invoked to achieve certain performance attributes.

An indirect approach to feasibility validation is to build a convincing case by analogy with already demonstrated applications of the projected technique. Such an approach may be adequate, provided that the application cited is truly representative of that proposed in a new system. It is important, however, that the comparison be quantitative rather than only qualitative so as to support the assumed performance resulting from the technology application.

A direct approach to validating the feasibility of a new physical implementation is to conduct experimental investigations of the techniques to be applied to demonstrate that the predicted performance characteristics can be achieved in practice. This approach is often referred to as "critical experiments," which are conducted early in the program to explore new implementation concepts.

The resources available for carrying out the validation process in the needs analysis phase are likely to be quite limited, since the commitment to initiate the actual development of the system has not yet been made. Accordingly, the quality of the validation process will depend critically on the experience and ingenuity of the systems engineering staff. The experience factor is especially important here because of the dependence of the work on knowledge of the operational environment, of the predecessor system, of analyses and studies previously performed, of the technological base, and of the methods of systems analysis and systems engineering.

Importance of Feasibility Demonstration. In defining a basis for developing a new system, the needs analysis phase not only demonstrates the existence of an important unfulfilled operational need but also provides evidence that satisfying the need is feasible. Such evidence is obtained by visualizing a realistic system concept that has the characteristics required to meet the operational objectives. This process illustrates a basic systems engineering principle that establishing realistic system requirements must include the simultaneous consideration of a system concept that could meet those requirements. This principle contradicts the widely held notion that requirements, derived from needs, should be established prior to consideration of any system concept that can fulfill those requirements.

6.7 SUMMARY

Originating a New System

Objectives of the needs analysis phase are to identify a valid operational need for a new system and to develop a feasible approach to meeting that need. This needs-driven system development approach is characteristic of most defense and other government programs and typically stems from a deficiency in current system capabilities. This type of development requires a feasible and affordable technical approach.

The other major type of approach is the technology-driven system development approach. This approach is characteristic of most commercial system development and stems from a major technological opportunity to better meet a need. This type of development requires demonstration of practicality and marketability.

Activities comprising the needs analysis phase are the following:

- *Operations Analysis*—understanding the needs for a new system;
- *Functional Analysis*—deriving functions required to accomplish operations;
- *Feasibility Definition*—visualizing a feasible implementation approach; and
- *Needs Validation*—demonstrating cost-effectiveness.

Operations Analysis

Studies and analyses are conducted to generate and understand the operational needs of the system. These studies feed the development of an objectives tree—describing the hierarchy of system expectations and outcomes.

Functional Analysis

Initial system functions are identified and organized that will achieve operational objectives. These functions are vetted through analysis and presentation to users and stakeholders.

Feasibility Definition

The system development approach is decided upon, articulated to stakeholders, and approximately costed. Moreover, an early feasible concept is articulated. Finally, developing operational requirements commences.

Needs Validation

The vetted set of operational needs is now validated by operational effectiveness analysis, usually at multiple levels within the analysis pyramid. System concepts that satisfy the operational needs are evaluated with agreed-upon MOE and reflect the entire system life cycle.

PROBLEMS

6.1 Describe and define the principal outputs (products) of the needs analysis phase. List and define the primary systems engineering activities that contribute to these products.

6.2 Identify the relationships between operational objectives and functional requirements for the case of a new commuter aircraft. Cite three operational objectives and the functional requirements that are needed to realize these objectives. (Use qualitative measures only.)

6.3 Referring to Figure 6.2, which illustrates the application of the systems engineering method to the needs analysis phase, select one of the four sections of the diagram and write a description of the processes pictured in the diagram. Explain the nature and significance of the two processes represented by circles and of each internal and external interaction depicted by arrows. The description should be several times more detailed than the definition of the step in the subsection describing the systems engineering method in needs analysis.

6.4 What is meant by "MOE"? For the effectiveness analysis of a sport utility vehicle (SUV), list what you think would be the 10 most important characteristics that should be exercised and measured in the analysis.

6.5 For six of the MOE of the SUV (see Problem 6.4), describe an operational scenario for obtaining a measure of its effectiveness.

6.6 Assume that you have a business in garden care equipment and are planning to develop one or two models of lawn tractors to serve suburban homeowners. Consider the needs of the majority of such potential customers and write at least six operational requirements that express these needs. Remember the qualities of good requirements as you do so. Draw a context diagram for a lawn tractor.

6.7 Given the results of Problem 6.6, describe how you would perform an analysis of alternatives to gain an understanding of the functional requirements and optional features that could fit the tractor to individual needs. Describe the

MOE you would use and the alternative architectures you would analyze. Describe the pros and cons for a single model as opposed to two models of different sizes and powers.

FURTHER READING

B. Blanchard and W. Fabrycky. *System Engineering and Analysis*, Fourth Edition. Prentice Hall, 2006, Chapter 3.

A. D. Hall. *A Methodology for Systems Engineering*. Van Nostrand, 1962, Chapter 6.

International Council on Systems Engineering. *Systems Engineering Handbook. A Guide for System Life Cycle Process and Activities*, Version 3.2, July 2010.

N. B. Reilly. *Successful Systems for Engineers and Managers*. Van Nostrand Reinhold, 1993, Chapter 4.

A. P. Sage and J. E. Armstrong, Jr. *Introduction to Systems Engineering*. Wiley, 2000, Chapter 3.

R. Stevens, P. Brook, K. Jackson, and S. Arnold. *Systems Engineering: Coping with Complexity*. Prentice Hall, 1998, Chapter 2.

7

CONCEPT EXPLORATION

7.1 DEVELOPING THE SYSTEM REQUIREMENTS

Chapter 6 discussed the process of needs analysis, which is intended to provide a well-documented justification for initiating the development of a new system. The process also produces a set of operational requirements (or objectives) that describe what the new system must be designed to do. Assuming that those responsible for authorizing the initiation of a system development have been persuaded that these preliminary requirements are reasonable and attainable within the constraints imposed by time, money, and other external constraints, the conditions have been achieved for taking the next step in the development of a new system.

The principal objective of the concept exploration phase, as defined here, is to convert the operationally oriented view of the system derived in the needs analysis phase into an engineering-oriented view required in the concept definition and subsequent phases of development. This conversion is necessary to provide an explicit and quantifiable basis for selecting an acceptable functional and physical system concept, and then for guiding its evolution into a physical model of the system. It must be

Systems Engineering Principles and Practice, Second Edition. Alexander Kossiakoff, William N. Sweet, Samuel J. Seymour, and Steven M. Biemer
© 2011 by John Wiley & Sons, Inc. Published 2011 by John Wiley & Sons, Inc.

remembered, however, that the performance requirements are an interpretation, not a replacement of operational requirements.

As in the case of operational requirements, the derivation of system performance requirements must also simultaneously consider system concepts that could meet them. However, to ensure that the performance requirements are sufficiently broad to avoid unintentionally restricting the range of possible system configurations, it is necessary to conceive not one, but to explore a variety of candidate concepts.

New systems that strive for a major advance in capability over their predecessors, or depend on the realization of a technological advance, require a considerable amount of exploratory research and development (R & D) before a well-founded set of performance requirements can be established. The same is true for systems that operate in highly complex environments and whose characteristics are not fully understood. For these cases, an objective of the concept exploration phase is to acquire the needed knowledge through applied R & D. This objective may sometimes take several years to accomplish, and occasionally, these efforts prove that some of the initial operational objectives are impracticable to achieve and require major revision.

For the above reasons it, is appropriate that this chapter, which deals with the development of system requirements, is entitled "Concept Exploration." Its intent is to describe the typical activities that take place in this phase of system development and to explain their whys and hows.

The discussion that follows is generally applicable to all types of complex systems. For information systems, in which software performs virtually all the functionality, the section on software concept development in Chapter 11 discusses software system architecture and its design and should also be consulted.

Place of Concept Exploration Phase in the System Life Cycle

The place of the concept exploration phase in the overall system development process is shown in Figure 7.1. It is seen that the top-level system operational requirements

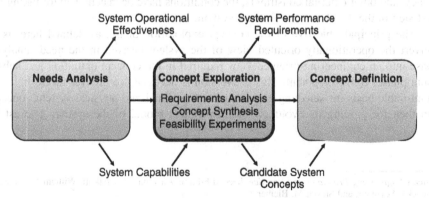

Figure 7.1. Concept exploration phase in a system life cycle.

come from needs analysis, which establishes that the needs are justified and that a development program is feasible within prescribed bounds. The outputs of the concept exploration phase are a set of system performance requirements down to the subsystem level and a number of potential system design concepts that analysis indicates to be capable of fulfilling those requirements.

While the formally defined concept exploration phase has a well-defined beginning and end, many of the supporting activities do not. For example, the exploratory development of advanced technological approaches or the quantitative characterization of complex system environments often begins before and extends beyond the formal terms of this phase, being supported by independent research and development (IRAD) or other nonproject funds. Additionally, considerable preliminary concept definition activity usually takes place well before the formal beginning of this phase.

The specific content of the concept exploration phase depends on many factors, particularly the relationship between the customer and the supplier or developer, and whether the development is needs driven or technology driven. If the system developer and supplier are different from the customer, as is frequently the case in needs-driven system developments, the concept exploration phase is conducted in part by the customer's own organization or with the assistance of a systems engineering agent engaged by the customer. The focus is on the development of performance requirements that accurately state the customer's needs in terms that one or more suppliers could respond to with specific product concepts. In the case of a technology-driven system development, the concept exploration phase is often conducted by the system developer and is focused on ensuring that all viable alternative courses of action are considered before deciding whether or not to pursue the development of a new system. In both cases, a primary objective is to derive a set of performance requirements that can serve as the basis of the projected system development and that have been demonstrated to ensure that the system product will meet a valid operational need.

For many acquisition programs, the period between the approval of a new system start and the availability of budgeted funds is often used to sponsor exploratory contractor efforts to advance technologies related to the anticipated system development.

System Materialization Status

The needs analysis phase was devoted to defining a valid set of operational objectives to be achieved by a new system, while a feasible system concept was visualized only as necessary to demonstrate that there was at least one possible way to meet the projected need. The term "visualize" is meant to connote the conceptualization of the general functions and physical embodiment of the subject in the case of needs analysis at the subsystem level.

Thus, in the concept exploration phase, one starts with a vision based generally on the above feasible concept. The degree of system materialization addressed in this phase has progressed to the next level, namely, the definition of the functions that the system and its subsystems must perform to achieve the operational objectives, and to the

TABLE 7.1. Status of System Materialization of the Concept Exploration Phase

| Level | Concept development | | | Engineering development | | |
	Needs analysis	Concept exploration	Concept definition	Advanced development	Engineering design	Integration and evaluation
System	Define system capabilities and effectiveness	Identify, explore, and synthesize concepts	Define selected concept with specifications	Validate concept		Test and evaluate
Subsystem		Define requirements and ensure feasibility	Define functional and physical architecture	Validate subsystems		Integrate and test
Component			Allocate functions to components	Define specifications	Design and test	Integrate and test
Subcomponent		Visualize		Allocate functions to subcomponents	Design	
Part					Make or buy	

Phase

visualization of the system's component configuration, as illustrated in Table 7.1 (an overlay of Table 4.1).

Systems Engineering Method in Concept Exploration

The activities in the concept exploration phase and their interrelationships are the result of the application of the systems engineering method (see Chapter 4). A brief summary of these activities is listed below; the names of the four generic steps in the method are shown in parentheses.

Operational Requirements Analysis (Requirements Analysis). Typical activities include
- analyzing the stated operational requirements in terms of their objectives; and
- restating or amplifying, as required, to provide specificity, independence, and consistency among different objectives, to assure compatibility with other related systems, and to provide such other information as may be needed for completeness.

Performance Requirements Formulation (Functional Definition). Typical activities include
- translating operational requirements into system and subsystem functions and
- formulating the performance parameters required to meet the stated operational requirements.

Implementation Concept Exploration (Physical Definition). Typical activities include
- exploring a range of feasible implementation technologies and concepts offering a variety of potentially advantageous options,
- developing functional descriptions and identifying the associated system components for the most promising cases, and
- defining a necessary and sufficient set of performance characteristics reflecting the functions essential to meeting the system's operational requirements.

Performance Requirements Validation (Design Validation). Typical activities include
- conducting effectiveness analyses to define a set of performance requirements that accommodate the full range of desirable system concepts; and
- validating the conformity of these requirements with the stated operational objectives and refining the requirements if necessary.

The interrelationships among the activities in the above steps in the systems engineering method are depicted in the flow diagram of Figure 7.2.

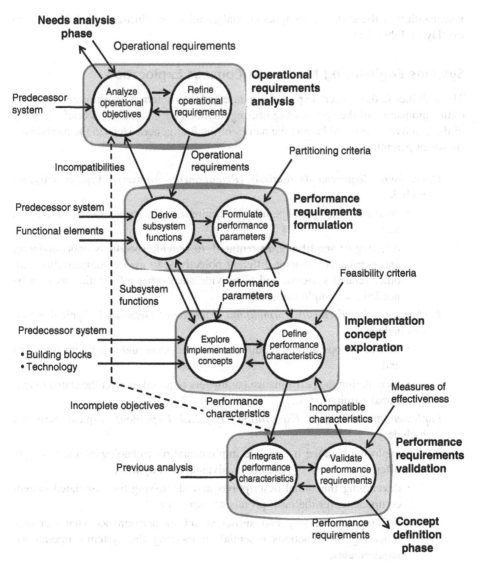

Figure 7.2. Concept exploration phase flow diagram.

7.2 OPERATIONAL REQUIREMENTS ANALYSIS

As in all phases of the system development process, the first task is to understand thoroughly, and, if necessary, to clarify and extend, the system requirements defined in the previous phase (in this case the operational requirements). In so doing, it is important to be alert for and to avoid shortcomings that are often present in the operational requirements as initially stated. We use a general process, known as *requirements*

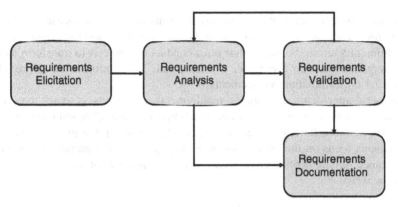

Figure 7.3. Simple requirements development process.

analysis, to identify and discover performance requirements, to synthesize and minimize initial sets of requirements, and finally, to validate the final set of requirements. This requirements analysis process, as mentioned in Chapter 4, occurs at each phase. However, the majority of effort occurs in the concept exploration phase where operational requirements are transformed into system performance requirements with measurable thresholds of performance. These system performance requirements tend to be the basis for contractual agreements between the customer and the developer and therefore need to be accurate and concise.

Figure 7.3 depicts the general process for developing requirements. Of course, this would be tailored to a specific application. The first activity involves the creation of a set of requirements. It is rare this occurs out of whole cloth—typically, a source of needs exists. In the concept exploration phase, a set of operational needs and requirements have been established. However, those needs and requirements are typically expressed in the language and context of an operator or user. These must be translated into a set of system-specific requirements describing its performance.

Requirements Elicitation

When analysts are developing operational requirements, they rely heavily on input from users and operators, typically through market surveys and interviews. When analysts are developing performance requirements, they rely on both people and studies. Initially, the customer (or buying agent within an organization) is able to provide thresholds of affordability and levels of performance that are desirable. But subject matter experts (SMEs) can also provide performance parameters as a function of technology levels, cost, and manufacturability. Previous studies and system development efforts can also assist in determining performance requirements. And finally, a requirements analyst performs a system effectiveness analysis to provide insight into the level of performance needed. All of these sources provide the analyst with an initial set of performance requirements.

Many times, however, this set of requirements contains inconsistencies, or even dichotomies. Further, many requirements are redundant, especially when they come from different sources. So the analyst must conduct a synthesis to transform an initial set of requirements into a concise, consistent set. More information is provided in Section 7.3 on formulating requirements.

A useful approach to developing requirements of any type is to ask the six interrogatives: who, what, where, why, when, and how. Of course, different types of requirements focus on different interrogatives, as described in Chapter 6. Operational requirements focus on the "why," defining the objectives and purpose of the system. Performance requirements focus on the "what," defining what the system should do (and how well).

Requirements Analysis

This activity starts with an initial set of requirements from the elicitation stage. Individual requirements as well as the set as a whole are analyzed for various attributes and characteristics. Some characteristics are desirable, such as "feasible" and "verifiable." Other characteristics are not, such as "vague" or "inconsistent."

For each requirement, a set of tests (or questions) is applied to determine whether the requirement is valid. And while many tests have been developed by numerous organizations, we present a set of tests that at least form a baseline. These tests are specific to the development of system performance requirements.

1. Is the requirement traceable to a user need or operational requirement?
2. Is the requirement redundant with any other requirement?
3. Is the requirement consistent with other requirements? (Requirements should not contradict each other or force the engineer to an infeasible solution.)
4. Is the requirement unambiguous and not subject to interpretation?
5. Is the requirement technologically feasible?
6. Is the requirement affordable?
7. Is the requirement verifiable?

If the answer to any of the questions above is "no," then the requirement needs to be revised, or possibly omitted. In addition, other requirements may need to be revised after performing this test.

In addition to individual requirements tests, a collective set of tests is also performed (usually after the individual tests have been performed on each requirement).

1. Does the set of requirements cover all of the user needs and operational requirements?
2. Is the set of requirements feasible in terms of cost, schedule, and technology?
3. Can the set of requirements be verified as a whole?

Both types of tests may need to be iterated before a final set of performance requirements exists.

Requirements Validation

Once a set of performance requirements is available, the set needs to be validated. This may be accomplished formally or informally. Formal validation means using an independent organization to apply various validation methods to validate the set of requirements against operational situations (i.e., scenarios and use cases) and to determine whether the requirements embodied within a system concept could achieve the user needs and objectives. Informal validation at this point means reviewing the set of requirements with the customer and/or users to determine the extent and comprehensiveness of the requirements. Section 7.5 provides further detail on the requirements validation process.

Requirements Documentation

A final, important activity is the documentation of the performance requirements. This is typically accomplished through the use of an automated tool, such as DOORS. Many tools exist that manage requirements, especially large, complex requirements hierarchies. As system complexity increases, the number and types of requirements tends to grow, and using simple spreadsheet software may not be sufficient to manage requirements databases.

Characteristics of Well-Stated Requirements

As mentioned above, the requirements analysis process leads to a concise set of performance requirements. This section examines the challenges associated specifically with translating operational requirements to performance requirements.

Since operational requirements are first formulated as a result of studies and analyses performed outside a formal project structure, they tend to be less complete and rigorously structured than requirements prepared in the subsequent managed phases of the development and are mainly oriented to justifying the initiation of a system development. Accordingly, in order to provide a valid basis for the definition of system performance requirements, their analysis must be particularly exacting and mindful of frequently encountered deficiencies, such as lack of specificity, dependence on a single assumed technical approach, incomplete operational constraints, lack of traceability to fundamental needs, and requirements not adequately prioritized. Each of these is briefly discussed in the succeeding paragraphs.

In an effort to cover all expected operating conditions (and to "sell" the project), operational requirements are often overly broad and vague where they should be specific. In the case of most complex systems, it is necessary to supplement the basic requirements with a set of well-defined operational scenarios that represent the range of conditions that the system is required to meet.

The opposite problem occurs if operational requirements are stated so as to be dependent on a specific assumed system configuration. To enable consideration of alternative system approaches, such requirements need to be restated to be independent of specific or "point" designs.

Often, operational requirements are complete only in regard to the active operational functions of the system and do not cover all the constraints and external interactions that the system must comply with during its production, transportation, installation, and operational maintenance. To ensure that these interactions are treated as fully as possible at this stage of development, it is necessary to perform a life cycle analysis and to provide scenarios that represent these interactions.

All requirements must be associated with and traceable to fulfilling the operational objectives of the user. This includes understanding who will be using the system and how it will be operated. Compliance with this guideline helps to minimize unnecessary or extraneous requirements. It also serves as a good communication link between the customer and developer when particular requirements subsequently lead to complex design problems or difficult technical trade-offs.

The essential needs of the customer must be given top priority. If the needs analysis phase has been done correctly, requirements stemming from these needs will be clearly understood by all concerned. When design conflicts occur later in development, a review of these primary objectives can often provide useful guidance for making a decision.

Beyond the above primary or essential requirements, there are always those capabilities that are desirable if they prove to be readily achievable and affordable. Requirements that are essential should be separately distinguished from those that are desirable but not truly necessary for the success of the primary mission. Often, preferences of the customer come through as hard and fast requirements, when they are meant to be desirable features. Examples of desirable requirements are those that provide an additional performance capability or design margin. There should be some indication of cost and risk associated with each desirable requirement so that an informed prioritization can be made. The discrimination between essential and desirable requirements and their prioritization is a key systems engineering function.

The Triumvirate of Conceptual Design

Above, we mentioned the use of the six primitive interrogatives in developing requirements. We also discussed that operational requirements' focus on "why" and functional requirements' focus on the "what" (along with performance requirements' focus on its associated interrogative, "how much"). So, if the two sets of requirements focus on why and what, then where does the analyst go to understand the other four primitive interrogatives? The answer lies with what we call the triumvirate of conceptual design, illustrated in Figure 7.4.

Three products are needed to describe the six interrogatives that collectively could be considered a system concept. The requirements (all three types we have addressed in detail up to this point) address why and what. A new product, the operational concept, sometimes referred to as a concept of operations (CONOPS), addresses how and who. And a description of the operational context, sometimes referred to as scenarios, addresses where and when. Of course, there is a significant overlap between the three, and often two or more of these products are combined into a single document.

Operational Concept (CONOPS)

Although the two terms are often used synonymously, in truth, an operational concept is a broader description of a capability that encompasses multiple systems. It tends to

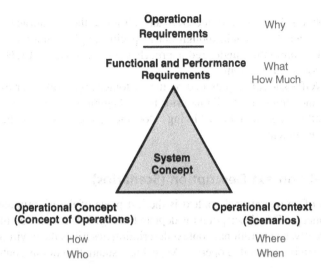

Figure 7.4. Triumvirate of conceptual design.

describe how a large collection of systems will operate. Examples would include an operational concept for the U.S. Transportation System (or even a subsystem of the whole system). In this case, "system" does not refer to a single system but a collection of systems. Another example would be an operational concept for an oil refinery—again referring to how a collection of systems would operate together. When referring to a single system, the term CONOPS is generally used. A further distinction relates to scenarios. An operational concept is sufficiently broad to be scenario independent. A CONOPS tends to relate to a single scenario or a set of related scenarios.

Operational concepts are useful since requirements should avoid prescribing how they should be fulfilled. Requirements documents risk inadvertently barring an especially favorable solution. However, a set of operational requirements alone is often insufficient to constrain the system solutions to the types desired. For example, the operational requirements for defending an airplane against terrorist attack could conceivably be met by counterweapons, passenger surveillance, or sensor technology. In a particular program initiative, the requirements would be constrained by adding a CONOPS, which would describe the general type of counterweapons that are to be considered. This extension of the operational requirement adds constraints, which express the customer's expectation for the anticipated system development.

The term *CONOPS* is quite general. The components of a CONOPS usually include

1. mission descriptions, with success criteria;
2. relationships with other systems or entities;
3. information sources and destinations; and
4. other relationships or constraints.

The CONOPS should be considered as an addition to the operational requirements. It defines the general approach, though not a specific implementation, to the desired system, thereby eliminating undesired approaches. In this way, the CONOPS clarifies the intended goal of the system.

The CONOPS should be prepared by the customer organization or by an agent of the customer and should be available prior to the beginning of the concept definition phase. Thereafter, it should be a "living" document, together with the operational requirements document.

Operational Context Description (Scenarios)

A description of an operational context is the last piece of the triumvirate in defining the system concept. This description (as depicted in Fig. 7.4) focuses on the where and when. Specifically, an operational context description describes the environment within which the system is expected to operate. A specific instantiation of this context is known as a scenario.

A scenario can be defined as "a sequence of events involving detailed plans of one or more participants and a description of the physical, social, economic, military, and political environment in which these events occur." With respect to system development, scenarios are typically projected into the future to provide designers and engineers a context for the system description and design.

Most scenarios include at least five elements:

1. *Mission Objectives:* a description of the overall mission with success criteria. The reader should notice this is the same as one of the components of a CONOPS. The mission can be of any type, for example, military, economic, social, or political.

2. *Friendly Parties:* a description of friendly parties and systems, and the relationships among those parties and systems.

3. *Threat Actions (and Plans):* a description of actions and objectives of threat forces. These threats need not be human; they could be natural (e.g., volcano eruption).

4. *Environment:* a description of the physical environment germane to the mission and system.

5. *Sequence of Events:* a description of individual events along a timeline. These event descriptions should not specify detailed system implementation details.

Scenarios come in all sizes and flavors. The type of scenario is determined by the system in questions and the problem being examined. Figure 7.5 shows different levels of scenarios that might be needed in a system development effort. During the early phases (needs analysis and concept exploration), the scenarios tend to be higher levels, near the top of the pyramid. As the development effort transitions to later phases, more detail is available as the design improves, and lower-level scenarios are used in engineering

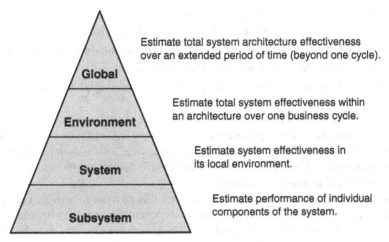

Estimate total system architecture effectiveness over an extended period of time (beyond one cycle).

Estimate total system effectiveness within an architecture over one business cycle.

Estimate system effectiveness in its local environment.

Estimate performance of individual components of the system.

Figure 7.5. Hierarchy of scenarios.

analyses. High-level scenarios continue to be used throughout to estimate the overall system effectiveness as the design matures.

Analysis of Alternatives

The needs analysis phase is usually conducted without the benefit of a well-organized and funded effort. In such cases, the operational requirements that are formulated during this phase are necessarily a preliminary and incomplete definition of the full mission objectives. Therefore, an essential part of the concept exploration phase is to develop the operational requirements into a complete and self-consistent framework as a basis for developing an effective operational system.

For the above reason, before initiating a major program, one or more studies are generally carried out to refine the operational requirements by modeling the interaction of operational scenarios. One of the common designations for such studies is "analysis of alternatives" because they involve the definition of a range of alternative system approaches to the general operational mission, and a comparative evaluation of their operational effectiveness. Such analyses define the realistic limits of expected operational effectiveness for the postulated operational situation and provide the framework for a set of complete, consistent, and realistic operational requirements.

Guidelines for Defining Alternative Concepts. As noted in the next section, conceiving new candidate approaches to satisfying a set of requirements is an inductive process and hence requires a leap of the imagination. For such a process, it is helpful to postulate some guidelines for selecting alternatives:

1. Start with the existing (predecessor) system as a baseline.
2. Partition the system into its major subsystems.

3. Postulate alternatives that replace one or more of the subsystems essential to the mission with an advanced, less costly, or otherwise superior version.
4. Vary the chosen subsystems (or superior version) singly or in combination.
5. Consider modified architectures, if appropriate.
6. Continue until you have a total of four to six meaningful alternatives.

Effectiveness Simulation. Where the analysis of alternatives involves complex systems, the analysis often requires the use of a computer simulation that measures the effectiveness of a model of a system concept in dealing with a model scenario of the system environment. Chapter 9 contains a brief description of the character and application of system effectiveness simulation.

The advantage of computer simulation is that it is possible to provide controls that vary the behavior of a selected system and environmental parameters in order to study their effect on the overall system behavior. This feature is especially valuable in characterizing the effect of operational and performance requirements on the system architecture necessary to satisfy them, and in turn, establishing practical bounds on the requirements. A range of solutions of varying capability and cost can be considered. Every particular application has its own key variables that can be called into play.

7.3 PERFORMANCE REQUIREMENTS FORMULATION

As noted previously, in the course of developing a new system, it is necessary to transform the system operational requirements, which are stated as required outcomes of system action, into a set of system performance requirements, which are stated in terms of engineering characteristics. This step is essential to permit subsequent stages of system development to be based on and evaluated in engineering rather than operational terms. Thus, system functional performance requirements represent the transition from operational to engineering terms of reference.

Derivation of Subsystem Functions

In deriving performance requirements from operational objectives, it is first necessary to identify the major functions that the system must perform to carry out the prescribed operational actions. That means, for example, that if a system is needed to transport passengers to such destinations as they may wish along existing roadways, its functional elements must include, among others, a source of power, a structure to house the passengers, a power-transmitting interface with the roadway, and operator-activated controls of locomotion and direction. Expressed in functional terms (verb–object), these elements might be called "power vehicle," "house passengers," "transmit power to roadway," "control locomotion," and "control direction."

As described in Chapter 6, a beginning in this process has already been made in the preceding phase. However, a more definitive process is needed to establish specific performance parameters. Correspondingly, as seen in Table 4.1, during this phase, the functional definition needs to be carried a step further, that is, to a definition of subsystem functions, and to the visualization of the functional and associated physical components, which collectively can provide these subsystem functions.

The Nondeterministic Nature of System Development

The derivation of performance requirements from desired operational outcomes is far from straightforward. This is because, like other steps in the system materialization process, the design approach is inductive rather than deductive, and hence not directly reversible. In going from the more general operational requirements to the more specifically defining system performance requirements, it is necessary to fill in many details that were not explicitly called out in the operational requirements. This can obviously be done in a variety of ways, meaning that more than one system configuration can, in principle, satisfy a given set of system requirements. This is also why in the system development process the selection of the "best" system design at a given level of materialization is accomplished by trade-off analysis, using a predefined set of evaluation criteria.

The above process is exactly the same as that used in inductive reasoning. For example, in designing a new automobile to achieve an operational goal of 600 mi on a tank of gasoline, one could presumably make its engine extremely efficient, or give it a very large gasoline tank, or make the body very light, or some combination of these characteristics. Which combination of these design approaches is selected would depend on the introduction of other factors, such as relative cost, development risk, passenger capacity, safety, and many others.

This process can also be understood by considering a deductive operation, as, for example, performance analysis. Given a specific system design, the system's performance may be deduced unambiguously from the characteristics of its components by first breaking down component functions, then by calculating their individual performance parameters, and finally by aggregating these into measures of the performance of the system as a whole. The reverse of this deductive process is, therefore, inductive and consequently nondeterministic.

One can see from the preceding discussion that, given a set of operational requirements, there is no direct (deductive) method of inferring a corresponding unique set of system performance characteristics that are necessary and sufficient to specify the requirements for a system to satisfy the operational needs. Instead, one must rely on experience-based heuristics, and to a large extent, on a trial and error approach. This is accomplished through a process in which a variety of different system configurations are tentatively defined, their performance characteristics are deduced by analysis or data collection, and these are subjected to effectiveness analysis to establish those characteristics required to meet the operational requirements. The above process is described in greater detail in the next section.

Functional Exploration and Allocation

The exploration of potential system configurations is performed at both the functional and physical levels. The range of different functional approaches that produce behavior suitable to meet the system operational requirements is generally much more limited than the possibilities for different physical implementations. However, there are often several significantly different ways of obtaining the called for operational actions. It is important that the performance characteristics of these different functional approaches be considered in setting the bounds on system performance requirements.

As noted earlier in Figure 7.2, one of the outputs of this step is the allocation of operational functions to individual subsystems. This is important in order to set the stage for the next step, in which the basic physical building block components may be visualized as part of the exploration of implementation concepts. These two steps are very tightly bound through iterative loops, as shown in the figure. Two important inputs to the functional allocation process are the predecessor system and functional building blocks. In most cases, the functions performed by the subsystems of the predecessor system will largely carry over to the new system. Accordingly, the predecessor system is especially useful as a point of departure in defining a functional architecture for the new system. And since each functional building block is associated with both a set of performance characteristics and a particular type of physical component, the building blocks can be used to establish the selection and interconnection of elementary functions and the associated components needed to provide the prescribed subsystem functions.

To aid in the process of identifying those system functions responsible for its operational characteristics, recall from Chapter 3 that functional media can be classed into four basic types: signals, data, material, and energy. The process addresses the following series of questions:

1. Are there operational objectives that require sensing or communications? If so, this means that signal input, processing, and output functions must be involved.
2. Does the system require information to control its operation? If so, how are data generated, processed, stored, or otherwise used?
3. Does system operation involve structures or machinery to house, support, or process materials? If so, what operations contain, support, process, or manipulate material elements?
4. Does the system require energy to activate, move, power, or otherwise provide necessary motion or heat?

Furthermore, functions can be divided again into three categories: input, transformative, and output. Input functions relate to the processes of sensing and inputting signals, data, material, and energy into the system. Output functions relate to the processes of interpreting, displaying, synthesizing, and outputting signals, data, material, and energy out of the system. Transformative functions relate to the processes of transforming the inputs to the outputs of the four types of functional media. Of course, for complex systems, the number of transformative functions may be quite large, and has successive "sequences" of transformations. Figure 7.6 depicts the concept of this two-dimensional construct, function category versus functional media.

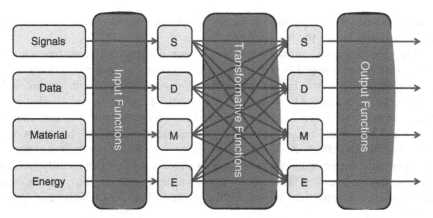

Figure 7.6. Function category versus functional media.

In constructing an initial function list, it helps to identify inputs and outputs (as described in Chapter 3). This list directly leads the engineer to a list of input and output functions. The transformative functions may be easier to identify when examining them in the light of a system's inputs and outputs.

As an example, while acknowledging it is not a complex system, consider a common coffeemaker (without any frills). By observation, an analyst can identify the necessary inputs:

- *Signals:* user commands (which we will simply identify as "on" and "off")
- *Data:* none
- *Materials:* fresh coffee grinds, filter, and water
- *Energy:* electricity
- *Forces:* mechanical support

Outputs can also be easily identified:

- *Signals:* status (which we will simply identify as on and off)
- *Data:* none
- *Materials:* brewed coffee, used filter, used coffee grinds
- *Energy:* heat
- *Forces:* none

Identifying inputs and outputs assists the analyst in identifying functions. Input functions will directly proceed from the input list (deductive reasoning). Output functions will directly proceed from the output list (deductive as well). The transformative functions will be more difficult to identify since doing so relies on inductive reasoning.

However, we now have a guide to this inductive process: we know that we must transform the six inputs into the five outputs.

This line of inquiry normally reveals all operationally significant functions and permits them to be grouped in relation to specific operational objectives. Further, this grouping naturally tends to bring together the elements of different subsystems, which are the first-level building blocks of the system itself. The above strategy is also appropriate even if the basic configuration is derived from a predecessor system because its generic and systematic approach tends to reveal elements that might otherwise be overlooked. In the coffeemaker case, we can focus on transforming the input materials and signals into output materials and signals. In other words, we can identify functions by answering the question "How do we transform fresh coffee grinds, a filter, water, and an on/off command into brewed coffee, a used filter, used coffee grinds, and a status?"

Keeping the list of functions minimal and high level, and using the verb–object syntax, an example list pertaining to the coffeemaker could be

Input Functions
1. Accept user command (on/off)
2. Receive coffee materials
3. Distribute electricity
4. Distribute weight

Transformative Functions
5. Heat water
6. Mix hot water with coffee grinds
7. Filter out coffee grinds
8. Warm brewed coffee

Output Functions
9. Provide status
10. Facilitate removal of materials
11. Dissipate heat

Can you map the inputs and outputs to one or more functions? Can you identify how the inputs are transformed into the outputs? Since a coffeemaker is a very simple system, the number of transformative functions was low. But keep in mind that regardless of a system's complexity, a top-level function list with about 5–12 functions can always be identified. So a complex system may have a large hierarchy of functions, but any system can be aggregated into an appropriate set of top-level functions.

Formulation of Performance Characteristics. As noted above, the objective of the concept exploration phase is to derive a set of system performance characteristics that are both necessary and sufficient. This means that a system possessing them will satisfy the following criteria:

1. A system that meets the system operational requirements and is technically feasible and affordable will comply with the performance characteristics.
2. A system that possesses these characteristics will meet the system operational requirements and can be designed to be technically feasible and affordable.

The condition that the set of performance requirements must be necessary as well as sufficient is essential to ensure that they do not inadvertently exclude a system concept that may be especially advantageous compared to others just because it may take an unusual approach to a particular system function. This often happens when the performance requirements are derived in part from a predecessor system and carry over features that are not essential to its operational behavior. It also happens when there is a preconceived notion of how a particular operational action should be translated into a system function.

For the above reasons, the definition of performance characteristics needs to be an exploratory and iterative process, as shown in Figure 7.2. In particular, if there are alternative functional approaches to an operational action, they should all be reflected in the performance characteristics, at least until some may be eliminated in the implementation and validation steps in the process.

Incompatible Operational Requirements. It should be noted that a given set of operational requirements does not always lead to feasible performance characteristics. In Chapter 6, the automobile was mentioned as a system that was required to undergo significant changes because of government-imposed regulations concerning safety, fuel economy, and pollution control. Initially, these areas of regulation were independently developed. Each set of requirements was imposed solely on the basis of a particular need, with little regard for either the associated engineering problems or other competing needs. When these regulations were subjected to engineering analysis, it was shown that they were not collectively feasible within the practical technology available at that time. Also, the investment in development and production would result in a per-unit cost far in excess of the then current automobile prices. The basic reason for these problems was that the available pollution controls necessary to meet emission requirements resulted in lowered fuel economy, while the weight reduction necessary to meet the required fuel economy defeated the safety requirements. In other words, the three independent sets of operational requirements turned out to be incompatible because no one had initially considered their combined impact on the design. Note that in this instance, an analysis of the requirements did not depend on a detailed design study since simply examining the design concepts readily revealed the conflicts.

Example: Concepts for a New Aircraft. An instructive example of concept exploration is illustrated by the acquisition of a new commercial aircraft. Assume for this discussion that an airline company serves short to medium domestic routes using two-engine propeller-driven aircraft. Many of the airports it serves have relatively short runways. This arrangement has worked well for a number of years. The problem that has become more and more apparent is that because of increasing maintenance and fuel expenses, the cost per passenger mile has increased to the point where the business is marginally profitable. The company is therefore considering a major change in its aircraft. In essence, the airline's need is to lower the cost per passenger mile to some acceptable value and to maintain its competitive edge in short-route service.

The company approaches several aircraft manufacturers for a preliminary discussion of a new or modified airplane to meet its needs. The discussions indicate that there are several options available. Three such options are the following:

1. A stretched aircraft body and increased power. Engines of the appropriate form and fit exist for such a configuration. This option permits a quick, relatively low-cost upgrade, which increases the number of passengers per aircraft, thereby lowering the overall cost per passenger mile.
2. A new, larger, four-engine propeller aircraft, using state-of-the-art technology. This option offers a good profit return in the near term. It is reasonably low risk, but the total useful life of the aircraft is not well-known, and growth potential is limited.
3. A jet-powered aircraft that is capable of takeoff and landing at most, but not necessarily all, of the current airports being served. This option permits a significant increase in passengers per airplane and opens up the possibility of competing for new, longer routes. This is also the most expensive option. Because of the inherent lower maintenance and fuel costs of jet engines relative to propeller engines, operating costs for this aircraft are attractive, but some existing routes will be lost.

It is evident that the final choice will require considerable expertise and should be based on a competition among interested manufacturers. The airline engages the services of an engineering consulting company to help its staff prepare a set of aircraft performance requirements that can serve as a basis for competitive bids and to assist in the selection process.

In exploring the above and related options, the alternative functional approaches are considered first. These appear to center on the choice between staying with propeller engines, an option that retains the basic features of the present aircraft, or moving to jet engines, which offer considerable operating economies. However, the latter is a major departure from the current system and will also affect its operational capabilities. To permit this choice to be left open to the bidders, the performance requirements such as runway length, cruising speed, and cruising altitude will need to be sufficiently broad to accommodate these two quite different functional approaches.

Requirements Formulation by Integrated Product Teams (IPTs)

As noted earlier, the responsibility for defining the performance requirements of a new product is that of the customer, or in the case of government programs, that of the acquisition agency. However, the organization of the process and its primary participants varies greatly with the nature of the product, the magnitude of the development, and the customer auspices.

As in the case of all acquisition practices, the Department of Defense (DoD) has had the most experience with various methods for organizing the acquisition process. A recent practice introduced by DoD is the use of IPTs throughout the acquisition process. IPTs are intended to bring a number of benefits to the process:

1. They bring senior industry participants into the system conceptual design process at the earliest opportunity, thereby educating them in the operational needs and injecting their ideas during the formative stages of the development.
2. They bring together the different disciplines and specialty engineering viewpoints throughout the development.
3. They capitalize on the motivational advantages of team collaboration and consensus building.
4. They bring advanced technology and COTS knowledge to bear on system design approaches.

As in the case of any organization, the success of this approach is highly dependent on the experience and interpersonal skills of the participants, as well as on the leadership qualities of the persons responsible for team organization. And perhaps even more important is the systems engineering experience of the team leaders and members. Without this, the majority of the team members, who tend to be specialists, will not be able to communicate effectively and hence the IPT will not achieve its objectives.

7.4 IMPLEMENTATION OF CONCEPT EXPLORATION

The previous section discussed the exploration of alternative functional approaches—concepts in which the nature of the activities involved differs from one case to the next. The physical implementation of such concepts involves the examination of different technological approaches, generally offering a more diverse source of alternatives. As in the case of examining alternative functional concepts, the objective of exploring implementation concepts is to consider a sufficient variety of approaches to support the definition of a set of system performance requirements that are feasible of realization in practice and do not inadvertently preclude the application of an otherwise desirable concept. To that end, the exploration of system concepts needs to be broadly based.

Alternative Implementation Concepts

The predecessor system, where one exists, forms one end of the spectrum to be explored. Given the operational deficiencies of the predecessor system to meet projected needs, modifications to the current system concept should first be explored with a view to eliminating these deficiencies. Such concepts have the advantage of being relatively easier to assess from the standpoint of performance, development risk, and cost than are radically different approaches. They can also generally be implemented faster, more cheaply, and with less risk than innovative concepts. On the other hand, they are likely to have severely limited growth potential.

The other end of the spectrum is represented by innovative technical approaches featuring advanced technology. For example, the application of powerful, modern microprocessors might permit extensive automation of presently employed manual operations. These concepts are generally riskier and more expensive to implement but offer large incremental improvements or cost reduction and greater growth potential. In between are intermediate or hybrid concepts, including those defined in the needs analysis phase for demonstrating the feasibility of meeting the proposed system needs.

Many techniques exist for developing new and innovative concepts. Perhaps the oldest is brainstorming, individually and within a group. Within the concept of brainstorming, several modern methods, or variations, to the old fashioned, largely unstructured brainstorming process have risen. One of our favorite techniques, which engineers may not be familiar with (but nonengineering practitioners may be), is Mind Maps. This particular technique uses visual images to assist in the brainstorming of new ideas. A simple Web search will point the reader to multiple Web sites describing the technique.

The natural temptation to focus quickly on a single concept or "point design" approach can easily preclude the identification of other potentially advantageous approaches based on fundamentally different concepts. Accordingly, several concepts spanning a range of possible design approaches should be defined and investigated. At this stage, it is important to encourage creative thinking. It is permissible, even sometimes desirable, to include some concepts that do not meet all of the requirements; otherwise, a superior alternative may be passed by because it fails to meet what may turn out to be a relatively arbitrary requirement. Just as in the needs analysis phase, negotiations with the customer regarding which requirements are really necessary and which are not can often make a significant difference in cost and risk factors while having minimal impact on performance.

Example: Concept Exploration for a New Aircraft. Returning to the example introduced in the previous section, it will be recalled that two principal functional options were explored to meet the need of the airline company: a propeller-driven and a jet-driven aircraft. It remains to explore alternative physical implementations of each of these options. As is usually the case, these are more numerous than the basic functional alternatives.

In the period since the airline's present fleet was acquired, a host of technological advances have occurred. For example, automation has become more widespread, especially in autopilots and navigation systems. Changes in safety requirements, such as for deicing provisions, must also be examined to identify those performance characteristics that should be called out. In exploring alternative implementations, the main features of each candidate system must first be analyzed to see if they are conceptually achievable. At this stage of development, a detailed design analysis is usually not possible because the concept is not yet sufficiently formulated. However, based on previous experience and engineering judgment, someone, usually the systems engineer, must decide whether or not the concept as proposed is likely to be achievable within the given bounds of time, cost, and risk.

There are numerous other options and variations of the above examples. It is noted that all the cited options have pros and cons, which typically leave the customer with no obvious choice. Note also that the option to use jet aircraft may partially violate the operational requirement that short-route capability be maintained. However, as noted earlier, it is not at all unusual at this stage to consider options that do not meet all the initial requirements to ensure that no desirable option is overlooked. The airline may decide that the loss of some routes is more than compensated for by the advantages to the overall system of using jet aircraft.

It is also important to note that the entire system life cycle must be considered in exploring alternatives. For example, while the jet option offers a number of performance advantages, it will require a substantial investment in training and logistic support facilities. Thus, assessment of these supporting functions must be included in formulating system requirements. In order to be a "smart buyer," the airline needs to have a staff well versed in aircraft characteristics, as well as in the business of running an airline, and access to consultants or engineering services organizations capable of carrying out the analyses involved in developing the requisite set of performance requirements.

Preferred System. Although in most cases it is best to refrain from picking a superior system concept prematurely, there are instances where it is permissible for the requirements definition effort to identify a so-called preferred system, in addition to considering a number of other viable system alternatives. Preference for a system or subsystem may be set forth when significant advanced development work has taken place and has produced very promising results in anticipation of future upgrades to the current system. Such work is often conducted or sponsored by the customer. Another justifying factor may be when there has been a recent major technological breakthrough, which promises high gains in performance at an acceptable risk. The idea of a preferred system approach is that subsystem analysis can start building on this concept, thereby saving time and cost. Of course, further analysis may show the favored approach not to be as desirable as predicted.

Technology Development

Whether the origin of a new system is needs driven or technology driven, the great majority of new systems have been brought into being, directly or indirectly, as a result of technological growth. In the process of exploring potential concepts for the satisfaction of a newly established need, a primary input is derived from what is called the technology base, which means the sum total of the then existing technology. It is, therefore, important for systems engineers to understand the nature and sources of technological advances that may be pertinent to a proposed system development.

System-oriented exploratory R & D can be distinguished according to whether it relates to new needs-driven or technology-driven systems. The former is mainly directed to gaining a firm understanding of the operational environment and the factors underlying the increased need for the new system.

The latter is usually focused on extending and quantifying the knowledge base for the new technology and its application to the new system objectives. In both instances, the objective is to generate a firm technical base for the projected system development, thus clarifying the criteria for selecting specific implementation concepts and transforming unknown characteristics and relationships into knowns.

Both industry and government support numerous programs of R & D on components, devices, materials, and fabrication techniques, which offer significant gains in performance or cost. For instance, most large automobile manufacturers have ongoing programs to develop more efficient engines, electrically powered vehicles, automated fuel controls, lighter and stronger bodies, and a host of other improvements that are calculated to enhance their future competitive position. In recent years, the greatest amount of technology growth has been in the electronics industry, especially computers and communication equipment, which in turn has driven the explosive growth of information systems and automation generally.

In government-sponsored R & D, there is also a continuing large-scale effort, mainly among government contractors, laboratories, and universities, directed toward the development of technologies of direct interest to the government. These cover many diverse applications, and their scope is almost as broad as that of commercial R & D. As has been noted previously, defense contractors are permitted to charge a percentage of their revenues from government contracts to IRAD as allowable overhead. A large fraction of such funds is devoted to activities that relate to potential new system developments. In addition, there is a specific category in the Congressional Research, Development, Test and Evaluation (RDT&E) appropriation, designated Research and Exploratory Development, which funds specific R & D proposals to the military services. Such projects are not intended to directly support specific new system developments but do have to be justified as contributing to existing mission areas.

Performance Characteristics

The derivation of performance characteristics by the exploration of implementation concepts can be thought of as consisting of a combination of two analytical processes: performance analysis and effectiveness analysis. Performance analysis derives a set of performance parameters that characterize each candidate concept. Effectiveness analysis determines whether or not a candidate concept meets the operational requirements and, if not, how the concept needs to be changed to do so. It employs an effectiveness model that is used to evaluate the performance of a conceptual system design in terms of a selected set of criteria or measures of effectiveness. This is a similar model to that used in the previous phase and to the one employed in the next step, the validation of performance requirements. The main difference in its use in the above applications is the level of detail and rigor.

Performance Analysis. The performance analysis part of the process is used to derive a set of relevant performance characteristics for each candidate system concept that has been found to satisfy the effectiveness criteria. The issue of relevancy arises because a full description of any complex system will involve many parameters, some of which may not be directly related to its primary mission. For example, some features, such as the ability of an aircraft search radar device to track some particular coded beacon transponder, might be included only to facilitate system test or calibration. Therefore, the performance analysis process must extract from the identified system characteristics only those that directly affect the system's operational effectiveness. At the same time, care must be taken to include all characteristics that can impact effectiveness under one or another particular operating condition.

The problem of irrelevant characteristics is especially likely to occur when the concept for a particular subsystem has been derived from the design of an existing subsystem employed in a different application. For example, a relatively high value of the maximum rate of train or elevation for a radar antenna assembly might not be relevant to the application now being examined. Thus, the derived model should not reflect this requirement unless it is a determining factor in the overall subsystem design concept. In short, as stated previously, the defined set of characteristics must be both necessary and sufficient to facilitate a valid determination of effectiveness for each candidate system concept.

Constraints. At this phase of the project, the emphasis will naturally be focused on active system performance characteristics and functions to achieve them. However, it is essential that other relevant performance characteristics not be overlooked, especially the interfaces and interactions with other systems or parts of systems, which will invariably place constraints on the new system. These constraints may affect physical form and fit, weight and power, schedules (e.g., a launch date), mandated software tools, operating frequencies, operator training, and so on. While constraints of this type will be dealt with in great detail later in the development process, it is not too soon to recognize their impact during the process of requirements definition. The immediate benefit of early attention to such problems is that conflicting concepts can be filtered out, leaving more time for analysis of the more promising approaches.

To accomplish the above objectives, it is necessary to consider the complete system life cycle. To a large extent, the constraints on the system will not depend on the specific system architecture. For example, environmental conditions of temperature, humidity, shock vibration, and so forth, for a great part of the system life cycle are often the same for any candidate system concept. Omission of any constraints such as these may result in serious deficiencies in the system design, which would adversely impact performance and operability.

7.5 PERFORMANCE REQUIREMENTS VALIDATION

Having derived the operationally significant performance characteristics for several feasible alternative concepts, all of which appear to be capable of meeting the system operational requirements, the next step is to refine and integrate them into a singular set to serve as a basis for the preparation of formal system performance requirements.

As stated earlier, these performance requirements, stated in engineering units, provide an unambiguous basis for the ensuing phases of system development, up to the stage where the actual system can be tested in a realistic environment.

The operations involved in the refinement and validation of system performance requirements can be thought of as two tightly coupled processes—an integration process, which compares and combines the performance characteristics of the feasible alternative concepts, and an effectiveness analysis process, which evaluates the validity of the integrated characteristics in terms of the operational requirements.

Performance Characteristics Integration

The integration process serves to select and refine those characteristics of the different system concepts examined in the exploration process that are necessary and sufficient to define a system that will possess the essential operational characteristics. Regardless of the analytical tools that may be available, this process requires the highest level of systems engineering judgment.

This and other processes in this phase can benefit greatly by the participation of systems engineers with experience with the predecessor system, which has been mentioned a number of times previously. The knowledge and database that comes with that system is an invaluable source of information for developing new requirements and concepts. In many cases, some of the key engineers and managers who directed its development may still be available to contribute to the development of new requirements and concepts. They may not only be aware of the current deficiencies but are likely to have considered various improvements. Additionally, they are probably aware of what the customer really wants, based on their knowledge of operational factors over a number of years. Just one key systems engineer with this background can provide significant help. Experienced people of this type will also have an educated "gut feel" about the viability of the requirements and concepts that are being considered. Their help, at least as consultants, will not alleviate the need for requirements analysis, but it may quickly point the effort in the right direction and avoid blind alleys that might otherwise be pursued.

Performance Characteristics Validation

The final steps in the process are to validate the derived performance characteristics against the operational requirements and constraints and to convert them into the form of a requirements document. Ideally, the performance characteristics derived from the refinement step will have been obtained from concepts validated in the implementation concept exploration process. However, it is likely that the effort to remove irrelevant or redundant characteristics in the integration step, and to add external constraints not present in the effectiveness model, will have significantly altered the resultant set of characteristics. Hence, it is essential to subject them once more to an effectiveness analysis to verify their compliance with the operational requirements. The effectiveness model in the above step should generally be more rigorous and detailed than models used in previous steps so as to ensure that the final product does not contain deficiencies due to omission of important evaluation criteria.

The above processes operate in closed-loop fashion until a self-consistent set of *system performance characteristics* that meets the following objectives is obtained:

1. They define what the system must do, and how well, but not how the system should do it.
2. They define characteristics in engineering terms that can be verified by analytical means or experimental tests, so as to constitute a basis for ensuing engineering phases of system development.
3. They completely and accurately reflect the system operational requirements and constraints, including external interfaces and interactions, so that if a system possesses the stated characteristics, it will satisfy the operational requirements.

Requirements Documentation

To convert the system performance characteristics into a requirements document involves skillful organization and editing. Since the system performance requirements will be used as the primary basis for the ensuing concept definition phase and its successors, it is most important that this document be clear, consistent, and complete. However, it is equally important to recognize that it is not carved in stone but is a living document, which will continue to evolve and improve as the system is developed and tested.

In a need-driven system development in which it is intended to compete the concept definition phase among a number of bidders, the system performance requirements are a primary component of the competitive solicitation, along with a complete statement of all other conditions and constraints. Such a solicitation is often circulated in draft form among potential bidders to help ensure its completeness and clarity.

In a technology-driven system development in which the same commercial company that will carry out the definition and subsequent phases conducts the exploratory phase, the end product typically serves as a basis for deciding whether or not to authorize and fund a concept definition phase preliminary to engineering development. For this purpose, the requirements document typically includes a thorough description of the most attractive alternative concepts investigated, evidence of their feasibility, market studies validating the need for a new system, and estimates of development, production, and market introduction costs.

7.6 SUMMARY

Developing the System Requirements

The objectives of the concept exploration phase (as defined here) are to explore alternative concepts to derive common characteristics and to convert the operationally oriented system view into an engineering-oriented view. Outputs of concept exploration are (1) system performance requirements, (2) a system architecture down to the subsystem level, and (3) alternative system concepts.

Activities that comprise concept exploration are the following:

- *Operational Requirements Analysis*—ensuring completeness and consistency;
- *Implementation Concept Exploration*—refining functional characteristics;
- *Performance Requirements Formulation*—deriving functions and parameters; and
- *Performance Requirements Validation*—ensuring operational validity.

Operational Requirements Analysis

Requirements development involves four basic steps: elicitation, analysis, validation, and documentation. These steps will, done correctly, lead to a robust set of well-articulated requirements.

Generating operational-level requirements usually involves analyses of alternative concepts, typically involving effectiveness models and simulations. In order to conduct these important analyses, three components are necessary: an initial set of operational requirements, an operational concept for the system in question, and the operational context—a set of operational scenarios depicting the environment.

Performance Requirements Formulation

System development is a nondeterministic process in that it requires an iterative inductive reasoning process, and many possible solutions can satisfy a set of operational requirements. The predecessor system can be of great assistance as it will help define the system functional architecture and the performance of functional building blocks.

Implementation Concept Exploration

Exploration of alternative implementation concepts should

- avoid the "point design syndrome";
- address a broad spectrum of alternatives;
- consider the adaptation of a predecessor system technology;
- consider innovative approaches using advanced technology; and
- assess the performance, risk, cost, and growth potential of each alternative.

Technology development is also an important component of system development. Industry and government support major R & D programs that lead to new technologies. This foundation of technology is typically referred to as the "technical base" and is the source of many innovative concepts.

System performance requirements are developed through analyses to establish the performance parameters of each concept. These requirements are then assessed for conformance with operational requirements and constraints. Sources of these constraints include (1) system operator, maintenance, and test considerations; (2) requirements for interfacing with other systems; (3) externally determined operational environments; and (4) fabrication, transportation, and storage environments.

When completed, system performance requirements define what the system should do, but not how it should do it. They present system characteristics in engineering terms—a necessary and sufficient set reflecting operational requirements and constraints.

Performance Requirements Validation

Performance requirements validation involves two interrelated activities: (1) integration of requirements derived from alternative system concepts and (2) effectiveness analyses to demonstrate satisfaction of the operational requirements. Performance requirements are defined in a living document; requirements are reviewed and updated throughout the system life cycle.

PROBLEMS

7.1 Explain why it is necessary to examine a number of alternative system concepts prior to defining a set of system performance requirements for the purpose of competitive system acquisition. What are the likely results of failing to examine a sufficient range of such concepts?

7.2 To meet future pollution standards, several automobile manufacturers are developing cars powered by electricity. Which major components of gasoline-powered automobiles would you expect to be retained with minor changes? Which ones would probably be substantially changed? Which would be new? (Do not consider components not directly associated with the automobile's primary functions, such as entertainment, automatic cruise control, power seats and windows, and air bags.)

7.3 List the characteristics of a set of well-stated operational requirements, that is, the qualities that you would look for in analyzing their adequacy. For each, state what could be the result if a requirement did not have these characteristics.

7.4 In the section of performance requirements formulation, the process of system development is stated to be "nondeterministic." Explain in your own words what is meant by this term. Describe an example of another common process that is nondeterministic.

7.5 Derive the principal functions of a DVD player by following the checklist shown in the subsection Functional Exploration and Allocation. How does each function relate to the operational requirements of the DVD player?

7.6 IPTs are stated to have four main benefits. What specific activities would you expect systems engineers to perform in realizing each of these benefits?

7.7 What role does exploratory R & D conducted prior to the establishment of a formal system acquisition program play in advancing the objective of a system acquisition program? What are the main differences between the organization and funding of R & D programs and system development programs?

7.8 In considering potential system concepts to meet the operational requirements for a new system, there is frequently a particular concept that appears to be an obvious solution to the system requirements. Knowing that premature focusing on a "point solution" is a poor systems engineering practice, describe two approaches for identifying a range of alternative system concepts for consideration.

7.9 (a) Develop a set of operational requirements for a simple lawn tractor. Limit yourself to no more than 15 operational requirements.

  (b) Develop a set of performance requirements for the same lawn tractor. Limit yourself to no more than 30 performance requirements.

  (c) Based on your experience, write a short paper defining the process of transforming operational requirements to performance requirements.

  (d) How would you go about validating the requirements in (b)?

FURTHER READING

B. Blanchard and W. Fabrycky. *System Engineering and Analysis*, Fourth Edition. Prentice Hall, 2006, Chapter 3.

W. P. Chase. *Management of Systems Engineering*. John Wiley, 1974, Chapters 4 and 5.

H. Eisner. *Essentials of Project and Systems Engineering Management*. Wiley, 1997, Chapter 8.

D. K. Hitchins. *Systems Engineering: A 21st Century Systems Methodology*. John Wiley & Sons, 2007, Chapters 5 and 8.

International Council on Systems Engineering. *Systems Engineering Handbook. A Guide for System Life Cycle Processes and Activities*. Version 3.2, July 2010.

K. Kendall and J. Kendall. *Systems Analysis and Design*, Sixth Edition. Prentice Hall, 2003, Chapters 4 and 5.

A. M. Law. *Simulation, Modeling & Analysis*, Fourth Edition. McGraw-Hill, 2007, Chapters 1, 2, and 5.

H. Lykins, S. Friedenthal, and A. Meilich. Adapting UML for an object-oriented systems engineering method (OOSEM). *Proceedings of 10th International Symposium INCOSE*, July 2000.

A. Meilich and M. Rickels. An application of object-oriented systems engineering to an army command and control system: A new approach to integration of systems and software requirements and design. *Proceedings of Ninth International Symposium INCOSE*, June 1999.

E. Rechtin. *Systems Architecting: Creating and Building Complex Systems*. Prentice Hall, 1991, Chapter 1.

N. B. Reilly. *Successful Systems for Engineers and Managers*. Van Nostrand Reinhold, 1993, Chapters 4 and 6.

A. P. Sage and J. E. Armstrong, Jr. *Introduction to Systems Engineering*. Wiley, 2000.Chapter 4.

R. Stevens, P. Brook, K. Jackson, and S. Arnold. *Systems Engineering, Coping with Complexity*. Prentice Hall, 1998, Section 8.

8

CONCEPT DEFINITION

8.1 SELECTING THE SYSTEM CONCEPT

The concept definition phase of the system life cycle marks the beginning of a serious, dedicated effort to define the functional and physical characteristics of a new system (or major upgrade of an existing system) that is proposed to meet an operational need defined in the preceding conceptual phases. It marks a commitment to characterize the system in sufficient detail to enable its operational performance, time of development, and life cycle cost to be predicted in quantitative terms. As illustrated in Chapter 4 (Figure 4.6), the level of effort in the concept definition phase is sharply greater than in previous phases, as system designers and engineering specialists are added to the systems engineers and analysts who largely staffed the preceding phases. In most needs-driven system developments, this phase is conducted by several competing developers, based on performance requirements developed in the preceding phases by or for the customer. The output of this phase is the selection, from a number of alternative system concepts, of a specific configuration that will constitute the baseline for development and engineering. From this phase on, the system development consists of implementing

Systems Engineering Principles and Practice, Second Edition. Alexander Kossiakoff, William N. Sweet, Samuel J. Seymour, and Steven M. Biemer
© 2011 by John Wiley & Sons, Inc. Published 2011 by John Wiley & Sons, Inc.

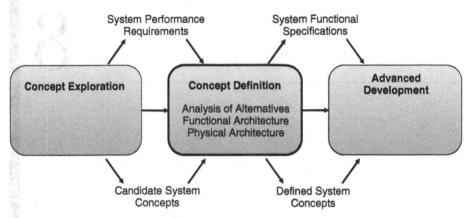

Figure 8.1. Concept definition phase in system life cycle.

the selected system concept (with modifications as necessary) in hardware and software, and engineering it for production and operational use.

With the advent and formal definition of systems architecting, this phase has been known in some sources as the system architecture phase. While this may not be entirely appropriate, systems architecting, as it is now defined and understood, is a major activity within this phase. The specifics of systems architecting are discussed in Section 8.8.

Place of the Concept Definition Phase in the System Life Cycle

The place of the concept definition phase in the overall system development is shown in Figure 8.1. It constitutes the last phase of the concept development stage and leads to the initiation of the engineering development stage, beginning with the advanced development phase. Its inputs are system performance requirements, the technology base that includes a number of feasible system concepts, and the contractual and organizational framework in which the system development is to be cast. Its outputs are system functional specifications, a defined system concept, and a detailed plan for the ensuing engineering program. The planning outputs of this phase are usually specified to include the systems engineering management plan (SEMP), which defines in detail the systems engineering approach to be followed, the project work breakdown structure (WBS), cost estimates for development and production, test plans, and such other supporting material as may be directed (see Chapter 5).

When the customer is the government, laws specify that all acquisition programs be conducted competitively, except in unusual circumstances. The competition frequently occurs during the concept definition phase. It customarily begins with a formal solicitation, which contains the system requirements, usually at the level of total system functionality, performance, and compatibility. Based on this solicitation, competing contractors carry out a proposal preparation effort, which embodies the concept definition phase of the program. The system concept and approach proposed by the successful bidder (or in some cases more than one) then becomes the baseline for the ensuing system development.

In the development of a commercial product, the concept definition phase generally begins after the conclusion of a feasibility study, which established a valid need for the product and the feasibility of meeting this need by one or more technical approaches. It is the point at which the company has decided to commit significant resources to define the product to a degree where a further decision can be made whether or not to proceed to full-scale development. Except for the formality and requirements for detailed documentation, the general technical activities during this phase for commercial and government programs are similar. One or several design concepts may be pursued, depending on the perceived importance of the objective and available funds.

Design Materialization Status

The previous phase was concerned with system design only to the level necessary to define a set of performance requirements that could be realized with a feasible system design, and that would not rule out other advantageous design concepts. For that purpose, it was sufficient to define functions at the subsystem level and only visualize the type of components that would be needed to implement the concept.

In order to define a system to the level where its operational performance, development effort, and production cost can be estimated with any degree of confidence (by analogy with previously developed systems), the conceptual design must be carried one level further. Thus, in the concept definition phase, the design focus is on components, the fundamental building blocks of systems. As indicated in Table 8.1, which is an overlay of Table 4.1, the focus in this phase is on the selection and functional definition of the system components and the definition of their configuration into subsystems.

Performance of the above tasks is primarily a systems engineering responsibility since they address technical issues that often cut across both technical disciplines and organizational boundaries. However, the functional definition task can be effectively carried out only if the component implementation used to achieve each prescribed function is reasonably well understood and is sufficiently visualized to serve as the basis for risk assessment and costing, which cannot be carried out solely at the functional level. Accordingly, as with many systems engineering tasks, consultation with and advice from experienced design specialists are almost always required, especially in cases where advanced techniques may be used to extend subsystem performance beyond previously achieved levels.

Systems Engineering Method in Concept Definition

The activities in the concept definition phase are discussed in the following sections in terms of the four steps of the systems engineering method (see Chapter 4), followed by a description of the planning of the ensuing system development effort and the formulation of system functional requirements. The four steps, as applied to this phase, are summarized below (generic names in parentheses):

TABLE 8.1. Status of System Materialization of Concept Definition Phase

	Phase					
	Concept development			Engineering development		
Level	Needs analysis	Concept exploration	Concept definition	Advanced development	Engineering design	Integration and evaluation
System	Define system capabilities and effectiveness	Identify, explore, and synthesize concepts	Define selected concept with specifications	Validate concept		Test and evaluate
Subsystem		Define requirements and ensure feasibility	Define functional and physical architecture	Validate subsystems		Integrate and test
Component			Allocate functions to components	Define specifications	Design and test	Integrate and test
Subcomponent		Visualize		Allocate functions to subcomponents	Design	
Part					Make or buy	

Performance Requirements Analysis (Requirement Analysis). Typical activities include

- analyzing the system performance requirements and relating them to operational objectives and to the entire life cycle scenario, and
- refining the requirements as necessary to include unstated constraints and quantifying qualitative requirements where possible.

Functional Analysis and Formulation (Functional Definition). Typical activities include

- allocating subsystem functions to the component level in terms of system functional elements and defining element interactions,
- developing functional architectural products, and
- formulating preliminary functional requirements corresponding to the assigned functions.

Concept Selection (Physical Definition). Typical activities include

- synthesizing alternative technological approaches and component configurations designed to performance requirements;
- developing physical architectural products; and
- conducting trade-off studies among performance, risk, cost, and schedule to select the preferred system concept, defined in terms of components and architectures.

Concept Validation (Design Validation). Typical activities include

- conducting system analyses and simulations to confirm that the selected concept meets requirements and is superior to its competitors, and
- refining the concept as may be necessary.

The application of the systems engineering method to the concept definition phase is illustrated in Figure 8.2, which is an elaboration of the generic diagram of Figure 4.12. Inputs are shown to come from the previous (requirements definition) phase in the form of system performance requirements and competitive design concepts. In addition, there are important external inputs in the form of technology, system building blocks (components), tools, models, and an experience knowledge base. Outputs include system functional requirements, a defined system concept, and (not shown in the diagram) detailed plans for the ensuing engineering stage of system development.

8.2 PERFORMANCE REQUIREMENTS ANALYSIS

As noted in Chapter 4, each phase of development must begin with a detailed analysis of all of the requirements and other terms of reference on which the ensuing program is to be predicated. In terms of problem solving, this is equivalent to first achieving a complete understanding of the problem to be solved.

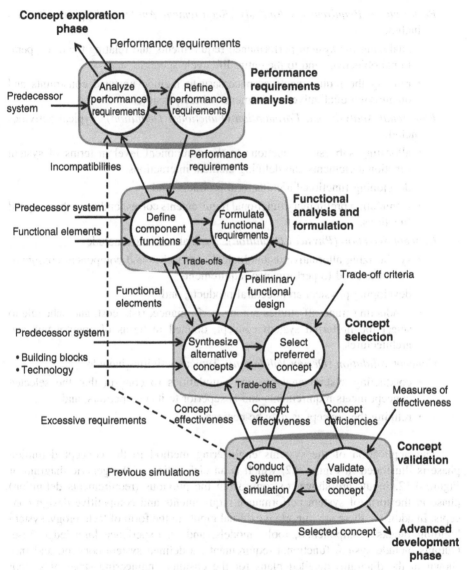

Figure 8.2. Concept definition phase flow diagram.

Analysis of Stated Performance Requirements

Requirements analysis in the concept definition phase is especially important because system performance requirements as initially stated often represent an imperfect interpretation of the user's actual needs. Even though the previous phases may have been thoroughly carried out, the derivation of a set of performance requirements for a complex system is necessarily an imprecise and often subjective process, not to mention iterative. In particular, the stated requirements tend to be influenced by personal and often not well-founded presumptions of what will turn out to be hard or easy to achieve. This may result in some performance requirements being unnecessarily stringent because they are believed to be readily achievable (a presumption that may turn out to be invalid). It is therefore essential that both the basis for the requirements and their underlying assumptions be clearly understood. Following this, steps can be taken to refine the requirements as necessary to support the definition of a truly viable system concept. The estimated relative difficulty of achieving the requirements will help to guide resource allocation during development.

The task of understanding the source of the given performance requirements in terms of user needs is the particular province of systems engineering. This task requires as intimate an acquaintance with the operational environment and with system users as circumstances may permit. In the case of complex operational systems, such an understanding can best be derived through years of work in the field.

Categories of System Requirements. In discussing the subject of requirements analysis, attention is usually focused on what functions the system must perform and how well. We have named these types of requirements, *functional* and *performance*. Such requirements are generally well defined. There are, however, other types of requirements that may be equally important but may be much more poorly defined, or even omitted up to this point. These include the following:

1. *Compatibility Requirements:* how the system is to interface with its operating site, its logistics support, and with other systems.
2. *Reliability, Maintainability, Availability (RMA) Requirements:* how reliable the system must be to fulfill its purpose, how it will be maintained, and what support facilities will be required.
3. *Environmental Requirements:* what extremes of the physical environment must the system be built to withstand throughout its lifetime.

RMA requirements, when explicitly stated, tend to be arbitrary and often not well defined. For the other two categories, requirements are often largely confined to the system's operational mode and leave out the conditions of shipping, storage, transit, assembling, and supporting the system. In these circumstances, it is necessary to investigate in detail the entire life of the system, from product delivery to the end of its operating life and its disposition.

System Life Cycle Scenario. To understand all of the situations that the system will encounter during its lifetime, it is necessary to develop a model or scenario that identifies all of the different circumstances to which the system will be exposed. These will include at least

1. storage of the system and/or its components,
2. transportation of the system to its operational site,

3. assembly and readying the system for operation,
4. extended deployment in the field,
5. operation of the system,
6. routine and emergency maintenance,
7. system modification and upgrading, and
8. system disposition.

The model of these phases of the system's use must be sufficiently detailed to reveal any interactions between the system and its environment that will affect its design. For example, the maintenance of the system will require a supply of spare parts, special test equipment, special test points, and other provisions that need to be recognized.

The model also needs to contain information for life cycle costing. Only by visualizing the complete life of the projected system can valid requirements and associated costs be developed.

Completion and Refinement of System Requirements

The development of a system life cycle model will almost always reveal that many important system requirements were not explicitly stated. This is likely to be true not only for the nonoperating phases of the system but also for its interaction with the physical environment. These environmental specifications are often derived from "boiler plate," especially in many military systems, rather than from a realistic model of the operating environment. In contrast, the desire to make use of standard commercial components may cause such specifications to be unduly relaxed or omitted entirely.

Probably the most important requirement that is often not stated is that of affordability. In competitive system developments, the projected system cost is one of the factors considered in selecting the winning proposal. Therefore, affordability must be considered as equivalent to other stated requirements, even though it may not be represented as such. It is, therefore, necessary to gain as much insight as practicable into what level of projected system cost development, production, and support will constitute an acceptable (or competitive) value.

Useful life is another system characteristic that is seldom stated as a requirement. To prevent early obsolescence, a system that uses high technology must be capable of periodic upgrading or modernization. To make such a process economically viable, the system must be designed with this objective in mind, making those subsystems or components that are susceptible to early obsolescence easy to modify or replace with newer technology.

In some programs, such upgrading or growth capability is explicitly provided for. This process is sometimes called "preplanned product improvement" (P^3I). In the majority of cases, however, especially when initial cost is a major concern, there is not a stated requirement for such capability. Nevertheless, it must be kept in mind as an important criterion for comparing alternative system concepts, since in practice, future changes in operating conditions and/or system environment (or product competition) will more often than not lead to increasing pressures for a system upgrade.

Unquantified Requirements. In order to be useful, a system requirement must be verifiable. This typically means measurable. Where the requirement is stated in nonquantifiable terms, the task of requirements analysis includes endowing it with as much quantification as possible. The following two examples are typical of such requirements.

A commonly unquantified area is that of user requirements, and especially the user–system interface. The overworked term "user friendly" does not translate readily into measurable form. Accordingly, it is important to gain a firsthand understanding of the user's needs and limitations. This, in turn, is complicated by the fact that there may be several users with different interfacing constraints and levels of training. There is also the maintenance interface, which has totally different requirements.

The interfaces between the system and other equipment at its operating site and with related systems are also often not stated in measurable terms. This may require a firsthand examination of the projected system environment, and even measurements of these interfaces, if necessary. For example, are there specifications for such parameters as available power or input signals that must be provided at the site?

Requirements and the Predecessor System. As noted previously, if there is a predecessor (current) system that performs the same or similar function as the projected system, as is usually the case, it is the single richest source of information on the requirements for the new system. It deserves detailed study by systems engineering at all stages of development, especially in the formative phases.

The predecessor system offers an excellent basis for understanding the exact nature of the deficiencies that led to the call for a new system. Since all its attributes are measurable, they can serve as a point of departure for quantifying the requirements for the new system. There is frequently documentation available that can provide a direct comparison to requirements for the new system.

The users of the predecessor system are usually the best source of information of what is needed in a new system. Thus, systems engineering should make the effort to gain a detailed firsthand understanding of system operation.

Operational Availability. There may or may not be a stated requirement for the date at which the system is to be ready for operational use. When there is, it is important to try to understand the priority of meeting this date relative to the importance of development cost, performance, and other system characteristics. This knowledge is needed because these factors are mutually interdependent, and their proper balance is essential to the success of the system development.

In any event, the time of availability is always important to the ultimate value of the system. This is because the growth of technology and competitive pressures operate continuously to shorten the new system's effective operational life. Thus, the time of operational availability must be considered a prime factor in the planning of a system development. In commercial developments, the first product to exploit a new technology often gains a lion's share of the market.

Determining Customer/User Needs. As noted previously, it is always necessary to clarify, extend, and verify the stated system requirements through contacts not only with the customer but also with present users of existing or similar systems.

In a competitive acquisition program, access to the customer may often be formally controlled. However, it should be used, insofar as possible, to clarify ambiguities and inconsistencies in the requirements as originally stated. This may be done directly, through correspondence, or at a bidders' conference, as appropriate.

A better opportunity to clarify system requirements is in the preproposal stage. In many large acquisition programs, a draft request for proposal (RFP) is circulated to prospective bidders for comment. During this period, it is usually possible to obtain a better understanding of the customer requirements than will be possible after the issuance of the RFP. This emphasizes the fact that the effort to respond to a system acquisition RFP must begin well before (months or years) its formal issuance.

In developing commercial systems, there is always an active and often an extended market survey to establish customer/user needs. In these cases, explicit system requirements may often not yet exist. As a prerequisite to the definition of a system concept and its associated performance requirements, it is therefore essential that systems engineering interact as directly as possible with potential customers and users of current systems to observe at first hand the system strengths, limitations, and associated operating procedures.

8.3 FUNCTIONAL ANALYSIS AND FORMULATION

It has been seen that in keeping with the inherent magnitude of designing a complex system, the systems engineering method divides the design task into two closely coupled steps: (1) analyzing and formulating the functional design of the system (what actions it needs to perform) and (2) selecting the most advantageous implementation of the system functions (how the actions can best be physically generated). The close coupling between these steps results from their mutual interdependence, which requires both visualization of the implementation step in formulating the functional design and iteration of the implementation step when alternative approaches are considered. Those familiar with software engineering will recognize these two steps as design and implementation, respectively.

Definition of Component Functions

The system materialization process in the concept definition phase is mainly concerned with the functional definition of system components (see Table 7.1). If the details of the concept exploration phase are available, the functional configuration at the system level has already been explored (recall the coffeemaker example in Chapter 7). If not, there will have almost always been exploratory studies preceding the formal start of concept definition that have laid out one or more candidate top-level concepts that can serve as a starting point for component functional design.

Functional Building Blocks. The general nature of the task of translating performance requirements into system functions can be illustrated by using the concept of system functional building blocks as summarized in Chapter 3. Extending the discussion in Chapter 7, the following steps are involved:

1. *Identification of Functional Media.* The type of medium (signals, data, materials, energy, and force) involved in each of the primary system functions can usually be readily associated with one of these five classes, using the criteria suggested in Chapter 7.
2. *Identification of Functional Elements.* Operations on each of the five classes of media are represented by five or six basic functional elements, listed in Chapter 3, each performing a significant function and found in a wide variety of system types. The system actions (functions) can be constructed from a selection of those functional building blocks.
3. *Relation of Performance Requirements to Element Attributes.* Each functional element possesses several key performance attributes (e.g., speed, accuracy, and capacity). If these can be related to the relevant system performance requirement(s), it confirms the correct selection of the functional element.
4. *Configuration of Functional Elements.* The functional elements selected to achieve the required performance characteristics must be interconnected and grouped into integrated subsystems. This may require adding interfacing (input/output) elements to achieve connectivity.
5. *Analysis and Integration of All External Interactions.* The given performance requirements often leave out important interactions of the system with its operational (or other) environment (e.g., external controls or energy source). These interactions need to be integrated into the total functional configuration.

It is not advisable to attempt to optimize at this stage. The initial formulation of the system functional design will need to be modified after the subsequent step of physical definition and the ensuing iteration.

Functional Interactions. The functional elements are inherently constituted to require a minimum of interconnections to other elements besides primary inputs and outputs. However, most of them depend on external controls and sources of energy, as well as being housed or supported by a material structure. Their grouping into subsystems should be such as to make each subsystem as self-sufficient as possible.

Minimizing critical functional interactions among different subsystems has two purposes. One is to aid the system development, engineering, integration, test, maintenance, and logistics support. The other is to facilitate making future changes in the system during its operational life to upgrade its effectiveness.

When several different ways to group functions (functional configurations) are comparably effective, these alternatives should be carried forward to the next step of the design process where a choice of the superior configuration may be more obvious.

Functional Block Diagramming Tools

Several formal tools and methods exist (and continue to be developed) for representing a system's functionality and their interactions. Commercial industry has used the functional flow diagram, formally referred to as the functional flow block diagram (FFBD), to represent not only functionality but also the flow of control (or any of the five basic elements). This diagramming technique can be used at multiple levels to form a hierarchy of functionality.

Recently developed is a method known as the integrated definition (IDEF) method. In fact, IDEF extends beyond functionality and now encompasses a range of capability descriptions for a system. Integrated definition zero (IDEF0) is the primary technique for representing system functionality. The basic construct is the functional entity, represented by a rectangle, as shown in Figure 8.3. Strict rules exist for identifying interfaces to and from a function. Sometimes, detail is included within the box, such as the listing of multiple functions performed by the entity; other times, the inside of the rectangle is left blank. Inputs always enter from the left; outputs exit to the right. Controls (separated from inputs) enter the function from the top, and mechanisms (or implementation) enter from the bottom.

One of the simplest diagramming techniques is the functional block diagram (FBD). This technique is similar to FFBDs, but without the flow structure, and IDEF0,

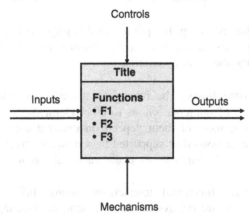

Figure 8.3. IDEF0 functional model structure.

but without the diagramming rules. Basically, each function is represented by a rectangle. Interfaces between functions are identified by directional arrows and are labeled to represent what is being passed between the functions. When a function interfaces with an external entity, the entity is represented in some fashion (e.g., rectangle and circle) and an interface arrow is provided.

Recall from Chapter 7 the example of the coffeemaker. Eleven functions were identified; they are relisted here:

Input Functions
- Accept user command (on/off)
- Receive coffee materials
- Distribute electricity
- Distribute weight

Transformative Functions
- Heat water
- Mix hot water with coffee grinds
- Filter out coffee grinds
- Warm brewed coffee

Output Functions
- Provide status
- Facilitate removal of materials
- Dissipate heat

Figure 8.4 represents an FBD using the 11 functions. Three external entities were also identified: the user, a power source (assumed to be an electrical outlet), and the environment. Notice that within the functions list, and the diagram, maintenance is not considered. This is due to the nature of household appliances in general, and coffeemakers in particular. They are not designed to be maintained. They are "expendable" or "throwaway."

Since it is difficult to avoid crossing lines, several mechanisms exist to distinguish between separate interface arrows. Color is probably the most prevalent. But other methods, such as dashed lines, are used as well. In the case of power, we have simply listed the functions that require power (e.g., "F5"). We have tried to be rather thorough in this example to help the reader think through the process of identifying functions and developing a functional structure for the system. Simplifying this diagram would not be difficult since we could omit several functions at this stage, as long as we did not forget about them later on. For example, function 10, "facilitate removal of materials," could be omitted at this stage, as long as the ultimate design does indeed allow the user to easily remove materials. Notice as well that we can categorize the functions into those handling the five basic elements:

Materials	Receive coffee materials
	Mix hot water with coffee grinds
	Filter out coffee grinds
	Facilitate removal of materials
Data	Provide status
Signals	Accept user commands
Energy	Distribute electricity
	Heat water
	Warm brewed coffee
	Dissipate heat
Force	Distribute weight

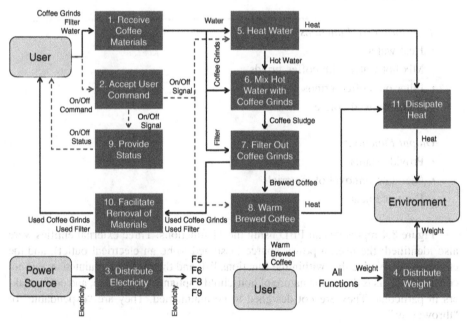

Figure 8.4. Functional block diagram of a standard coffeemaker.

This is not a "clean" categorization, since some functions input one type of element and convert it into another type. For example, function 2, "accept user commands," inputs a datum and converts it to signals. Subjective judgment is necessary.

Hardware–Software Allocation. The issue of whether a given function should be performed by hardware or software may seem like a question of implementation rather than function. However, system-level issues are almost always involved in such decisions, such as the effect on operator interfaces, test equipment, and widespread interaction with other system elements. Accordingly, the definition of functional building blocks makes a clear distinction between software elements (e.g., control system and control processing) and hardware elements (e.g., process signal and process data). For these reasons, the functional definition at the component level should include the allocation of all significant processing functions to either hardware or software. An important consideration in such decisions is provision for future growth potential to keep up with the rapidly advancing data processing technology.

In software-embedded systems, as defined in Chapter 11, software tends to be assigned most of the critical functions, especially those related to controls, because of its versatility. In software-intensive systems, in which virtually all the functionality is performed by software, functional allocation is not as straightforward because of the absence of commonly occurring functional elements. Chapter 11 describes the inherent differences between hardware and software and their effect on system design, and addresses the methods used in designing software system architectures.

To the extent that decisions may be involved in selecting functional elements, configuring them, or quantifying their functional characteristics, trade-offs should be made among the candidates using a set of predefined criteria. The principles and methods of trade-off analysis are described in Chapter 9.

Simulation

The analysis of the behavior of systems that have dynamic modes of response to events occurring in their environment often requires the construction of computer-driven models that simulate such behavior. The analysis of the motion of an aircraft, or for that matter of any vehicle, requires the use of a simulation that embodies its kinematic characteristics.

Simulations can be thought of as a form of experimental testing. They are used to obtain information critical to the design process in a much shorter time and at lesser cost than building and testing system components. In effect, simulations permit designers and analysts to gain an understanding of how a system will behave before the system exists in physical form. Simulations also permit designers to conduct "what-if" experiments by making selected changes in key parameters. Simulations are dynamic; that is, they represent time-dependent behavior. They are driven by a programmed set of inputs or scenarios, whose parameters may be varied to produce the particular responses to be studied, and may include input–output functional models of selected system elements. These characteristics are especially useful for conducting system trade-off studies.

In the concept definition phase, system simulation is particularly useful in the concept selection process, especially in cases where the dynamic behavior of the system is important. Simulation of the several alternative concepts permits the conduct of "experiments" that present the candidates with a range of critical potential challenges. The use of simulation results in scoring the candidates is generally more meaningful and persuasive than using judgment alone. Chapter 9 describes in greater detail some of the different types of simulation used in system development.

Formulation of Functional Specifications

One of the outputs of the concept definition phase is a set of system functional specifications to serve as an input to the advanced development phase. It is appropriate to formulate a preliminary set of functional specifications at this step in the process to lay the groundwork for more formal documents. This also serves as a check on the completeness and consistency of the functional analysis.

In stating functional specifications, it is essential to quantify them insofar as may be inferred from the performance and compatibility requirements. The quantification should be considered provisional at this time, to be iterated during the physical definition step and incorporated into the formal system functional specification document at the end of the concept definition phase. It is at this level in the system hierarchy that the physical configuration becomes clearly evident.

8.4 FUNCTIONAL ALLOCATION

The decisions in the process of concept definition center on the selection of a particular system configuration or concept and the definition of the functions it is to perform. These decisions do more to determine the ultimate performance, cost, and utility of the new system than those in any subsequent phase of the development. Further, in a competitive acquisition process, selection of who will develop the system is largely based on the evaluation of the proposed concept and the supporting documentation. For those reasons, the functional allocation process is of crucial importance.

The systems engineering method calls for such decisions to be made by a structured process that considers the relative merit of a number of alternatives before any one is selected. This process is called "trade-off studies" or "trade-off analysis" and is used in decision-making processes throughout system development. Trade-off analysis is most conspicuously employed during the concept definition phase, largely in the selection of the physical implementation of system components. As stated previously, Chapter 9 contains a description of the principles and methods of trade-off analysis.

Formulation of Alternative Concepts

The first step in selecting a preferred system concept is to formulate a set of alternative solutions, or in this case, system concepts. In the early development phases, the alternative construction begins by allocating the functions identified above to physical components of the system. In other words, we must determine how we will implement the functions above. Of course, this might entail decomposing the top-level functions in an FBD (or other functional representation) into lower-level functions. Many times, this activity provides insight into alternative methods of implementing each function.

As we identify system components, beginning with subsystems, we are constantly faced with the question of whether multiple functions can and should be implemented by a single physical component. The converse is also an issue: should a single function be implemented by multiple subsystems? Ideally, a one-to-one mapping is our goal. However, other factors may lead one to map multiple functions to a single component, or vice versa.

A specific allocation of functions to physical components, and the functional and physical interfaces that result from that allocation, is considered a single alternative. Other allocation schemes will result in different alternatives. The trade-offs mentioned above can occur at multiple levels, from the entire system to individual components. Many times, these trade-offs are part of the functional allocation process.

An important objective is to ensure that no potentially valuable opportunities are omitted. The following paragraphs discuss issues with developing alternatives.

The Predecessor System as a Baseline. As noted earlier, most system developments are aimed at extending the capabilities or increasing the efficiency of some function that is presently being inadequately performed by an existing system. In cases where the functions of the current system are the same or similar to those of the new system, the current system provides a natural point of departure for system concept definition. Where the main driving force comes from serious deficiencies of limited portions of the current system, an obvious (partial) set of alternative approaches would begin with a minimum modification of the system, restricted to those subsystems or major components that are clearly deficient. Other alternatives would progressively modify or replace other subsystems that may be made obsolescent by modern technology. The general configuration of the system would be retained.

In cases where there are new and improved technological advances at the component level, or when there are standard commercial off-the-shelf components that could be applied to the new system, the impetus for change to a new system would be technology-driven. In this case, a commonly used approach is to introduce improvements sequentially over time as modifications to the current system configuration.

Even when there are reasons against retaining any parts of the current system, as, for example, when moving from a conventional, manually controlled process to an automated and higher-speed operation, the current system's general functional configuration, component selection, materials of construction, special features, and other characteristics usually provide a useful point of departure for alternative concepts.

Technological Advances. As noted in Chapter 6, some new system developments are driven more by advances in technology than by operational deficiencies in the previous system. These advances may arise either in exploratory research and development programs aimed at particular application areas, such as development of advanced jet engines, or may come from broadly applicable technology such as high-speed computing and communication devices.

Such advances are often incorporated into an existing system to achieve specific performance improvements. However, if their impact is major, the possibility of a radical departure from the previous configuration should be included among the alternatives. Beyond a certain point, the existing framework may overly constrain the achievable benefits and should therefore be abandoned. Thus, when advanced technology is involved, a wide range of choices for change should be examined.

Original Concepts. In relatively rare instances, a really different concept is advanced to meet an operational need, especially when the need had not been previously met. In such instances, there is not likely to be a previous system to use for comparison, so that different types of alternatives would need to be examined. Often, various versions of the new concept can be considered, differing in the degree of reliance on new and unproven technology in exchange for projected performance and cost.

Modeling of Alternatives

For comparing alternative concepts, each must be represented by a model that possesses the key attributes on which the relative values of the alternatives will be judged. As a minimum, an FFBD of each should be constructed, and a pictorial or other physical description produced for providing a more realistic view of the system candidate.

Both the above modeling and the simulation of alternative concepts will contribute important context to the selection process and associated trade-offs.

8.5 CONCEPT SELECTION

The objective of trade-off studies in the concept definition phase is to assess the relative "goodness" of alternative system concepts with respect to

- operational performance and compatibility,
- program cost,
- program schedule, and
- risk in achieving each of the above.

The results are judged not only by the *degree* to which each characteristic is expected to be achieved but also by the *balance* among them. Such a judgment is of necessity highly program dependent because of the differing priorities that may be placed on the above characteristics.

Design Margins. In a competitive program, there is always a tendency to maximize system performance so as to gain an edge over competing system proposals. This often results in pushing the system design to a point where various design margins are reduced to a bare minimum. The term "design margin" refers to the amount that a given system parameter can deviate from its nominal value without producing unacceptable behavior of the system as a whole. A reduction in design margins is inevitably reflected in tighter restrictions on the environmentally induced changes in component characteristics during system operation and/or on the fabrication tolerances imposed in the production process. Either can lead to higher program risk, cost, or both. Accordingly, the issue of design margins should be explicitly addressed as an important criterion when selecting a preferred system concept.

System Performance, Cost, and Schedule. To the extent that stated performance requirements are quantified, are found to be an accurate expression of operational needs, and are within current system capabilities, they may be considered a minimum baseline for the system. However, where they are found to stress the state of the art, or to be desirable rather than truly essential, they need to be considered elastic and capable of being traded off against cost, schedule, risk, or other factors. Unstated requirements found to be significant should always be included among the variables.

Program cost must be derived from the system life cycle cost, which in turn must be derived from a model of the complete system life cycle. The appropriate relative weighting of the near-term versus long-term costs depends on the financial constraints of the acquisition strategy. Specific cost drivers should be identified wherever possible.

The appropriate weighting of schedule requirements is very program dependent and may be difficult to establish. There is an inherent tendency, especially in government and other programs where competition among contractors is especially strong, to estimate both cost and schedule of a new acquisition on the optimistic side, making no provision for the unforeseen delays that always occur in new system developments and are often caused by "unk-unks," as discussed in Chapter 4. This optimism factor also applies to the estimation of system performance and technical risk. Overall, it tends to slant the trade-off process toward the selection of advanced concepts and optimistic schedules over more conservative ones.

Program Risks. The assessment of risk is another primary systems engineering task. It involves estimating the probability that a given technical approach will *not* succeed in achieving the intended objective at an affordable cost. Such risk is present in every previously untried approach. In the development of new complex systems, there are many areas in which risk of failure must be explicitly considered and measures taken to avoid such risks or to reduce their potential impact to manageable levels.

Chapter 5, which devotes a section to the subject of risk management, shows that program risk can be considered to consist of two factors: (1) probability of failure—the probability that the system will fail to achieve an essential program objective, and (2) criticality of failure—the impact of the failure on the success of the program. Thus, the seriousness of each risk can be qualitatively considered as a combination of the probability of the failure weighted by its criticality to the system. For the purposes of this chapter, the following are examples of conditions that may result in a significant probability of program failure:

- A leading-edge unproven technology is to be applied.
- A major increase in performance is required.
- A major decrease in cost is required for the same performance.
- A significantly more severe operating environment is postulated.
- An unduly short development schedule is imposed.

Selection Strategy. The preceding discussion shows that the principal criteria involved in selecting a preferred system concept are complex, semiquantitative at best, and involve comparisons of incommensurables. This means that the evaluation of the relative merits of alternatives must be such as to expose and illuminate their most critical characteristics and to allow the maximum exercise of judgment throughout the evaluation process.

Two additional guidelines for conducting complex trade-off analyses may be useful: (1) to conserve analytical effort, use a staged approach to the selection process, in which only the most likely winners are subjected to the full system evaluation; and (2) to retain the visibility of the complete evaluation profile of each concept (against each critical measure of effectiveness) until the final selection, rather than combining the components into a single figure of merit, a practice that is often employed but that tends to submerge significant differences.

In pursuing a staged approach, the following suggestions can serve as a checklist, to be applied where appropriate:

1. For the first stage of evaluation, make sure that a sufficient number of alternative approaches are considered to address all needs and to explore all relevant technical opportunities.

2. If the number of alternatives is larger than can be individually evaluated in detail, perform a preliminary comparison to winnow out the "outliers." This is equivalent to qualifying the candidates. But be careful not to discard prematurely any candidates that present a new and unique technological opportunity, unless they are inherently incapable of qualifying.

3. For the next stage of evaluation, examine the list of performance and compatibility requirements and select a subset of the most critical ones that are also the most likely to reject unsuitable system concepts. Include consideration of growth capability and design margins as appropriate.

4. For each candidate concept, evaluate its expected compliance with each selected criterion. In the case of partial noncompliance, attempt to adjust the concept where possible to satisfy the criteria. Estimate the resultant performance, cost, risk, and schedule. In the event of conspicuous imbalance in the above, attempt to modify further the concept to achieve an acceptable balance for all requirements.

5. Assign weighting factors or priorities to the evaluation criteria, including cost, risk, and schedule, and apply to the ranking of each concept. Avoid concepts that do not have a sound balance of the above factors.

6. For each evaluation criterion, rank order the several candidate concepts.

7. Look for and eliminate clear losers.

8. Unless there is a single clear winner, perform a significantly more detailed comparison among the two or three potential winners. To this end, develop a life cycle model for each concept, along with a WBS, and a risk abatement plan.

In making the final system concept selection, review the evaluation profile of the merit of each candidate concept against each critical measure of effectiveness to ensure that the choice has no major weaknesses. Check for the sensitivity of the result to a reasonable variation of the weighting of individual criteria.

As stated previously, use each of the above suggestions only where it may be appropriate to the particular selection process. Chapter 9 devotes a section to the fundamentals of trade-off analysis, with an example of their application.

8.6 CONCEPT VALIDATION

The task of designing a model of the system environment to serve as the basis for concept validation builds on the set of parameters initially established for use in the trade-off studies of the selection process.

Modeling the System and Its Environment

Since the degree of system definition at this stage is largely functional, its validation must rely primarily on analysis rather than on testing. The rapid growth of computer modeling and simulation in recent years is providing powerful tools for the validation of complex system concepts.

System Effectiveness Models. In complex operational systems, system effectiveness models are developed in the needs analysis and concept exploration phases to provide a fuller understanding of the effectiveness of existing systems in performing their missions and in identifying deficiencies that need to be remedied. These are most often computer simulations that include provisions for varying key parameters to establish the sensitivity of overall performance to environmental and system parameter variations and to determine the nature and extent of system changes needed to offset any identified deficiencies (see also Chapter 9).

In the concept definition phase, the construction of system effectiveness models by the system developer depends on whether or not the models used in the previous phases are available, as in the case where the developer is also the customer. In that case, the models can be readily extended to conform to the selected system concept for the validation process. If not, the construction of the model becomes part of the concept definition task. For this and other reasons, the preparation for the competitive effort often begins months (and sometimes years) before the start of the formal competition.

Computer models are also capable of validating a host of subsystem or component-level technical design features. Areas such as aerodynamic design, microwave antennae, hydrodynamics, heat transfer, and many others can be modeled for analysis through the use of special computer codes. Advances in computer capabilities have made such modeling more and more accurate in predicting system behavior for purposes of design and evaluation.

Critical Experiments. When a proposed system concept relies on technical approaches that have not been previously proven in similar applications, its feasibility must be demonstrated. Often this cannot be done credibly through analysis alone and must be subjected to experimental verification. This is difficult to accommodate in the limited time and constrained resources of a competitive acquisition, but must nevertheless be undertaken to support the proposed system concept.

The term "critical experiment" is appropriate in such instances because it is related to the specific purpose of substantiating a critical feature of the design. It purposely stresses the proposed design feature to its extreme limits to ensure that it is not just marginally satisfactory. The term "experiment" rather than "test" is appropriate because it is performed for the purpose of obtaining sufficient data to understand thoroughly the behavior of the system element, rather than merely to measure whether or not the element operates within certain limits. By the same token, extensive data analyses are also performed to illuminate the system behavior.

Analysis of Validation Results

The analysis of the results of system validation simulations can produce three different types of unsatisfactory findings that require remedial action: (1) deficiencies in the assumed characteristics of the system being modeled, (2) deficiencies in the test model, or (3) excessively stringent system requirements. It is the purpose of the analysis process to attribute the results of the simulation to one or more of the above causes. Beyond these findings, the analysis should also indicate what kind and degree of changes would eliminate the discrepancies. This latter finding usually requires a series of simulations or analyses that test the effect of alternative remedial actions.

The feedback resulting from the validation analysis results in an iterative process in which the system model design and environmental model are refined as necessary to bring the system model in compliance with the requirements.

Iteration of System Concepts and Requirements

The above description of the validation process implies that only one concept was found to be superior in the concept trade-off evaluation, and that this concept was then validated against the full system requirements. Not infrequently, two and sometimes more concepts turn out to be nearly equal in preliminary rankings. In that case, each should be evaluated against the full requirements to see if the more rigorous comparison produces a clear discriminator for selecting the preferred concept.

The system requirements should always be regarded as flexible up to a point. If the validation or trade-off results show that one or more stated requirements appear to be responsible for unduly driving up system complexity, cost, or risk, they should be subjected to critical analysis, and if appropriate, highlighted for discussions with the customer by program management.

8.7 SYSTEM DEVELOPMENT PLANNING

A major product of the concept definition phase is a set of plans that define how the engineering program is to be managed. Among these are the WBS, the life cycle model, the SEMP or its equivalent, system development schedules, the operational (or integrated logistic) support plan, and such others as may be specified by the contracting agency to provide all participants with clear objectives and timescales for accomplishing their respective tasks.

Of the above plans, systems engineering has prime responsibility only for the SEMP. However, it is also deeply involved in all the others by having to provide a detailed description and ongoing assessment of the development process to those who are directly responsible for the other technical management documents. For example, systems engineers are often asked to review initial estimates of the time and effort required to perform a particular engineering task, and based on their appraisal of the associated technical risks, to recommend approval or modification as appropriate.

WBS

The WBS, which was described in Chapter 5, is one of the essential development planning vehicles. The WBS provides a hierarchical framework designed to accommodate all the tasks that need to be accomplished during the entire life of the project. The topmost level represents the project as a whole; the next contains the system product itself, and the principal supporting and management categories. Succeeding levels subdivide the total effort into successively smaller work elements. This subdivision is continued until the complexity and cost of each work element or task are reduced to the point that the task can be directly planned, costed, scheduled, and controlled. The process must ensure that no necessary task is overlooked and that realistic cost and schedule estimates can be made.

The specific form of the WBS is dependent on the nature of the project and is often stipulated in the contract for the system development, especially if the government is the customer. Government programs have had to comply with standards, which define a specific hierarchical structure that provides a logical framework and a place for every aspect of a system product, often with a high degree of detail.

As an example of a typical WBS structure, the system project is at level 1, and the next level (level 2) is broken down into five types of activities, abbreviated from the more detailed descriptions in Chapter 5:

1. *System Product*, including the total effort of developing, producing, and integrating the system itself, together with any auxiliary equipment required for its operation. It includes all of the design, engineering, and fabrication of the system, as well as the testing of its components (unit test).
2. *System Support* (also referred to as "integrated logistics support"), involving provision of equipment, facilities, and services necessary for the development

and operation of the system product. It includes all equipment, facilities, and training for both development and system operations.

3. *System Test*, beginning at the integration test level, unit tests of individual components being part of the effort of developing the system product. It includes integration and testing of subsystems and of the total system.

4. *Project Management*, covering the project planning and control effort throughout the program.

5. *Systems Engineering*, covering all aspects of systems engineering support.

The WBS is by its nature an evolving document. As noted previously, it begins in the concept exploration phase, when only the topmost level can be identified. It is in the concept definition phase, when the system components and architecture have been defined, that serious costing and scheduling may be undertaken. Thereafter, the WBS must evolve along with the development and engineering of the system components and progressive discovery and resolution of problems. Thus, at any time, the WBS should reflect the latest knowledge of the program tasks and their status, and should constitute a reliable basis for program planning.

As noted in Chapter 5, the WBS is structured so that every task is identified at the appropriate place within the WBS hierarchy. Systems engineering plays an important role in helping the project manager to structure the WBS so as to achieve this objective.

SEMP

Chapter 5 described the nature and purpose of the planning of the systems engineering tasks that are to be performed in the course of developing a system. In many system acquisition programs, such a plan is referred to as the SEMP and is a required deliverable as part of a proposal for a system development program.

The SEMP is a detailed plan showing how the key systems engineering activities are to be conducted. It typically covers three main activities:

1. *Development Program Management*—including organization, scheduling, and risk management;

2. *Systems Engineering Process*—including requirements, functional analysis, and trade-offs; and

3. *Engineering Specialty Integration*—including reliability, maintainability, producibility, safety, and human factors.

Life Cycle Cost Estimating

The provision of a credible cost estimate for development, production, and (usually) operational support of the proposed new system is a required product of the concept definition phase. While systems engineering is not primarily responsible for this task, it has an essential role in providing key items of information to those who are.

The only basis for deriving costs for a new task is through the identification of a similar and successfully completed task whose costs are known. To this end, the system concept must be decomposed into elements analogous to existing components. Since the concept at this stage is still mainly functional, the systems engineer must visualize the likely physical embodiment of these functions. Once this is done, and any unusual features are identified, those experienced in cost estimating can usually make a reasonable estimate of the prospective costs.

The main guides for deriving system costs are the WBS, the life cycle model, and costing models. The WBS, which spells out all the tasks to be performed during system development, is the chief reference for deriving development costs.

The costs of developing new or modified components are usually derived from estimates provided by those who expect to do the development—whether subcontractors or in-house. Special care must be taken to assure that these estimates reflect an assessment of the associated development risk that is neither unduly optimistic nor overly cautious. These estimates should be reviewed critically by systems engineering to provide a check on the above factors.

The costs for component production, assembly, and testing are usually derived using a cost model developed for this purpose. The cost model is based on the accumulated experience of the developing organization and is updated after each new program. The actual costing is usually done by cost estimating specialists. However, these specialists must rely heavily on the vision of the system elements as provided by systems engineers and the design engineers responsible for component development.

The preparation of cost estimates must not only be as expertly performed as possible, but it must also be documented so as to be credible to management and to the customer. In a competitive acquisition program, the magnitude and credibility of the cost estimates, especially development costs that are the most immediate, weigh heavily in the evaluation.

The "Selling" of the System Development Proposal

The selection of a feasible and affordable concept in the concept definition phase is a necessary but not sufficient step to assure that the engineering of that concept into an operational system will be undertaken. Progression to the engineering development stage requires a management decision to devote much larger resources to the project than have as yet been expended in the conceptual phases. Whether the concept is to be part of a competitive proposal for a formal acquisition program or is to be presented informally to in-house management, there are always other ways to spend the money required to develop the proposed system. Accordingly, such a decision requires compelling evidence that the result will be well worth the cost and time to be expended.

To accomplish its purpose, the concept definition phase must produce persuasive evidence in favor of proceeding with the development of the proposed system. This requires that the reasons for selecting the proposed concept are clear and compelling, that the feasibility of the approach is persuasively demonstrated, and that the plan for carrying out the system development is thoroughly thought out and documented. The end result must be to instill a high degree of confidence that the new system will achieve the required performance within the estimated cost and time and be superior to other potential system approaches.

In developing such a case, it must be remembered that those making the decision to proceed are not likely to be technical experts, so that the evidence will have to be couched in terms that intelligent laymen can understand. This is a very difficult constraint, which must nevertheless be observed. Translating and condensing design specialist jargon and test data into a form that is readily understood, and is clearly relevant to the issues of concept feasibility, risk, and cost, is a very important responsibility that is commonly also assigned to systems engineering.

In this task of selling the system concept and development plan, the following general approach is recommended:

1. Show the shortfalls in existing systems and the need to be filled by the proposed system.
2. Demonstrate that the proposed concept was selected after a thorough examination of alternatives. Illustrate the alternatives and indicate which main features of the selected system drove the decision.
3. Fully discuss program risks and the proposed means for their management. Describe results of critical experiments designed to reveal problems and identify solutions, especially in the application of new technology.
4. Display evidence of careful planning of the development and production program. Documents such as the WBS, SEMP, TEMP, and other formal plans give evidence of such planning.
5. Present evidence of the organization's experience and previous successes in system developments of a similar nature, and the carryover of key staff to the proposed system.
6. Present the derivation of the life cycle costing for the project and the level of confidence in the conservatism of the estimates.
7. Provide further justification as indicated by the specific evaluation criteria listed in the system requirements. Discuss environmental impact analysis if that is an issue.

8.8 SYSTEMS ARCHITECTING

When we think of the word "architecture," something like Figure 8.5 comes to mind. For many people, architecture refers to buildings, and an architect is someone who designs buildings. Over two decades ago, though, a professor at the University of Southern California challenged that notion. He reasoned that as systems grew in complexity, the top-level design, or more accurately the conceptual design of a system, as defined at the time, was insufficient to guide engineers and designers to accurate and efficient designs. He looked to the field of architecture to understand how complex systems (i.e., buildings) could be created and developed, and (as far as we understand) coined the term "systems architecting." That man was Eberhardt Rechtin.

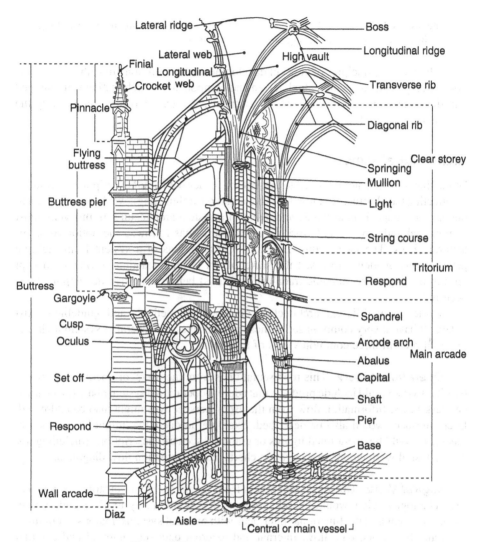

Figure 8.5. Traditional view of architecture.

The Institute of Electrical and Electronics Engineers (IEEE) Std 610.12 defines an architecture as "the structure of components, their relationships, and the principles and guidelines governing their design and evolution over time." This applies to complex systems, such as aircraft, power plants, and spacecraft, as much as buildings. Therefore, Rechtin's premise was to apply the principles from the field of architecture to systems engineering, not as a replacement, but as part of developing a system.

Dr. Rechtin defined the term *systems architecting* in this way:

> The essence of architecting is structuring. Structuring can mean bringing form to function, bringing order out of chaos, or converting the partially formed ideas of a client

into a workable conceptual model. The key techniques are balancing the needs, fitting the interfaces, and compromising among the extremes.

Read closely, the principles of concept development and definition are within his definition. Twenty years ago, conceptual design and components of architecting were lumped into the phrase "preliminary design." Fortunately, that term has been replaced by the more extensive "architecting."

Architectural Views

While this section is not intended to present the reader with a full description of systems architecting (see Further Reading for more detail on architecting), we do want to present the basic concepts behind the development of a system architecture. In this vein, most commercial and government work on architectures has followed the notion of architectural views. The idea is this. Develop representations of a system from multiple perspectives, or views, to assist the stakeholders in understanding a system concept (and in making those valuable trade-off decisions) before extensive development has occurred.

While many different architecture development methods and guidelines exist today, all have a very common set of these perspectives. In general, a system architecture will present three common views of a system.

Operational View. This representation is from the users' or operators' perspective. This view would include products that address operational system phases, scenarios, and task flows. Information flow from the users' perspectives might also be addressed. User interfaces would also be described. Example products that might be included in this view would be operational figures or graphics, scenario descriptions (including use cases), task flow diagrams, organization charts, and information flow diagrams.

Logical View. This representation is from the manager's or customer's perspective. The logical view would include products that define the system's boundary with its environment and the functional interfaces with external systems, major system functions and behaviors, data flow, internal and external data sets, internal and external users, and internal functional interfaces. Example products for this view would be FFBDs, context diagrams, N2 diagrams, IDEF0 diagrams, data flow diagrams, and various stakeholder-specific products (including business-related products).

Physical View. This representation is from the designers' perspective. This view would include products that define the physical system boundary, the system's physical components and how they interface and interact together, the internal databases and data structures, the information technology (IT) infrastructure of the system and the external IT infrastructure with which the system interfaces, and the standards in force in its development. Example products include physical block diagrams down to a fairly high level of detail, database topologies, interface control documents (ICDs), and standards.

Different architectural guidelines and standards may use different names, but all three of these perspectives are included in every architectural description.

A common question from someone just introduced to the concept of systems architecting is "What is the difference between architecting and designing?" A convenient method of answering that question is to delineate the uses of an architecture versus a design.

A system architecture is used

- to discover and refine operational and functional requirements,
- to drive the system to a specific use or purpose,
- to discriminate between options, and
- to resolve make/buy decisions.

A system design is used

- to develop system components,
- to build and integrate system components, and
- to understand configuration changes as the system is modified.

The nature of these uses means there is a difference between architecting and engineering. Systems architecting is largely an inductive process that focuses on functionality and behavior. Consequently, architecting deals with unmeasurable parameters and characteristics as much if not more than measureable ones. The toolset is largely unquantitative and imprecise—diagramming is a large component of the architect's toolset. Heuristics typically guide an architect's decisions rather than algorithms.

Design engineering can be contrasted with architecting since it relies on deductive processes. Engineering focuses on form and physical decomposition and integration. Consequently, design engineering deals with measurable quantities, characteristics, and attributes. Thus, analytical tools derived from physics are the engineer's primary tools.

Given these characteristics of the two fields (which should certainly not be considered loosely coupled), the architect tends to be active in the early phases of the system development life cycle. The architect tends to be rather dormant during the detailed design, fabrication, and unit testing phases. Integration and system testing will see the architect emerge again to ensure requirements and top-level architectures are being followed. In contrast, the design engineer's activity peaks during the architect's dormant phases, though he is by no means completely inactive during the early and late phases of system development.

Architecting in the Engineering Hierarchy. With the differences between architecting and engineering, it is obvious the two activities are separate. An obvious question then arises: who works for whom? Although there are exceptions, our role of systems architecting leads to the management structure where the architect works for the systems engineer. Systems architecting is a subset of systems engineering. This is different from the role and place of the traditional architect—which is typically at the top. When a new building is designed, developed, and constructed, the architect plays the primary role in the building's design and continues with that prominent role throughout development and construction. In system development, the systems engineer holds the prominent technical position and the architect works for the systems engineer.

Architecture Frameworks

As mentioned, architectures are used extensively now in large, complex system development programs. The architect and his team have a large latitude in developing and integrating products. This initially led to architectures that were technically accurate but diverse in their structure. In order to standardize the architecture development effort and the products associated with architectures, many organizations developed and mandated the use of architecture frameworks.

An architecture framework is a set of standards that prescribes a structured approach, products, and principles for developing a system architecture. Two early frameworks that emerged were the Command, Control, Communications, Computers, Intelligence, Surveillance and Reconnaissance (C4ISR) Architecture Framework mandated by the U.S. Department of Defense (DoD) and The Open Group Architecture Framework (TOGAF) developed for commercial organizations.

Other frameworks have emerged recently as well, and some that have been around for decades are being recognized as architecture frameworks, though that particular title was not applied until recently (e.g., the Zachman Framework). The early frameworks were focused on individual systems and their architectures. Newer versions, however, have expanded into the field of enterprise architecture, a subset of enterprise engineering or enterprise systems engineering (see Chapter 3 for a discussion of enterprise systems engineering). All of the current versions, including the Department of Defense Architecture Framework (DODAF) and TOGAF, have enterprise editions of their frameworks.

Many architecture frameworks that can be applied to system development exist, even if the primary purpose is enterprise architecting. Below is a selected list of architecture frameworks:

- DODAF
- TOGAF
- The Zachman Framework
- Ministry of Defense Architecture Framework (MODAF)
- Federal Enterprise Architecture Framework (FEAF)
- NATO Architecture Framework (NAF)
- Treasury Enterprise Architecture Framework (TEAF)
- Integrated Architecture Framework (IAF)
- Purdue Enterprise Reference Architecture Framework (PERAF)

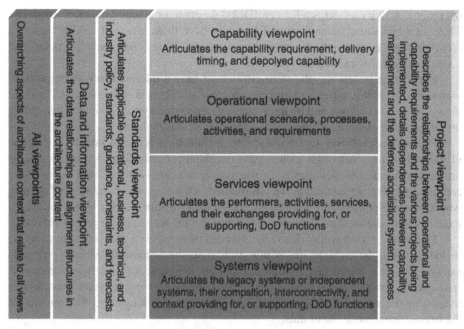

Figure 8.6. DODAF version 2.0 viewpoints.

DODAF. Although by no means more important or "better" than any other framework, we discuss the basic products of the DODAF to illustrate the basic components of a framework.

The DOD framework, like all frameworks mentioned, is divided into a series of perspectives, or viewpoints. Figure 8.6 depicts these viewpoints using a figure from the DODAF description. The viewpoints can be observed in three bundles. The first consists of four viewpoints that describe the overall system and its environment: capability, operational, services, and systems. The second bundle consists of the underlying principles, infrastructure, and standards: all data and information and standards. The final bundle is a single viewpoint focusing on the system development project.

Version 2 of this framework is easily scalable from the system level to the enterprise level, where multiple systems are under development and would be integrated into a legacy system architecture. In fact, each of the three major system-level architecture frameworks, DODAF, MODAF, and TOGAF, are now compatible with enterprise development efforts. Furthermore, with the addition a services viewpoint, service-oriented architectures are now possible within the DODAF framework.

Within each viewpoint, a set of views is defined. A total of 52 views are defined by DODAF, organized within the eight viewpoints. For each view, a variety of methods and techniques are available to represent the view. For example, one view within the operational viewpoint is the operational activity model. This view can be represented by a variety of models, such as the FFBD. Other models can be used to represent the operational activity model, such as an IDEF0 diagram, or a combination of diagrams. Thus, an architecture framework will typically have three layers of entities: a set of *viewpoints* that compose the *framework*, a set of *views* that define each viewpoint, and a set of *models* that can represent the view.

Every large system development effort must have a minimum set of architecture views. Rarely will a system architecture contain all 52 architecture views. Pertinent views are decided beforehand by the systems engineer and system architect, depending on the intended communication and the appropriate stakeholders.

The key to developing successful system architectures is to understand the purpose of the architecture. Although each system development effort is different, depending on the magnitude and complexity of the system, all architectures have at least one common purpose: to communicate information. Choosing which framework to use, which viewpoints within the framework, which views within the viewpoint, and which models within the view all depends on the purpose the architect is trying to achieve.

The existing frameworks define the superset of viewpoints and views that may be included within the architecture. Within each view, the framework typically suggests candidate models, which can be used to represent the view. A hallmark of the current frameworks, however, is the flexibility inherent within each view. If the architect desires to use a model not included in the candidate list, he can—as long as he does not violate the overall framework constraints.

For example, many of the current frameworks were initially defined using traditional, structured analysis models (e.g., IDEF0, FFBD, data flow diagrams) to define their views. However, engineers familiar with object-oriented (OO) models began to use a combination of OO and structured analysis models to represent views. As the trend increased, the organizations responsible for the common architecture frameworks revised the available models to include OO models that can represent the views. Section 8.9 discusses two languages that implement OO models.

8.9   SYSTEM MODELING LANGUAGES: UNIFIED MODELING LANGUAGE (UML) AND SYSTEMS MODELING LANGUAGE (SysML)

All architecture frameworks use models to represent aspects, perspectives, and views of the system. Traditional models, like standard block diagramming techniques, are based on the top-down decomposition of a system. These methods are typically functionally based and are formed into a hierarchy of models representing attributes of the system in increasing levels of detail. In the 1970s, when software engineering was expanding at a significant rate, a formal modeling construct emerged and was called "structured analysis and design" (SAAD). The term has been applied to systems in general and is not restricted to software systems only.

Models that have been in use for decades resemble many of the SAAD constructs, and they have been grouped into what we call *traditional hierarchical methods*, or simply *traditional systems modeling*. This book uses many of the traditional models to represent aspects of systems. This informal modeling language has evolved into an excellent educational language for communicating principles and techniques.

After the advent of SAAD, a new set of modeling languages has emerged, based on object-oriented analysis and design (OOAD) principles. This analysis and design method is primarily bottoms-up in approach and focuses on entities, as opposed to functions, though the two are closely related. In the 1990s, a new modeling language that incorporated OOAD principles and techniques was formalized: the UML.

UML

It was noted that in developing a complex system, it is essential to create high-level models of its structure and behavior to gain an understanding of how it may be configured to meet its requirements. In the development of OOAD methodology, several of the principal practitioners separately developed such models. In the mid-1990s, three of them (Booch, Rumbaugh, and Jacobson), developed a common modeling terminology they called the "UML." This language has been adopted as a standard by the software community and is widely used throughout industry and government. It is supported by sophisticated tools produced by several major software tool developers.

Whereas structured methodology employs three complementary views of a system, UML provides OO analysts and designers with 13 different ways to diagram different system characteristics. They may be divided into six static or structural diagrams and seven dynamic or behavioral diagrams. Figure 8.7 also lists the two sets of diagrams.

Structural diagrams represent different views of system entity relationships:

- *Class Diagrams* show a set of classes, their relationships, and their interfaces.
- *Object Diagrams* show a set of instances of classes and their relationships.
- *Component Diagrams* are typically used to illustrate the structure of, and relationships among, physical objects.
- *Deployment Diagrams* show a static view of the physical components of the system.
- *Composite Structure Diagrams* provide a runtime decomposition of classes.
- *Package Diagrams* present a hierarchy of components.

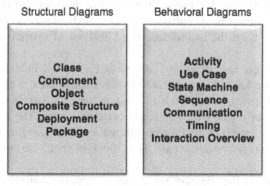

Figure 8.7. UML models.

Behavioral diagrams represent different views of system dynamic characteristics.

- *Use Case Diagrams* show interrelations among a set of use cases representing system functions that respond to interactions with external entities ("actors").
- *Sequence Diagrams* show the interactions among a set of objects in executing a system scenario, arranged in chronological order.
- *State Machine Diagrams* model the transition events and activities that change the state of the system.
- *Activity Diagrams* are flowcharts of activities within a portion of the system showing control flows between activities.
- *Communication Diagrams* define links between objects, focusing on their interactions.
- *Interaction Overview Diagrams* are a mix of sequence and activity diagrams.
- *Timing Diagrams* present interactions between objects with timing information.

UML class diagrams correspond approximately to entity relationship diagrams in structural analysis, while state chart diagrams correspond to state transition diagrams. Others, especially activity diagrams, are different views of functional flow diagrams.

The new language was quickly adopted by the software engineering community as the de facto standard for representing software concepts and software-intensive systems. Although the origins of the language are in the software world, recently, the language has been used successfully in developing systems that include both hardware and software.

UML is governed by the Object Management Group (OMG), a worldwide consortium. UML will continue to evolve with new releases and added complexity.

Rather than providing examples and explanations to all of the diagrams, we present some examples—several behavioral diagrams: the use case diagram, the activity diagram, and the sequence diagram; and one structural diagram: the class diagram.

Use Case Diagram. We present the use case diagram first due to its utility in defining a system's operation. In software, and in some hardware applications, use cases have been used to assist the identification and analysis of operational and functional requirements.

The form of a use case diagram is shown in Figure 8.8, modeling the interaction of an "actor" on the left side (represented by the stick figures) with a single use case (represented by an oval), which leads to a subordinate activity (a separate use case), while the other three interact with a second (external) actor. The arrows indicate the initiation of the use case, not the flow of information. For example, the librarian actor can initiate the "manage loans" use case. The "check-in book" use case may also initiate the same use case.

Each use case in the diagram represents a separate sequence of activities and events. UML defines a standard set of components for a use case, including

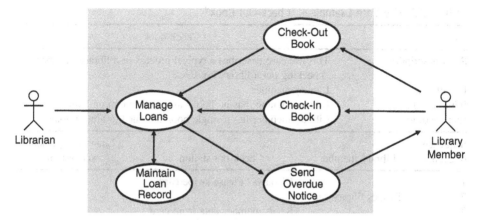

Figure 8.8. Use case diagram.

- title;
- short description;
- list of actors;
- initial (or pre-) conditions describing the state of the environment before the use case occurs (or is executed);
- end (or post-) conditions describing the state of the environment after the use case occurs (or has been executed); and
- sequence of events, a list of actions or events that occur in a defined sequence.

Table 8.2 displays an example use case description for "check-out book." The sequence of events lists the actions and activities that both actors and subsystems execute. In this case, the use case involves one actor and two subsystems—the check-out station and the loan management subsystem. This use case represents an automated check-out system at a library using the Universal Product Code (UPC) symbology.

Although not required, it can be beneficial to use columns to separate actions of each actor and subsystem, such as was done in Table 8.2. This allows the reader to easily determine who is performing the action and in what order (sometimes simultaneously). Use cases can, of course, be stylized or tailored to specific situations and may demonstrate the preferences of their authors. In other words, two engineers may come up with different use case sequences of events for the same use case. This may not represent a flaw or problem. In fact, a use case may have several different variants, known in UML as "scenarios." Unfortunately, the use of the term scenarios differs from our traditional definition provided earlier.

Activity Diagram. As another example of a behavior diagram, we turn to the activity diagram. Activity diagrams can represent any type of flow inherent in a system, including processes, operations, or control. The diagram accomplishes this through a

TABLE 8.2. Use Case Example—"Check-Out Book"

Title	Check book
Short description	This use case describes a typical process of a library member checking out a library book.
List of actors	Library member
Initial conditions	Library member has no books assigned to him on loan.
End conditions	Library member has a single book assigned to him on loan.

Sequence of events	Library member	Check-out station	Loan management subsystem
1		Displays "Please swipe card"	
2	Swipes library car		
3		Reads member data from card	
4		Sends request to confirm member is in good standing	
5			Checks database for member information
6			Confirms good standing
7		Receives confirmation	Sends confirmation
8		Displays "Place book UPC under scanner"	
9	Places book UPC symbol under scanner		
10		Scans book UPC	
11		Sends request to confirm book is available	
12			Checks database for book information
13			Confirms availability
14		Receives confirmation	Sends confirmation
15		Displays "Thank you! Book is due in two weeks."	Indicates book as "out"

sequence of activities and events. The sequence of activities and events is regulated via various control nodes. The basic components of the activity diagram are described below:

- *Action:* an elementary executable step within an activity (rectangle with rounded corners);
- *Activity Edge:* a connecting link between actions, and between actions and nodes (an arrow); activity edges are further divided into two types: object flows and control flows;
- *Object Flow:* an activity edge that transports objects (or object tokens);

- *Control Flow:* an activity edge that represents direction of control (also transports control tokens);
- *Pin:* a connecting link between action parameters and a flow (a box connected to an action and a flow); a pin accepts explicit inputs or produces explicit outputs from an action;
- *Initial Node:* the starting point for a control flow (solid circle);
- *Final Node:* the termination point for a control flow (solid circle within an open circle);
- *Decision Node:* a branch point for a flow in which each branch flow contains a condition that must be satisfied (diamond);
- *Merge Node:* a combination point in which multiple flows are merged into a single flow (diamond);
- *Fork Node:* a point at which a single flow is split into multiple concurrent flows (a solid line segment); and
- *Join Node:* a point in which multiple flows are synchronized and joined into a single flow (a solid line segment).

Figure 8.9 represents a simple activity diagram, which is analogous to a functional flow diagram, for our library book system. The diagram shows the activity path to split into two concurrent activities, one of which follows one of two logical paths, of returning or borrowing a library book.

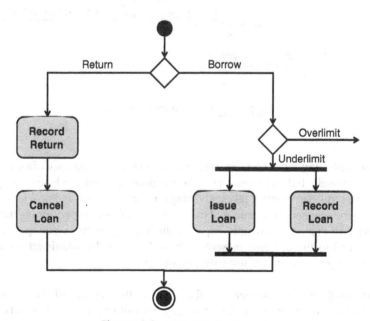

Figure 8.9. UML activity diagram.

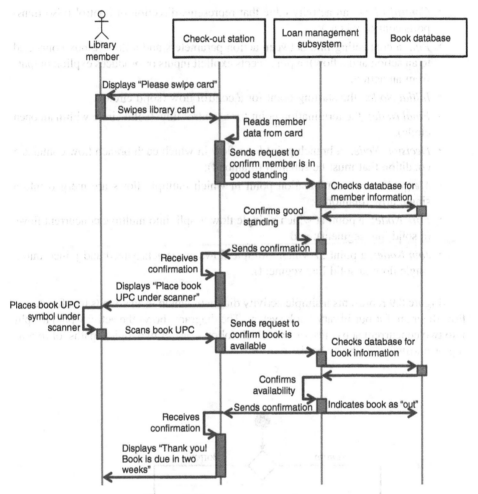

Figure 8.10. UML sequence diagram.

Sequence Diagram. Our last behavior diagram is the sequence diagram. These diagrams are usually linked to a use case where actions or events are listed in sequential formats. The sequence diagram takes advantage of this sequence and provides a visual depiction of the sequence of events, tied to the actor or subsystem performing the action.

Figure 8.10 depicts an example sequence diagram of the check-out operation. The diagram is tied to the use case presented above but provides additional information over what was presented in the use case description.

Class Diagram. At the heart of the UML is the concept of the class and is depicted in the class diagram. A class is simply a set of objects (which can be real or virtual) that have the same characteristics and semantics. In this case, an object can be

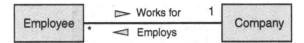

Figure 8.11. Example of a class association.

almost anything and, within the UML, can be represented in software. The class typically describes the structure and behavior of its objects.

Within a class definition, three primary components exist (among others):

- *Attributes:* the structural properties of the class;
- *Operations:* the behavior properties of the class; and
- *Responsibilities:* the obligations of the class.

Classes typically have relationships with other classes. The basic structural relationship is known as an *association*. Figure 8.11 depicts a simple association between the two classes, "employee" and "company." The line linking the two classes can have an arrow; however, if no arrow is present, then a bidirectional relationship is assumed. The nature of the association can also be provided by using a triangle. The association is then read like a sentence, "Employee *works for* company," and "Company *employs* employee." Finally, if the author wants to designate the association as a numerical relationship, he can use *multiplicity*. Multiplicity designates the numerical aspects of the association and can be expressed with specific numbers or a series of shorthand notations. For example, 0..2 means that any value between 0 and 2 can exist as part of the association. The star symbol, *, is used as a wildcard symbol, and can be thought of as "many." Thus, in our example, both the star and the number "1" are used to represent the fact that an employee works for only one company, and the company employs many employees.

Two other relationship types between classes are *generalization* and *dependency*. Generalization refers to a taxonomic relationship between a special, or specific, class and a general class. Figure 8.12 depicts a generalization relationship between the three classes, customer, corporate customer, and personal customer. In this case, both the corporate and the personal customers are specific class types belonging to the general class, customer. This relationship is depicted as an arrow with a large arrowhead. In this diagram, the class attributes and operations are provided for each.

When a generalization relationship is defined, the specific classes inherit the attributes and operations of the parent. Thus, the corporate customer class not only has its own specific attributes and operation but would also contain the attributes Name and Address, in addition to the operation, getCreditRating(). The same is true for the personal customer class.

Dependency is the third type of relationship and denotes the situation where one class requires the other for its specification or implementation. We should note that dependency is a relationship type that can be used among other elements within the UML, not just classes.

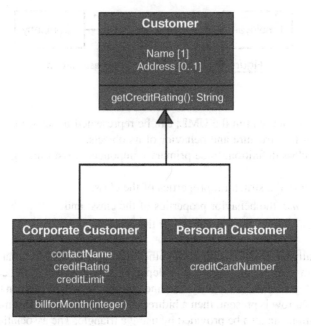

Figure 8.12. Example of a class generalization association.

Figure 8.13 includes the dependency association with our library example. The class diagram depicts several association types as presents a number of classes that would be defined as part of the library check-out system.

Systems Modeling Language (SysML)

Although UML has been applied to systems that include both hardware and software, it became evident that a variant form of UML, developed specifically for systems that combine software and hardware, could be used more effectively. Additionally, with the evolution of systems engineering, and specifically systems architecting, during the 1990s, a formal modeling language was recognized as beneficial to establish a consistent standard. The International Council on Systems Engineering (INCOSE) commissioned an effort in 2001 to develop a standard modeling language. Due to its popularity and flexibility, the new language was based on UML, specifically version 2.0. The OMG collaborated with this effort and established the Systems Engineering Domain Special Interest Group in 2001. Together, the two organizations developed and published the systems engineering extension to UML, called the SysML for short.

Perhaps the most important difference between UML and SysML is that a user of SysML need not be an expert in OOAD principles and techniques. SysML supports many traditional systems engineering principles, features, and models. Figure 8.14 presents the diagrams that serve as the basis for the language.

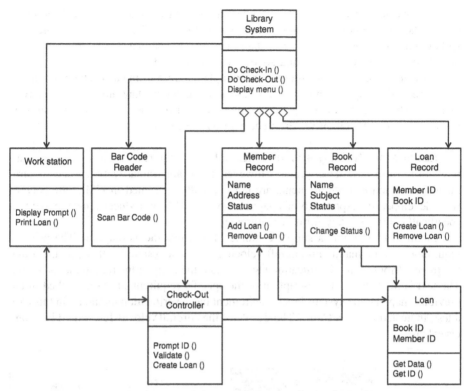

Figure 8.13. Class diagram of the library check-out system.

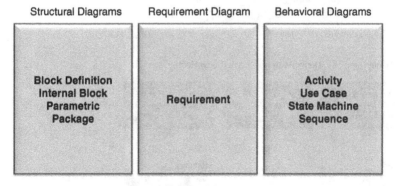

Figure 8.14. SysML models.

A new category, consisting of a single diagram of the same name, has been introduced: the requirements diagram. Only four of the 13 UML diagrams are included without changes: package, use case, state machine, and sequence. Diagrams that rely heavily on OO methodologies and approaches are omitted.

As with UML above, we present an example diagram from each category—in this case three—the requirements diagram, the internal block diagram, and the activity diagram. The latter two correspond closely to the UML class and activity diagrams; however, we will highlight the differences in our discussion.

Requirements Diagram. In UML, software requirements are primarily captured in the use case descriptions. However, these are primarily functional requirements; nonfunctional requirements are not explicitly presented in UML. Stereotypes were developed in response to this gap; however, SysML introduces a new model that specifically addresses any form of requirements.

Figure 8.15 presents a simple example of a requirements diagram. The primary requirement is the maximum aircraft velocity. This is a system-level requirement that has three attributes: an identification tag, text, and the units of the requirements metric. The text is the "classical" description of the specific requirement. As described in the previous chapters, the system-level requirement has a verification method—in this case a test, indicated by "TestCase." The details of the AircraftVelocityTest would be found elsewhere.

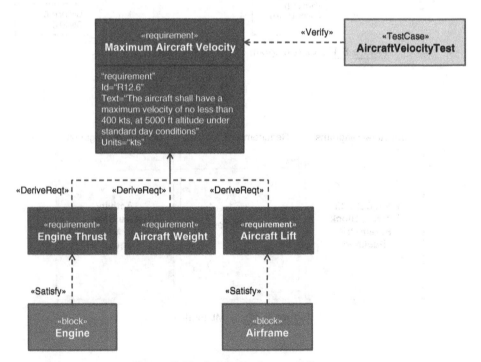

Figure 8.15. SysML requirements diagram.

This system-level requirement may lead to a set of derived requirements, typically associated with subsystems of the system. In the figure, three derived requirements are included: engine thrust, aircraft weight, and aircraft lift. These requirements would also have attributes and characteristics, although they are not shown in this particular diagram.

Finally, the satisfy relationship is depicted in the figure. This indicates a mechanism, or entity, that will satisfy the derived requirement. In the case of engine thrust, the engine subsystem is responsible for satisfying the derived requirement.

The requirements diagram is typically a series of rectangles that identify and associate many system-level requirements with subsystem-level requirements, their verification methods, derived requirements, and their satisfaction concepts. The latter allows the concept of mapping or tracing requirements to functional and physical entities.

As with operational, performance, and functional requirements, these diagrams are updated throughout the systems engineering method and the system development process. Linkages between components of the requirements model represented in this diagram, and the functional and physical models represented in other SysML diagrams, are crucial to successful systems engineering. Modern tools have been, and are being, developed to facilitate these linkages between model components.

Allocation. In SysML, a formal mechanism has been developed to enable the user to connect, or bind, elements of different models together. This mechanism is called allocation. SysML provides three types of allocations, although users can define others: behavior, structure, and object flow. The behavior allocation links, or allocates, behavior (represented in one or more of the behavioral diagrams) to a block that realizes this behavior. Recall that behavior is typically an activity or action. The structure allocation links, or allocates, logical structures with physical structures (and vice versa). This mechanism enables the engineer to link components of a logical definition of the system (typically represented by logical blocks) with components of a physical definition of the system (typically represented by physical blocks and packages). Finally, the object flow allocation connects an item flow (found in the structure diagram) with an object flow edge (found in the activity diagram). Allocation can be signified by a dashed arrow in many of the SysML diagrams.

Block Definition Diagram. In UML, the basic element is the *class*, with the *object* representing its instantiation. Because these terms are so closely identified with software development, SysML uses a different name to represent its basic element—the *block*. The structure and meaning of the block is almost identical to the class. A block contains attributes, may be associated with other blocks, and may also describe a set of activities that it conducts or behaviors it exhibits.

Blocks are used to represent the static structure of a system. They may represent either logical (or functional) elements or physical elements. The latter can also be divided into many types of physical manifestations—hardware, software, documentation, and so on. Figure 8.16 depicts an example block definition. The various components of a block definition are also depicted. This definition would be part of the block definition diagram (or sets of diagrams).

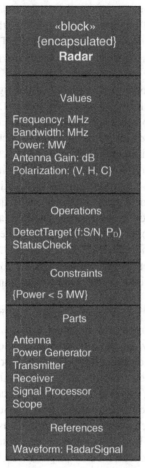

Figure 8.16. SysML block definition.

The block name is at the top. Values are the attributes or characteristics of the radar that are pertinent; the figure displays a sample set of attributes for this radar block. The next section down is the operations or the actions and behaviors of the block. In this example, the radar conducts only two types of operations, DetectTarget and StatusCheck. In reality, of course, common radars would perform many other operations. There may be constraints put on the operations or attributes of the block, so the next section lists any constraints. The block may also be defined with its subsystems or components, typically referred to as "parts." The example lists six basic subsystems of the radar. Finally, references (to other blocks) are provided.

Figure 8.17 depicts several types of block associations. Associations, similar to their counterparts in UML, represent relationships between blocks. Simple associations are depicted as lines connecting blocks. If direction is needed, then an arrow is placed

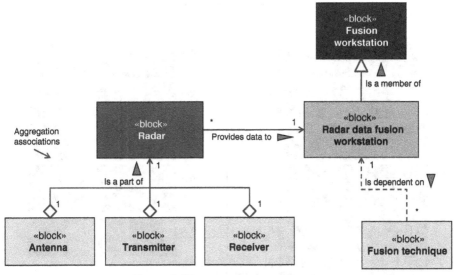

Figure 8.17. SysML block associations.

on one end—this type of association is called a *navigable association*. Special catego-ries are also available: aggregation associations represent blocks that are part of a whole; composition associations represent blocks that are part of a composite; depen-dency associations represent blocks that are dependent on other blocks; and generaliza-tion associations represent specialized blocks that are incorporated into a general block.

Activity Diagram. Of UML's behavioral diagrams, only one has been signifi-cantly expanded within SysML: the activity diagram. Four major extensions have been incorporated:

- Control flow has been extended with control operators.
- Modeling of continuous systems is now enabled using continuous object flows.
- Flows can have associated probabilities.
- Modeling rules for activities have been extended.

With these extensions, some existing functional modeling techniques can be imple-mented, such as the extended functional flow block diagram (EFFBD). Additionally, with the new extensions, a function tree can be represented quite easily, as shown in Figure 8.18a. This example uses the coffeemaker functions provided in Figure 8.4.

These functions can be arranged into a more traditional activity diagram, shown in Figure 8.18b. For clarity, the diagram does not include all 11 functions. The general control flow is indicated by the flow arrows and follows the general flow of Figure 8.4 (the FBD). Inputs and outputs are depicted by separate connectors—arrows with pins (or rectangles connected to the activity). These connectors are labeled with the entities

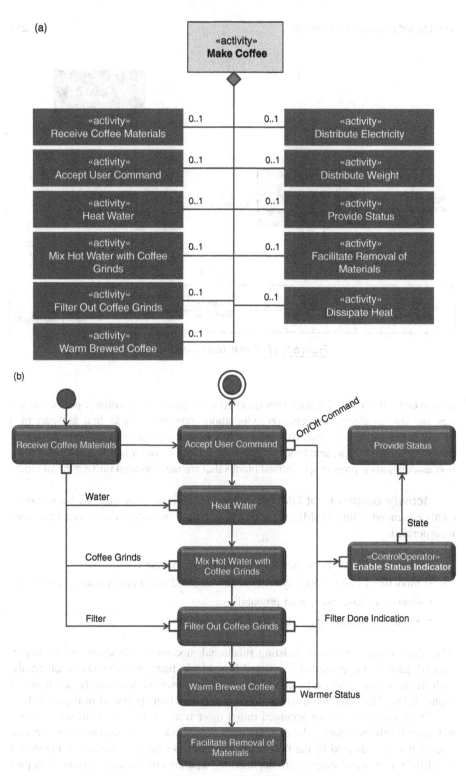

Figure 8.18. (a) SysML functional hierarchy tree. (b) SysML activity diagram.

passed across the interfaces. A control operator is also included to illustrate this type of special control mechanism. In this case, a control operator regulates what is passed to the *Display Status* activity, depending on the combination of its three inputs.

We have presented three SysML diagrams to illustrate some of the basic techniques of the language—one from each diagram category. Like UML, SysML offers the systems engineer and the systems architect with a flexible modeling kit with which to represent many aspects and perspectives of a system concept. Furthermore, it overcomes some of the inherent challenges within the UML when representing the more traditional methods of systems engineering, the requirements diagram being perhaps the most relevant example. With the advent of SysML, numerous commercial applications have risen to assist the engineer in developing, analyzing, and refining system concepts.

8.10   MODEL-BASED SYSTEMS ENGINEERING (MBSE)

With the advent of formal modeling languages, such as UML and SysML, and system architecture frameworks, such as DODAF and TOGAF, the ability of systems engineers to represent system requirements, behaviors, and structures has never been greater. Thus, exploring and defining system concepts have now been formalized and a new subset of systems engineering, systems architecting, has risen from obscurity to significance. In broad terms, the system architecture can be thought of as a model of the system, or at least the system concept. This is not to be confused with the fact that the term "model" is also used to denote the basic building blocks of a system architecture.

Soon after the first formal version of UML was released, OMG released the first version of their new model-driven architecture (MDA). This architecture was the first formal architecture framework that recognized the shift from a code-centric software development paradigm to an object-centric paradigm, enabled by the then de facto standard for software engineering model languages, UML. The MDA presented a set of standard principles, concepts, and model definitions that allowed for consistency in defining object models across the software community.

MDA delineated between the real system and its representation by a set of models. These models, in turn, would conform to a metamodel definition, which would in turn, conform to a meta-meta model definition. Several concepts, processes and techniques were presented in the literature using this concept, although the names differed: model-driven development, model-driven system design (MDSD), and model-driven engineering. They were all based on the basic concepts of focusing on a model and its metamodel to represent the system from the early stages of development through deployment and operations.

With the attempt to merge software and systems engineering processes and principles, model-driven development was applied several times to system development in various forms. In 2007, these attempts (along with their techniques and concepts) were grouped by INCOSE under the banner of MBSE. And with the release of the current versions of SysML, this approach has continued to increase in popularity.

The basic notion behind MBSE is that a model of the system is developed early in the process and evolves over the system development life cycle until the model becomes, in essence, the build-to baseline. Early in the life cycle, the models have low levels of fidelity and are used primarily for decision making (not unlike the system architecture in Section 8.8 above). As the system is developed, the level of fidelity increases until the models can be used for design. Finally, the models are transformed yet again into the build-to baseline. At each stage, similar to the standard systems engineering method introduced in Chapter 4, a subprocess is performed to evolve the set of system models. Baker introduced this subprocess for his approach (which he called MDSD). This subprocess is shown in Figure 8.19.

Additionally, Baker defined an early information model, or view, for an MDSD. This is provided in Figure 8.20 and is read similarly to a UML class diagram. The arrows represent the direction of the relationship, not the flow of information.

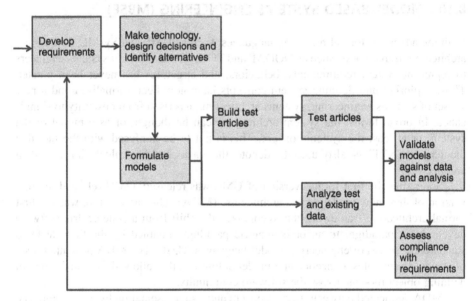

Figure 8.19. Baker's MDSD subprocesses.

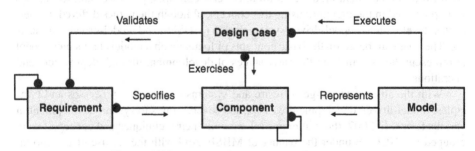

Figure 8.20. Baker's information model for MDSD.

Although this approach may sound familiar to the traditional systems engineering approach, several significant differences exist between the two. The foremost difference is the products of each. In traditional systems engineering (including either of the structured analysis or OO approaches), the primary products early in the system development life cycle are documents. Regardless of whether these documents are electronic or paper, they tend to be static representations of the system. With MBSE, the primary products are models, which can be executed to some extent. Thus, reviewing an MDSD (regardless of where one is along the life cycle) involves interrogating a set of models, which is an automated process. Reviewing traditional systems engineering products involves largely reading text and diagrams (although modern representations and displays greatly assist in this).

Of course, there is a price for this ability. Additional computing resources (applications, databases, hardware, visualization, and networking) are required to facilitate the MDSD effort. Currently, few of these resources are available, although more are in development and should be available to engineers soon. Furthermore, until projects are implemented using this approach, we do not yet have a rich lessons learned database.

With this inexperience in mind, INCOSE set about to identify and document the products which implemented this approach in part or whole. The INCOSE MBSE Focus Group published its finding in May 2007 and they identified five methodologies:

1. *Telelogic's Harmony®–SE.* This proprietary methodology is modeled after the products classical systems engineering "Vee" process, except that a requirements and model repository is established and updated during each step in the process. Additionally, a test data repository is also established and updated to track test cases and data. Several tools and applications have been developed or revised to facilitate the harmony methodology. Telelogic produces several of these (e.g., Rhapsody, Popkin, DOORS), although the methodology itself is application-neutral.

2. *INCOSE's Object-Oriented Systems Engineering Method (OOSEM).* This approach implements the model-based approach using SysML to support the specification, analysis, design, and verification of a system. The basic set of activities produces artifacts that can be refined and used in other applications. These activities and artifacts are listed below:
   a. Analyze stakeholder needs.
   b. Define system requirements.
   c. Define logical architecture.
   d. Synthesize candidate allocated architectures.
   e. Optimize and evaluate alternatives.
   f. Validate and verify the system.

3. *IBM's Rational Unified Process for Systems Engineering (RUP–SE).* The goal of the RUP–SE process was to apply the discipline and best practices found in the RUP and to apply them to the challenges of system specification, analysis, design, and development. Moreover, RUP–SE was developed specifically to

implement model-driven system development. This adaptation of the existing unified process focuses on four modeling levels: context, analysis, design, and implementation, each incorporating higher levels of fidelity than the previous. These first three model levels are then cross-indexed with six viewpoints: worker, logical, information, distribution, process, and geometric, to produce 17 architecture artifacts (the context/process pair does not produce an artifact, and the implementation model produces actual *physical* artifacts). These artifacts become the basis of the RUP–SE architecture framework.

4. *Vitech's MBSE Methodology.* This approach is based on four primary activities that are integrated through a common design repository:

   a. source requirements analysis,
   b. functional/behavior analysis,
   c. architecture/synthesis, and
   d. design validation and verification.

   This methodology requires a common information model to manage the syntax and semantics of artifacts. Vitech has defined a system definition language (SDL) for use with their process (which also can be used with their tool, CORE), although the process itself can use any information model language.

5. *Jet Propulsion Laboratory's (JPL) State Analysis (SA).* This last methodology leverages a model- and state-based control architecture to capture system requirements and design. This process distinguishes between a system's state and one's knowledge of that state. Generally, the knowledge of the system state is represented by more abstract concepts than the actual states themselves. How the system evolves from state to state is represented within a set of models. Finally, system control is also represented by models, although complete control is considered impossible due to system complexity.

The establishment and maturation of OO methods, systems modeling languages, and the proliferation of tools and applications implementing those methods and languages have led to an increased awareness of the benefits of using a model-driven approach in systems engineering. And although the approach does come with a price in increased resources, the benefits may indeed provide for an adequate return on investment. Case studies are slowly being offered as "proof" that this approach can indeed work. More time and experience is necessary before the community as a whole embraces MBSE; however, its basic principles are sound. And this methodology and approach is one more step in the convergence of software and systems engineering practices.

8.11 SYSTEM FUNCTIONAL SPECIFICATIONS

The concept definition phase is not complete until a formal basis is created to guide the follow-on engineering design stage. A linchpin of such a basis is a statement describing completely and concisely all the functions that the system must be designed to perform in order to fulfill its operational requirements. In major government acquisitions, such a statement is usually called the "system specification" or "A-Spec."

The system specification can be thought of as a textual and diagrammatic representation of the system concept. It does not, however, address specifically how the system is implemented to perform its functions but stipulates what functions are to be performed, with what precision, and under what conditions. In so doing, it is essential that the definitions be stated in measurable terms because the engineering implementation of those functions will rely on these definitions.

While the preparation of system specifications is logically a part of the concept definition phase, in a competitive acquisition process, it is usually prepared immediately after the selection process by the successful contractor team. In commercial product development, the process is not as formal but is similar in purpose.

The system specification document should address at least the following subjects:

System Definition
Mission and concept of operation system functions
Configuration and organization of system interfaces
Required Characteristics
Performance characteristics (hardware and software) and compatibility requirements
RMA requirements
Support Requirements
Shipping, handling, and storage training
Special facilities
Special Requirements
Security and safety human engineering

The leadership and much of the actual work involved in formulating the system specification document is the responsibility of systems engineering.

8.12 SUMMARY

Selecting the System Concept

Objectives of the concept definition phase are to select a preferred system configuration and to define system functional specifications, as well as a development schedule and cost.

Concept definition concludes the concept development stage, which lays the basis for the engineering development stage of the system life cycle. Defining a preferred concept also provides a baseline for development and engineering.

Activities that comprise concept definition are

- *Performance Requirements Analysis*—relating to operational objectives,
- *Functional Analysis and Formulation*—allocating functions to components,
- *Concept Selection*—choosing the preferred concept by trade-off analysis, and
- *Concept Validation*—confirming the validity and superiority of the chosen concept.

Performance Requirements Analysis

Performance requirements analysis must include ensuring compatibility with the system operating site and its logistics support. The analysis must also address reliability, maintainability, and support facilities, as well as environmental compatibility. A specific focus on the entire life cycle, from production to system disposition, must be kept. Finally, the analysis must resolve the definition of unquantified requirements.

Functional Analysis and Formulation

Functional system building blocks (Chapter 3) are useful for functional definition. The selection of a preferred concept is a systems engineering function, which formulates and compares evaluation of a range of alternative concepts.

Functional Allocation

Developing alternative concepts requires part art and part science. Certainly, the predecessor system can act as a baseline for further concepts (assuming a predecessor is available). Brainstorming and other team innovation techniques can assist in developing alternatives.

Concept Selection

System concepts are evaluated in terms of (1) operational performance and compatibility, (2) program cost and schedule, and (3) risks in achieving each of the above. Program risk can be considered to consist of a combination of two factors: likelihood that the system will fail to achieve its objectives and impact of the failure on the success of the program.

Program risks can result from a number of sources:

- unproven technology,
- difficult performance requirements,
- severe environments,
- inadequate funding or staffing, and
- an unduly short schedule.

Trade-off analysis is fundamental in all systematic decision making.

Concept Validation

In concept selection, trade-off analysis should be

- *Organized*—set up as a distinct process,
- *Exhaustive*—consider the full range of alternatives,
- *Semiquantitative*—use relative weightings of criteria,
- *Comprehensive*—consider all major characteristics, and
- *Documented*—describe the results fully.

Justification for the development of the selected concept should

- show the validity of the need to be met;
- state reasons for selecting the concept over the alternatives;
- describe program risks and means for containment;
- give evidence of detailed plans, such as WBS, SEMP, and so on;
- give evidence of previous experience and successes;
- present life cycle costing; and
- cover other relevant issues, such as environmental impact.

System Development Planning

The WBS is essential in a system development program and is organized in a hierarchical structure. It defines all of the constituent tasks in the program.

The SEMP (or equivalent) defines all systems engineering activities through the system life cycle.

Systems Architecting

Systems architecting is primarily the development and articulation of different perspectives, or viewpoints, of a system. Almost all system architectures have at least three perspectives:

- *Operational View*—a system representation from the user's or operator's perspective,
- *Logical View*—a system representation from the customer's or manager's perspective, and
- *Physical View*—a system representation from the designer's perspective.

Architecture frameworks define the structure and models used to develop and present a system architecture. These frameworks are meant to ensure consistency across programs in articulating the various perspectives.

System Modeling Languages: UML and SysML

The UML provides 13 system models to represent both structural and behavioral aspects of the system. Although UML was developed for software development applications, it has been successfully applied to software-intensive systems. The language differs from the traditional structured analysis approach by focusing on entities (represented by classes and objects) instead of functions and activities.

The SysML is an extension of UML that enables a more complete modeling of software/hardware systems and facilitates the top-down approach of traditional systems engineering. An emphasis on requirements to drive the development effort is inherent in SysML. To distinguish the two languages, SysML uses the block as its primary entity, in place of the class.

MBSE

The basic notion behind MBSE is that a model of the system is developed early in the process and evolves over the system development life cycle until the model becomes, in essence, the build-to baseline. Early in the life cycle, the models have low levels of fidelity and are used primarily for decision making (not unlike the system architecture in Section 8.8 above). As the system is developed, the level of fidelity increases until the models can be used for design. Finally, the models are transformed yet again into the build-to baseline.

System Functional Specifications

System functional specifications address the system functional description, its required characteristics, and the support requirements.

PROBLEMS

8.1 Describe three principal differences between system performance requirements, which are an input to the concept definition phase, and system functional specifications, which are an output (see Fig. 8.1).

8.2 Both the concept exploration and concept definition phases analyze several alternative system concepts. Explain the principal differences in the objectives of this process in the two phases and in the manner in which the analysis is performed.

8.3 Describe what is meant by the term "functional allocation" and illustrate its application to a personal computer. Draw a functional diagram of a personal computer using the functional elements described in Chapter 3 as building blocks. For each building block, describe what functions it performs, how it interacts with other building blocks, and how it relates to the external inputs and outputs of the computer system.

8.4 Under the subsection Program Risks, five examples are listed of conditions that may result in a significant probability of program failure. For each example, explain briefly what consequences of the condition may lead to a program failure.

8.5 In the subsection Selection Strategy, it is recommended that in comparing different concepts, the weighted evaluations of the individual criteria for each concept should not be collapsed into a single figure of merit for each concept (as is commonly done) but should be retained in the form of an evaluation "profile." Explain the rationale for this recommendation and illustrate it with a hypothetical example.

8.6 Discuss how you would use trade-off analysis to prioritize the efforts to be allocated to the mitigation of identified high and medium program risks.

8.7 The section The "Selling" of the System Development Proposal lists seven elements in a recommended approach to the authorities responsible for making the decision. Illustrate the utility of each element by explaining in each case what the authorities might conclude in the absence of a suitable discussion of the subject.

8.8 (a) Develop a top-level function list for an ATM system. Limit yourself to no more than 12 functions.

(b) Draw an FBD of the ATM using the functions in (a).

8.9 (a) Identify the functions of a common desktop computer.

(b) Identify the components of a common desktop computer.

(c) Allocate the functions in (a) to the components in (b).

8.10 Suppose you have completed the functional analysis and allocation activities within the concept definition phase of a system's development.

(a) Suppose that you have some functions that are allocated to multiple components (as opposed to a single component). What does that mean regarding your conceptual design? Is this a problem?

(b) Suppose that you have many functions that are allocated to a single component. What does that mean regarding your conceptual design? Is this a problem?

8.11 Convert the coffeemaker FBD in Figure 8.4 to an IDEF0 diagram.

8.12 Draw a physical block diagram of the coffeemaker represented in Figure 8.4. Within the diagram, use rectangles to represent physical components and label the interfaces between the components.

8.13 Draw a diagram that presents the associations and relationships between the following:

• the system,

• system architecture,

• architecture framework,

• viewpoint,

- view,
- modeling language, and
- model.

The diagram should include seven rectangles (one for each entity above) and labeled arrows that describe the relationships between the entities.

8.14 Convert the coffeemaker FBD in Figure 8.4 to a UML activity diagram.

8.15 Write a two-page essay comparing and contrasting the latest versions of DODAF and TOGAF.

8.16 Suppose you are the system architect for a new private business jet aircraft that is intended to seat eight executives. Suppose also that you have been asked to use DODAF as your architecture framework. Decide and explain which views you would include in your architecture. Of course, all of the views within DODAF will not be necessary for this type of system.

8.17 Build a matrix that maps UML models to DODAF views. In other words, which UML model(s) would be appropriate for each DODAF view? Hint: many DODAF views will be not applicable while others will have more than a single UML view. Please use a matrix or table.

8.18 Repeat Problem 8.17, but map SysML models to DODAF.

8.19 Repeat Problem 8.17, but map UML to TOGAF.

8.20 Research MBSE and write an essay comparing and contrasting MBSE with traditional systems engineering, as described in Chapters 1–8 of this book. What are the principles of MBSE? What is different? Can traditional systems engineering implement the basic principles without significant upgrades?

FURTHER READING

L. Baker, P. Clemente, B. Cohen, L. Permenter, B. Purves, and P. Salmon. *Foundational Concepts for Model Driven System Design*. INCOSE Model Driven Design Interest Group, INCOSE, July 2000.

L. Balmelli, D. Brown, M. Cantor, and M. Mott. Model-driven systems development. *IBM Systems Journal*, 2006, 45(3), 569–585.

B. Blanchard and W. Fabrycky. *System Engineering and Analysis*, Fourth Edition. Prentice Hall, 2006, Chapter 3.

F. P. Brooks, Jr. *The Mythical Man Month—Essays on Software Engineering*. Addison-Wesley, 1995.

W. P. Chase. *Management of Systems Engineering*. John Wiley, 1974, Chapters 3 and 4.

H. Chesnut. *Systems Engineering Methods*. John Wiley, 1967.

S. Dam. *DOD Architecture Framework: A Guide to Applying System Engineering to Develop Integrated, Executable Architectures*. SPEC, 2006.

Defense Acquisition University. *Systems Engineering Fundamentals*. DAU Press, 2001, Chapters 5 and 6.

Defense Acquisition University. *Risk Management Guide for DoD Acquisition*, Sixth Edition. DAU Press, 2006.

Department of Defense Web site. DoD Architecture Framework Version 2.02. http://cio-nii.defense.gov/sites/dodaf20.

H. Eisner. *Computer-Aided Systems Engineering*. Prentice Hall, 1988, Chapter 12.

J. A. Estefan. Survey of model-based systems engineering (MBSE) methodologies, INCOSE Technical Document INCOSE-TD-2007-003-02, Revision B, June 10, 2008.

M. Fowler. *UML Distilled: A Brief Guide to the Standard Object Modeling Language*, Third Edition. Addison-Wesley, 2004.

H. Hoffmann. SysML-based systems engineering using a model-driven development approach. Telelogic White Paper, Version 1, January 2008.

International Council on Systems Engineering. *Systems Engineering Handbook. A Guide for System Life Cycle Processes and Activities*. Version 3.2, July 2010.

J. Kasser. *A Framework for Understanding Systems Engineering*. The Right Requirement, 2007.

M. Maier and E. Rechtin. *The Art of Systems Architecting*. CRC Press, 2009.

The Open Group. *TOGAF Version 9 Enterprise Edition*, Document Number G091. The Open Group, 2009. http://www.opengroup.org/togaf/.

R. S. Pressman. *Software Engineering: A Practitioner's Approach*. McGraw Hill, 2001.

N. B. Reilly. *Successful Systems for Engineers and Managers*. Van Nostrand Reinhold, 1993, Chapter 12.

A. P. Sage and J. E. Armstrong, Jr. *Introduction to Systems Engineering*. Wiley, 2000, Chapter 3.

D. Schmidt. Model-driven engineering. *IEEE Computer*, 2006, 39(2), 25–31.

R. Stevens, P. Brook, K. Jackson, and S. Arnold. *Systems Engineering, Coping with Complexity*. Prentice Hall, 1998, Chapter 4.

9

DECISION ANALYSIS AND SUPPORT

The preceding chapters have described the multitude of decisions that systems engineers must make during the life cycle of a complex new system. It was seen that many of these involve highly complex technical factors and uncertain consequences, such as incomplete requirements, immature technology, funding limitations, and other technical and programmatic issues. Two of the strategies that have been devised to aid in the decision process are the application of the systems engineering method and the structuring of the system life cycle into a series of defined phases.

Decision making comes in a variety of forms and within numerous contexts. Moreover, everyone engages in decision making almost continuously from the time they wake up to the time they fall asleep. Put simply, not every decision is the same. Nor is there a one-size-fits-all process for making decisions. Certainly, the decision regarding what you will eat for breakfast is not on par with deciding where to locate a new nuclear power plant.

Decision making is not independent of its context. In this chapter, we will explore decisions typically made by systems engineers in the development of complex systems. Thus, our decisions will tend to contain complexity in their own right. They are the hard decisions that must be made. Typically, these decisions will be made under levels

Systems Engineering Principles and Practice, Second Edition. Alexander Kossiakoff, William N. Sweet, Samuel J. Seymour, and Steven M. Biemer
© 2011 by John Wiley & Sons, Inc. Published 2011 by John Wiley & Sons, Inc.

of uncertainty—the systems engineer will not have all of the information needed to make an optimal decision. Even with large quantities of information, the decision maker may not be able to process and integrate the information before a decision is required.

9.1 DECISION MAKING

Simple decision making typically requires nothing more than some basic information and intuition. For example, deciding what one will have for breakfast requires some information—what food is available, what cooking skill level is available, and how much time one has. The output of this simple decision is the food that is to be prepared. But complex decisions require more inputs, more outputs, and much more planning. Furthermore, information that is collected needs to be organized, integrated (or fused), and presented to decision makers in such a way as to provide adequate support to make "good" decisions.

Figure 9.1 depicts a simplified decision-making process for complex decisions. A more detailed process will be presented later in the chapter.

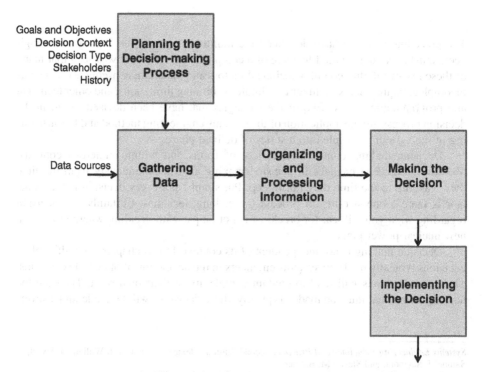

Figure 9.1. Basic decision-making process.

Obviously, this appears to be rather cumbersome. However, how much time, energy, and the level of resource commitment devoted to each stage will be dependent on the type, complexity, and scope of the decision required. Formal decisions, typical in large government acquisition programs, may take years, while component decisions for a relatively simple system may require only hours or less.

Each stage requires a finite amount of time. Even "making the decision" is not necessarily instantaneous. For example, if more than a single person must make and approve the decision, this stage may be quite lengthy. If consensus is required, then this stage may become quite involved, and would include political as well as technical and programmatic considerations. Government legislatures are good examples in understanding the resources required in each step. Planning, gathering, and organizing are usually completed by staffs and through public and private hearings. The stage, making the decision, is actually an involved process that includes political maneuvering, deal making, marketing, campaigning, and posturing. This stage has lasted months in many cases.

Regardless of the type of decision, or the forum within which the decision will be made, there are many factors that must be considered to initiate and complete the planning stage.

Factors in the Decision-Making Process

Complex decisions require an understanding of the multidimensionality of the process before an appropriate and useful decision can be made. The following factors need to be considered as part of the planning stage.

Goals and Objectives. Before making decisions, one needs to ask: what are the goals and objectives of the stakeholders? These will probably be different at different levels of the organization. The goals of a line supervisor will be different than a program manager. Which holds the higher priority? And what are the goals of management above the decision maker? The decision should be made to satisfy (as far as possible) the goals and objectives of the important stakeholders.

Decision Type. The decision maker needs to understand the type of decision required. Many bad decisions stem from a misunderstanding about the type required. Is the decision binary? Maybe the decision is concerned with a permission of some sort. In these cases, a simple yes/no decision is required. Other binary decisions may not be yes or no, but a choice between two alternatives, make or buy being a classic example. More complex decisions typically involve one or more choices among a set of alternatives. Lastly, the decision maker needs to understand who and what will be affected. Is the decision purely technical, or is there a personal element? Providing the wrong type of decision will certainly lead to significantly negative consequences.

In the same vein, understanding who needs to be included in the decision is vital. Is this decision to be made by an individual? Or is a consensus among a group required? Who needs to approve the decision before it is implemented? The answers to these questions influences when, and how, decisions will be made.

Decision Context. Understanding the scope of the decision is also essential to making a proper decision. A global (or enterprise-wide) decision will be much different than a system component decision. The consequences of a wrong decision will be far-reaching if the decision affects the enterprise, for example. Context involves understanding the problem or issue that led to a decision point. This will be difficult since context has many dimensions, leading to different goals and objectives for your decision maker:

- technical, involving physical entities, such as subsystem decisions;
- financial, involving investment instruments and quantities;
- personnel, involving people;
- process, involving business and technical procedures, methods, and techniques;
- programmatic, involving resource allocations (including time, space, and funding);
- temporal, meaning the time frame in which a decision is needed (this may be dynamic); and
- legacy, involving past decisions.

Stakeholders. Stakeholders can be defined as anyone (people or organizations) who will be affected by the results of the decision. Understanding who the stakeholders are with respect to a decision needs to be established before a decision is made. Many times, this does not occur—stakeholders are not recognized before a decision is made. Yet, once the decision is announced or implemented, we can be sure that all who are affected will make their opinion heard.

Legacy Decisions. Understanding what relevant decisions have been made in the past helps with both the context (described above) and the environment in which the current decision must be made. Consequences and stakeholders can be identified more readily if the decision maker has knowledge of the past.

Supporting Data. Finally, necessary supporting data for the decision need to be provided in a timely fashion. A coherent and timely data collection plan is needed to ensure proper information can be gathered to support the decision. Accuracy in data collected is dependent on the decision type and context. Many times, decisions are delayed unnecessarily because greater accuracy than needed was demanded before the decision maker would act.

Decision Framework

As mentioned above, understanding the type of decision needed is critical in planning for and executing any process. Several decision frameworks are available in the literature to assist in understanding the decision type. In Table 9.1, we present a framework that is a combination of several.

TABLE 9.1. Decision Framework

Type of Decision	Scope of Control			Technology needed
	Operational	Managerial	Strategic planning	
Structured	Known procedures algorithms	Policies Laws Trade-off analysis Logic	Historical analysis Goal-oriented task analysis	Information systems
Semistructured	Tailored procedures Heuristics	Tailored policies Heuristics Logic	Causality ROI analysis Probabilities	Decision support systems
Unstructured	Intuition Experimental	Intuition Experimental	Intuition Creativity Theory	Expert systems

There are many ways to categorize decisions. Our categorization focuses on three types of decisions: structured, semistructured, and unstructured.

Structured. These types of decisions tend to be routine, in that the context is well understood and the decision scope is known. Supporting information is usually available, and minimal organization or processing is necessary to make a good decision. In many cases, standards are available, either globally or within an organization, to provide solution methods. Structured decisions have typically been made in the past; thus, a decision maker has a historical record of similar or exact decisions made like the one he is facing.

Semistructured. These types of decisions fall outside of "routine." Although similar decisions may have been made, circumstances are different enough that past decisions are not a clear indicator of the right decision choice. Typically, guidance is available though, even when specific methods are not. Many systems engineering decisions fall within the category.

Unstructured. Unstructured decisions represent complex problems that are unique and typically one-time. Decisions regarding new technologies tend to fall into this category due to the lack of experience or knowledge of the situation. First-time decisions fall into this category. As experience grows and decisions are tested, they may transition from an unstructured decision to the semistructured category.

In addition to the type, the scope of control is important to recognize. Decisions within each scope are structured differently, have different stakeholders, and require different technologies to support.

Operational. This is the lowest scope of control that systems engineering is concerned about. Operational control is at the practitioner level—the engineers, analysts, architects, testers, and so on, who are performing the work. Many decisions at this scope of control involve structured or semistructured decisions. Heuristics, procedures, and algorithms are typically available to either describe in detail when and how decisions should be made or at least to provide guidelines to decision making. In rare cases, when new technologies are implemented, or a new field is explored, unstructured decisions may rise.

Managerial. This scope of control defines the primary level of systems engineering decision making—that of the chief engineer, the program manager, and of course, the systems engineer. This scope of control defines the management, mentoring, or coaching level of decisions. Typically, for semistructured decisions, policies, heuristics, and logical relationships are available to guide the systems engineer in these decisions.

Strategic Planning. This level of control represents an executive- or enterprise-level control. Semistructured decisions usually rely on causality concepts to guide decisions making. Additionally, investment decisions and decisions under uncertainty are typically made at this scope of control level.

Supporting Decisions

The level of technologies needed to support the three different decision types varies. For structured decisions, uncertainty is minimal. Databases and information systems are able to organize and present information clearly, enabling informed decisions. For semistructured decisions, however, simply organizing information is not sufficient. Decision support systems (DSS) are needed to analyze information, to fuse information from multiple sources, and to process information to discover trends and patterns.

Unstructured decisions require the most sophisticated level of technology, expert systems, sometimes called knowledge-based systems. Due to the high level of uncertainty and a lack of historical precedence and knowledge, sophisticated inference is required from these systems to provide knowledge to decision makers.

Formal Decision-Making Process

In 1976, Herbert Simon, in his landmark work on management decision science, provided a structured decision process for managers consisting of four phases. Table 9.2 is a depiction of this process.

This process is similar to the one in Figure 9.1 but provides a new perspective—the concept of modeling the decision. This concept refers to the activities of developing a model of the issue or problem at hand and predicting the outcome of each possible alternative choice available to the decision maker.

Developing a model of the decision means creating a model that represents the decision context and environment. If the decision refers to an engineering subsystem trade-off, then the model would be of the subsystem in question. Alternative configura-

TABLE 9.2. Simon's Decision Process

Phase I: Intelligence	Define problem
	Collect and synthesize data
Phase II: Design	Develop model
	Identify alternatives
	Evaluate alternatives
Phase III: Choice	Search choices
	Understand sensitivities
	Make decision(s)
Phase IV: Implementation	Implement change
	Resolve problem

tions, representing the different choices available, would be implemented in the model and various outcomes would be captured. These are then compared to enable the decision maker to make an informed choice.

Of course, models can be quite complex in scope and fidelity. Available resources typically provide the constraints on these two attributes. Engineers tend to desire a large scope and high fidelity, while the available resources constrain the feasibility of attaining these two desires. The balance needed is one responsibility of the systems engineer. Determining the balance between what is desired from a technical perspective with what is available from a programmatic perspective is a balance that few people beyond the systems engineer are able to strike.

Although we have used the term "model" in the previous chapters, it is important to realize that models come in all shapes and sizes. A spreadsheet can be a model of a decision. A complex digital simulation can also be an appropriate model. What type of model to develop to support decision making depends on many factors.

1. *Decision Time Frame.* How much time does the decision maker have to make the decision? If the answer is "not much," then simple models are the only available resource, unless more sophisticated models are already developed and ready for use.
2. *Resources.* Funding, personnel, skill level, and facilities/equipment are all constraints on one's ability to develop and exercise a model to support decisions.
3. *Problem Scope.* Clearly, simple decisions do not need complicated models. Complex decisions generally do. The scope of the problem will, in some respects, dictate the scope and fidelity of the model required. Problem scope itself has many factors as well: range of influence of the decision, number and type of stakeholders, number and complexity of entities involved in the decision space, and political constraints.
4. *Uncertainty.* The level of uncertainty in the information needed will also affect the model type. If large uncertainty exists, some representation of probabilistic reasoning must be included in the model.

5. *Stakeholder Objectives and Values.* Decisions are subjective by nature, even with objective data to support them. Stakeholders have values that will affect the decision and, in turn, will be affected by the decision. The systems engineer must determine how values will be represented. Some may, and should, be represented within the model. Others can, and should, be represented outside of the model. Keep in mind that a large part of stakeholder values involves their risk tolerance. Individuals and organizations have different tolerances for risk. The engineer will need to determine whether risk tolerance is embedded within the model or handled separately.

In summary, modeling is a powerful strategy for dealing with decisions in the face of complexity and uncertainty. In broad terms, modeling is used to focus on particular key attributes of a complex system and to illuminate their behavior and relationships apart from less important system characteristics. The objective is to reveal critical system issues by stripping away properties that are not immediately concerned with the issue under consideration.

9.2 MODELING THROUGHOUT SYSTEM DEVELOPMENT

Models have been referred to and illustrated throughout this book. The purpose of the next three sections is to provide a more organized and expanded picture of the use of modeling tools in support of systems engineering decision making and related activities. This discussion is intended to be a broad overview, with the goal of providing an awareness of the importance of modeling to the successful practice of systems engineering. The material is necessarily limited to a few selected examples to illustrate the most common forms of modeling. Further study of relevant modeling techniques is strongly recommended.

Specifically, the next three sections will describe three concepts:

- *Modeling:* describes a number of the most commonly used static representations employed in system development. Many of these can be of direct use to systems engineers, especially during the conceptual stage of development, and are worth the effort of becoming familiar with their usage.
- *Simulation:* discusses several types of dynamic system representations used in various stages of system development. Systems engineers should be knowledgeable with the uses, value, and limitations of simulations relevant to the system functional behavior, and should actively participate in the planning and management of the development of such simulations.
- *Trade-Off Analysis:* describes the modeling approach to the analysis of alternatives (AoA). Systems engineers should be expert in the use of trade-off analysis and should know how to critically evaluate analyses performed by others. This section also emphasizes the care that must be taken in interpreting the results of analyses based on various models of reality.

9.3 MODELING FOR DECISIONS

As stated above, we use models as a prime means of coping with complexity, to help in managing the large cost of developing, building, and testing complex systems. In this vein, a model has been defined as "a physical, mathematical, or otherwise logical representation of a system entity, phenomenon, or process." We use models to represent systems, or parts thereof, so we can examine their behavior under certain conditions. After observing the model's behavior within a range of conditions, and using those results as an estimate of the system's behavior, we can make intelligent decisions on a system development, production, and deployment. Furthermore, we can represent processes, both technical and business, via models to understand the potential impacts of implementing those processes within various environments and conditions. Again, we gain insight from the model's behavior to enable us to make a more informed decision.

Modeling only provides us with a representation of a system, its environment, and the business and technical processes surrounding that system's usage. The results of modeling provide only estimates of a system's behavior. Therefore, modeling is just one of the four principal decision aids, along with simulation, analysis, and experimentation. In many cases, no one technique is sufficient to reduce the uncertainty necessary to make good decisions.

Types of Models

A model of a system can be thought of as a simplified representation or abstraction of reality used to mimic the appearance or behavior of a system or system element. There is no universal standard classification of models. The one we shall use here was coined by Blanchard and Fabrycky, who define the following categories:

- *Schematic Models* are diagrams or charts representing a system element or process. An example is an organization chart or data flow diagram (DFD). This category is also referred to as "descriptive models."
- *Mathematical Models* use mathematical notation to represent a relationship or function. Examples are Newton's laws of motion, statistical distributions, and the differential equations modeling a system's dynamics.
- *Physical Models* directly reflect some or most of the physical characteristics of the actual system or system element under study. They may be scale models of vehicles such as airplanes or boats, or full-scale mock-ups, such as the front section of an automobile undergoing crash tests. In some cases, the physical model may be an actual part of a real system, as in the previous example, or an aircraft landing gear assembly undergoing drop tests. A globe of the earth showing the location of continents and oceans is another example, as is a ball and stick model of the structure of a molecule. Prototypes are also classified as physical models.

The above three categories of models are listed in the general order of increasing reality and decreasing abstraction, beginning with a system context diagram and ending with a production prototype. Blanchard and Fabrycky also define a category of "analog models," which are usually physical but not geometrical equivalents. For the purpose of this section, they will be included in the physical model category.

Schematic Models

Schematic models are an essential means of communication in systems engineering, as in all engineering disciplines. They are used to convey relationships in diagrammatic form using commonly understood symbology. Mechanical drawings or sketches model the component being designed; circuit diagrams and schematics model the design of the electronic product.

Schematic models are indispensable as a means for communication because they are easily and quickly drawn and changed when necessary. However, they are also the most abstract, containing a very limited view of the system or one of its elements. Hence, there is a risk of misinterpretation that must be reduced by specifying the meaning of any nonstandard and nonobvious terminology. Several types of schematic models are briefly described in the paragraphs below.

Cartoons. While not typically a systems engineering tool, cartoons are a form of pictorial model that illustrates some of the modeled object's distinguishing characteristics. First, it is a simplified depiction of the subject, often to an extreme degree. Second, it emphasizes and accentuates selected features, usually by exaggeration, to convey a particular idea. Figure 2.2, "The ideal missile design from the viewpoint of various specialists," makes a visual statement concerning the need for systems engineering better than words alone can convey. An illustration of a system concept of operations may well contain a cartoon of an operational scenario.

Architectural Models. A familiar example of the use of modeling in the design of a complex product is that employed by an architect for the construction of a home. Given a customer who intends to build a house to his or her own requirements, an architect is usually hired to translate the customer's desires into plans and specifications that will instruct the builder exactly what to build and, to a large extent, how. In this instance, the architect serves as the "home systems engineer," with the responsibility to design a home that balances the desires of the homeowner for utility and aesthetics with the constraints of affordability, schedule, and local building codes.

The architect begins with several sketches based on conversations with the customer, during which the architect seeks to explore and solidify the latter's general expectations of size and shape. These are pictorial models focused mainly on exterior appearance and orientation on the site. At the same time, the architect sketches a number of alternative floor plans to help the customer decide on the total size and approximate room arrangements. If the customer desires to visualize what the house would more nearly look like, the architect may have a scale model made from wood or cardboard. This would be classified as a physical model, resembling the shape of the proposed house. For homes with complex rooflines or unusual shapes, such a model may be a good investment.

The above models are used to communicate design information between the customer and the architect, using the form (pictorial) most understandable to the customer. The actual construction of the house is done by a number of specialists, as is the building of any complex system. There are carpenters, plumbers, electricians, masons, and so on, who must work from a much more specific and detailed information that they can understand and implement with appropriate building materials. This information is contained in drawings and specifications, such as wiring layouts, air conditioning routing, plumbing fixtures, and the like. The drawings are models, drawn to scale and dimensioned, using special industrial standard symbols for electrical, plumbing, and other fixtures. This type of model represents physical features, as do the pictorials of the house, but is more abstract in the use of symbols in place of pictures of components. The models serve to communicate detailed design information to the builders.

System Block Diagrams. Systems are, of course, far more complex than conventional structures. They also typically perform a number of functions in reacting to changes in their environment. Consequently, a variety of different types of models are required to describe and communicate their structure and behavior.

One of the most simple models is the "block diagram." Hierarchical block diagrams have the form of a tree, with its branch structure representing the relationship between components at successive layers of the system. The top level consists of a single block representing the system; the second level consists of blocks representing the subsystems; the third decomposes each subsystem into the components, and so on. At each level, lines connect the blocks to their parent block. Figure 9.2 shows a generic system block diagram of a system composed of three subsystems and eight components.

The block diagram is seen to be a very abstract model, focusing solely on the units of the system structure and their physical relationships. The simple rectangular blocks are strictly symbolic, with no attempt to depict the physical form of the system elements. However, the diagram does communicate very clearly an important type of relationship among the system elements, as well as identify the system's organizing

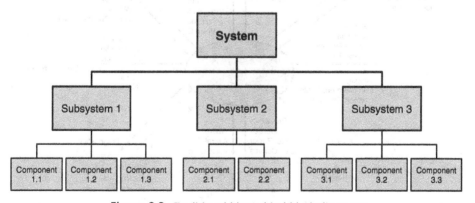

Figure 9.2. Traditional hierarchical block diagram.

principle. More complex interactions across the subsystems and components are left to more detailed diagrams and descriptions. The interactions among blocks may be represented by labeling the connecting lines.

System Context Diagrams. Another useful model in system design is the context diagram, which represents all external entities that may interact with a system, either directly or indirectly. We have already seen the context diagram in Figure 3.2. Such a diagram pictures the system at the center, with no details of its interior structure, surrounded by all its interacting systems, environments, and activities. The objective of a system context diagram is to focus attention on external factors and events that should be considered in developing a complete set of system requirements and constraints. In so doing, it is necessary to visualize not only the operational environment but also the stages leading up to operations, such as installation, integration, and operational evaluation.

Figure 9.3 shows a context diagram for the case of a passenger airliner. The model represents the external relationships between the airliner and various external entities. The system context diagram is a useful starting point for describing and defining the system's mission and operational environment, showing the interaction of a system with all external entities that may be relevant to its operation. It also provides a basis for formulating system operational scenarios that represent the different conditions under which it must be designed to operate. In commercial systems, the "enterprise diagram" also shows all the system's external inputs and outputs but also usually includes a representation of the related external entities.

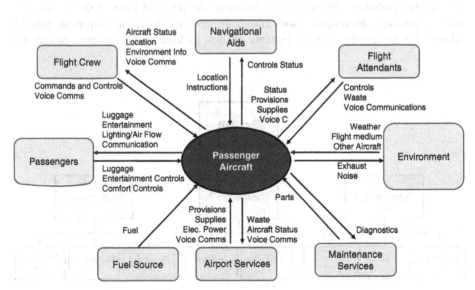

Figure 9.3. Context diagram of a passenger aircraft.

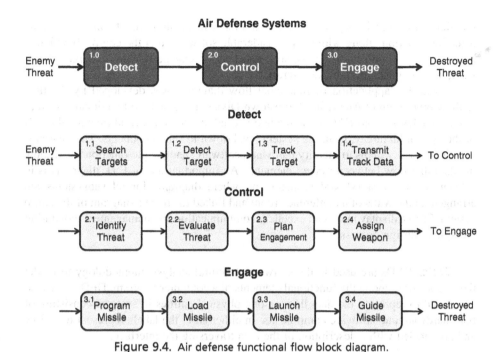

Figure 9.4. Air defense functional flow block diagram.

Functional Flow Block Diagrams (FFBDs). The models discussed previously deal primarily with static relationships within the system's physical structure. The more significant characteristics of systems and their components are related to how they behave in response to changes in the environment. Such behavior results from the functions that a system performs in response to certain environmental inputs and constraints. Hence, to model system behavior, it is necessary to model its principal functions, how they are derived, and how they are related to one another. The most common form of functional model is called the FFBD.

An example of an FFBD is shown in Figure 9.4. The figure shows the functional flow through an air defense system at the top-level functions of detect, control, and engage, and at the second-level functions that make up each of the above. Note the numbering system of the functional blocks that ties them together. Note also that the names in the blocks represent functions, not physical entities, and thus, all begin with a verb instead of a noun. The arrowheads on the lines between blocks in an FFBD indicate the flow of control and, in this case, also the flow of information. Keep in mind that flow of control does not necessarily equate with flow of information in all cases. The identity of the functions flowing between the blocks may be denoted on the FFBD as an optional feature but is not expected to be complete as it would be in a software DFD.

In the above example, the physical implementation of the functional blocks is not represented and may be subject to considerable variation. From the nature of the functions, however, it may be inferred that a radar installation may be involved in the detection function, along with very considerable software; that the control function is mostly software with operator displays; and that the engage function is largely hardware, such as guns, missiles, or aircraft.

A valuable application of functional flow diagrams was developed by the then Radio Corporation of America, Moorestown Division. Named the functional flow diagrams and descriptions (F^2D^2), the method is used to diagram several functional levels of the system hierarchy, from the system level down to subcomponents. The diagrams use distinctive symbols to identify hardware, software, and people functions, and show the data that flow between system elements. An important use of F^2D^2 diagrams is in a "war room" or storyboard arrangement, where diagrams for all subsystems are arranged on the walls of a conference room and linked to create a diagram of the entire system. Such a display makes an excellent communication and management tool during the system design process.

DFDs. DFDs are used in the software structural analysis methodology to model the interactions among the functional elements of a computer program. DFDs have also been used to represent the data flow among physical entities in systems consisting of both hardware and software components. In either case, the labels represent data flow and are labeled with a description of the data traversing the interface.

Integrated Definition Language 0 (IDEF0) Diagrams. IDEF0 is a standard representation of system activity models, similar to software DFDs, and was described in Chapter 8. Figure 8.3 depicts the rules for depicting an activity. IDEF0 is widely used in the modeling of complex information systems. As in FFBD and F^2D^2 diagrams, the functional blocks are rectangular and the sides of the activity boxes have a unique function. Processing inputs always enter from the left, controls from the top, and mechanisms or resources from the bottom; outputs exit on the right. The name of each block starts with a vowel and carries a label identifying its hierarchical location.

Functional Flow Process Diagrams (FFPD). The functional flow diagrams described earlier model the functional behavior of a system or a system product. Such diagrams are equally useful in modeling processes, including those involved in systems engineering. Examples of FFPDs are found in every chapter. The system life cycle model is a prime example of a process FFPD. In Chapter 4, Figures 4.1, 4.3, and 4.4 define the flow of system development through the defined stages and phases of the system life cycle. In Chapters 5–8, the first figures show the functional inputs and outputs between the corresponding life cycle phase and those immediately adjoining.

The systems engineering method is modeled in Chapter 4, Figure 4.10, and in greater detail in Figure 4.11. The functional blocks in this case are the principal processes that constitute the systems engineering method. Inside each block is a functional flow diagram that represents the functions performed by the block. The inputs coming from outside the blocks represent the external factors that contribute to the respective processes. Chapters 5–8 contain similar functional flow diagrams to illustrate the processes that take place during each phase of system development.

FFPDs are especially useful as training aids for production workers by resolving complex processes into their elementary components in terms readily understandable by the trainees. All process diagrams have a common basic structure, which consists of three elements: input → processing → output.

Trigonal System Models. In attempting to understand the functioning of complex systems, it is useful to resolve them into subsystems and components that individually are more simple to understand. A general method that works well in most cases is to resolve the system and each of its subsystems into three basic components:

1. sensing or inputting signals, data, or other media that the system element operates on;
2. processing the inputs to deduce an appropriate reaction to the inputs; and
3. acting on the basis of the instructions from the processing element to implement the system element's response to the input.

In an example of a system simulation described in the previous subsection, an air defense system was shown to be composed of three functions, namely, detect, control, and engage (see Fig. 9.4). The detect function is seen to correspond to the input portion, control (or analyze and control response) to the processing portion, and engage to the response action portion.

The input–processing–output segmentation can then be applied to each of the subsystems themselves. Thus, in the air defense system example, the detect function can be further resolved into the radar, which senses the reflection from the enemy airplane or missile, the radar signal processor, which resolves the target reflection from interfering clutter and jamming, and the automatic detection and track software, which correlates the signal with previous scans to form a track and calculate its coordinates and velocity vector for transmission to the control subsystem. The other two subsystems may be similarly resolved.

In many systems, there is more than a single input. For example, the automobile is powered by fuel but is steered by the driver. The input–processing–output analysis will produce two or more functional flows: tracing the fuel input will involve the fuel tank and fuel pump, which deliver the fuel, the engine, which converts (processes) the fuel into torque, and the wheels, which produce traction on the road surface to propel the car. A second set of components are associated with steering the car, in which the sensing and decision is accomplished by the driver, with the automobile executing the actual turn in response to steering wheel rotation.

Modeling Languages. The schematic models described above together were developed relatively independently. Thus, although they have been in use for several decades, they are used according to the experience of the engineer. However, these models do have certain attributes in common. They are, by and large, activity focused.

They communicate functionality of systems, whether that is of the form of activities, control, or data. Even block diagrams representing physical entities include interfaces among the entities showing flow of materials, energy, or data. Because of their age (basic block diagrams have been around for over 100 years), we tend to categorize these models as "functional" or "traditional."

When software engineering emerged as a significant discipline within system development, a new perspective was presented to the engineering community: object-oriented analysis (OOA). Rather than activity based, OOA presented concepts and models that were object based, where object is defined in very broad terms. Theoretically, anything can be an object. As described in Chapter 8, Unified Modeling Language (UML) is now a widely used modeling language for support of systems engineering and architecting.

Mathematical Models

Mathematical models are used to express system functionality and dependencies in the language of mathematics. They are most useful where system elements can be isolated for purposes of analysis and where their primary behavior can be represented by well-understood mathematical constructs. If the process being modeled contains random variables, simulation is likely to be a preferable approach. An important advantage of mathematical models is that they are widely understood. Their results have inherent credibility, provided that the approximations made can be shown to be of secondary importance. Mathematical models include a variety of forms that represent deterministic (not random) functions or processes. Equations, graphs, and spreadsheets, when applied to a specific system element or process, are common examples.

Approximate Calculations. Chapter 1 contains a section entitled The Power of Systems Engineering, which cites the critical importance of the use of approximate ("back of the envelope") calculations to the practice of systems engineering. The ability to perform "sanity checks" on the results of complex calculations or experiments is of inestimable value in avoiding costly mistakes in system development.

Approximate calculations represent the use of mathematical models, which are abstract representations of selected functional characteristics of the system element being studied. Such models capture the dominant variables that determine the main features of the outcome, omitting higher-order effects that would unduly complicate the mathematics. Thus, they facilitate the understanding of the primary functionality of the system element.

As with any model, the results of approximate calculations must be interpreted with full knowledge of their limitations due to the omission of variables that may be significant. If the sanity check deviates significantly from the result being checked, the approximations and other assumptions should be examined before questioning the original result.

In developing the skill to use approximate calculations, the systems engineer must make the judgment as to how far to go into the technical fundamentals in each specific case. One alternative is to be satisfied with an interrogation of the designers who made the original analysis. Another is to ask an expert in the discipline to make an independent check. A third is to apply the systems engineer's own knowledge, to augment it by reference to a handbook or text, and to carry out the approximate calculation personally.

The appropriate choice among these alternatives is, of course, situation dependent. However, it is advisable that in selected critical technical areas, the systems engineer becomes sufficiently familiar with the fundamentals to feel comfortable in making independent judgments. Developing such skills is part of the systems engineer's special role of integrating multidisciplinary efforts, assessing system risks, and deciding the areas that require analysis, development, or experimentation.

Elementary Relationships. In every field of engineering and physics, there are some elementary relationships with which the systems engineer should be aware, or familiar. Newton's laws are applicable in all vehicular systems. In the case of structural elements under stress, it is often useful to refer to relationships involving strength and elastic properties of beams, cylinders, and other simple structures. With electronic components, the systems engineer should be familiar with the elementary properties of electronic circuits. There are "rules of thumb" in most technical fields, which are usually based on elementary mathematical relationships.

Statistical Distributions. Every engineer is familiar with the Gaussian (normal) distribution function characteristic of random noise and other simple natural effects. Some other distribution functions that are of interest include the Rayleigh distribution, which is valuable in analyzing signals returned from radar clutter, the Poisson distribution, the exponential distribution, and the binomial distribution; all of these obey simple mathematical equations.

Graphs. Models representing empirical relationships that do not correspond to explicit mathematical equations are usually depicted by graphs. Figure 2.1a in Chapter 2 is a graph illustrating the typical relationship between performance and the cost to develop it. Such models are mainly used to communicate qualitative concepts, although test data plotted in the form of a graph can show a quantitative relationship. Bar charts, such as one showing the variations in production by month, or the cost of alternative products, are also models that serve to communicate relationships in a more effective manner than by a list of numbers.

Physical Models

Physical models directly reflect some or most of the physical characteristics of an actual system or system element under study. In that sense, they are the least abstract and therefore the most easily understood type of modeling. Physical models, however, are by definition simplifications of the modeled articles. They may embody only a part of the total product; they may be scaled-down versions or developmental prototypes. Such models have multiple uses throughout the development cycle, as illustrated by the examples described next.

Scale Models. These are (usually) small-scale versions of a building, vehicle, or other system, often used to represent the external appearance of a product. An example of the engineering use of scale models is the testing of a model of an air vehicle in a wind tunnel or of a submersible in a water tunnel or tow tank.

Mock-Ups. Full-scale versions of vehicles, parts of a building, or other structures are used in later stages of development of systems containing accommodation for operators and other personnel. These provide realistic representations of human–system interfaces to validate or possibly to modify their design prior to a detailed design of the interfaces.

Prototypes. Previous chapters have discussed the construction and testing of development, engineering, and product prototypes, as appropriate to the system in hand. These also represent physical models of the system, although they possess most of the properties of the operational system. However, strictly speaking, they are still models.

Computer-based tools are being increasingly used in place of physical models such as mock-ups and even prototypes. Such tools can detect physical interferences and permit many engineering tasks formerly done with physical models to be accomplished with computer models.

9.4 SIMULATION

System simulation is a general type of modeling that deals with the dynamic behavior of a system or its components. It uses a numerical computation technique for conducting experiments with a software model of a physical system, function, or process. Because simulation can embody the physical features of the system, it is inherently less abstract than many forms of modeling discussed in the previous section. On the other hand, the development of a simulation can be a task of considerable magnitude.

In the development of a new complex system, simulations are used at nearly every step of the way. In the early phases, the characteristics of the system have not yet been determined and can only be explored by modeling and simulation. In the later phases, estimates of their dynamic behavior can usually be obtained earlier and more economically by using simulations than by conducting tests with hardware and prototypes. Even when engineering prototypes are available, field tests can be augmented by using simulations to explore system behavior under a greater variety of conditions. Simulations are also used extensively to generate synthetic system environmental inputs for test purposes. Thus, in every phase of system development, simulations must be considered as potential development tools.

There are many different types of simulations and one must differentiate static from dynamic simulations, deterministic from stochastic (containing random variables), and discrete from continuous. For the purposes of relating simulations to their application to systems engineering, this section groups simulations into four categories: operational, physical, environmental, and virtual reality simulation. All of these are either wholly or partly software based because of the versatility of software to perform an almost infinite variety of functions.

Computer-based tools also perform simulations at a component or subcomponent level, which will be referred to as engineering simulation.

Operational Simulation

In system development, operational simulations are primarily used in the conceptual development stage to help define operational and performance requirements, explore alternative system concepts, and help select a preferred concept. They are dynamic, stochastic, and discrete event simulations. This category includes simulations of operational systems capable of exploring a wide range of scenarios, as well as system variants.

Games

The domain of analyzing operational mission areas is known as operations analysis. This field seeks to study operational situations characteristic of a type of commerce, warfare, or other broad activity and to develop strategies that are most suitable to achieving successful results. An important tool of operations analysis is the use of games to evaluate experimentally the utility of different operational approaches. The military is one of the organizations that relies on games, called war games, to explore operational considerations.

Computer-aided games are examples of operational simulations involving people who control a simulated system (blue team) in its engagement with the simulated adversary (red team), with referees observing both sides of the action and evaluating the results (white team). In business games, the two sides represent competitors. In other games, the two teams can represent adversaries.

The behavior of the system(s) involved in a game is usually based on that of existing operational systems, with such extensions as may be expected to be possible in the next generation of the system. These may be implemented by variable parameters to explore the effect of different system features on their operational capabilities.

Gaming has several benefits. First, it enables the participants to gain a clearer understanding of the operational factors involved in various missions, as well as of their interaction with different features of the system, which translates into experience in operational decision making. Second, by varying key system features, the participants can explore system improvements that may be expected to enhance their effectiveness. Third, through variation in operational strategy, it may be possible to develop improved operational processes, procedures, and methods. Fourth, analysis of the game results may provide a basis for developing a more clearly stated and prioritized set of operational requirements for an improved system than could be derived otherwise.

Commercial games are utilized by large corporations to identify and assess business strategies over a single and multiple business cycles within a set of plausible economic scenarios. Although these games do not typically predict technological

breakthroughs, they can identify "breakthrough" technologies that could lead to paradigm shifts in an industry.

Military organizations conduct a variety of games for multiple purposes such as
assessing new systems within a combat situation, analyzing a new concept for transport-
ing people and material, or evaluating a new technology to detect stealthy targets.
The games are facilitated by large screen displays and a bank of computers. The geographic displays are realistic, derived from detailed maps of the globe available on the
Internet and from military sources. A complex game may last from a day to several
weeks. The experience is highly enlightening to all participants. Short of actual operational experience, such games are the best means for acquiring an appreciation of the
operational environment and mission needs, which are important ingredients in systems
engineering.

Lastly, government organizations and alliances conduct geopolitical games to
assess international engagement strategies. These types of games tend to be complex
as the dimensions of interactions can become quite large. For example, understanding
national reactions to a country's policy actions involves diplomatic, intelligence, military, and economic (DIME) ramifications. Also, because interactions are complex,
standard simulation types may not be adequate to capture the realm of actions that a
nation might take. Therefore, sophisticated simulations are developed specifically to
model various components of a national entity. These components are known as *agents*.

System Effectiveness Simulation

During the concept exploration and concept definition phases of system development,
the effort is focused on the comparative evaluation of different system capabilities and
architectures. The objective is first to define the appropriate system performance
requirements and then to select the preferred system concept to serve as the basis for
development. A principal vehicle for making these decisions is the use of computer
system effectiveness simulations, especially in the critical activity of selecting a preferred system concept during concept definition. At this early point in the system life
cycle, there is neither time nor resources to build and test all elements of the system.
Further, a well-designed simulation can be used to support the claimed superiority of
the system concept recommended to the customer. Modern computer display techniques
can present system operation in realistic scenarios.

The design of a simulation of a complex system that is capable of providing a basis
for comparing the effectiveness of candidate concepts is a prime systems engineering
task. The simulation itself is likely to be complex in order to reflect all the critical
performance factors. The evaluation of system performance also requires the design
and construction of a simulation of the operational environment that realistically challenges the operational system's capabilities. Both need to be variable to explore different operational scenarios, as well as different system features.

A functional block diagram of a typical system effectiveness simulation is illustrated in Figure 9.5. The subject of the simulation is an air defense system, which is
represented by the large rectangle in the center containing the principal subsystems
detect, control, and engage. At the left is the simulation of the enemy force, which

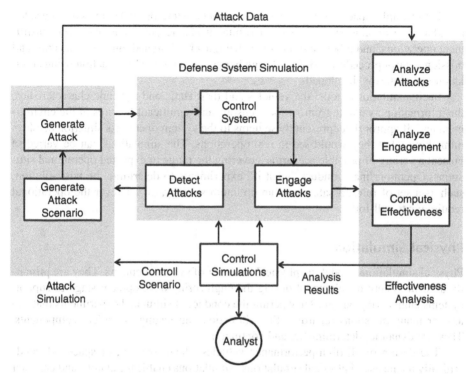

Figure 9.5. System effectiveness simulation.

contains a scenario generator and an attack generator. At the right is the analysis subsystem, which assesses the results of the engagement against an expected outcome or against results from other engagements. The operator interface, shown at the bottom, is equipped to modify the attacking numbers and tactics and also to modify the performance of these system elements to determine the effects on system effectiveness.

The size and direction of system effectiveness variations resulting from changes in the system model should be subjected to sanity checks before acceptance. Such checks involve greatly simplified calculations of the system performance and are best carried out by analysts not directly responsible for either the design or the simulation.

Mission Simulation

The objective of the simulations referred to as mission simulations is focused on the development of the operational modes of systems rather than on the development of the systems themselves. Examples of such simulations include the conduct of air traffic control, the optimum trajectories of space missions, automobile traffic management, and other complex operations.

For example, space missions to explore planets, asteroids, and comets are preceded by exhaustive simulations of the launch, orbital mechanics, terminal maneuvers, instrument operations, and other vital functions that must be designed into the spacecraft and mission control procedures. Before design begins, an analytical foundation using simulation techniques is developed.

Such simulations model the vehicles and their static and dynamic characteristics, the information available from various sensors, and significant features of the environment and, if appropriate, present these items to the system operator's situation displays mimicking what they would see in real operations. The simulations can be varied to present a variety of possible scenarios, covering the range of expected operational situations. Operators may conduct "what if" experiments to determine the best solution, such as a set of rules, a safe route, an optimum strategy, or whatever the operational requirements call for.

Physical Simulation

Physical simulations model the physical behavior of system elements. They are primarily used in system development during the engineering development stage to support systems engineering design. They permit the conduct of simulated experiments that can answer many questions regarding the fabrication and testing of critical components. They are dynamic, deterministic, and continuous.

The design of all high-performance vehicles—land, sea, air, or space—depends critically on the use of physical simulations. Simulations enable the analyst and designer to represent the equations of motion of the vehicle, the action of external forces, such as lift and drag, and the action of controls, whether manual or automated. As many experiments as may be necessary to study the effects of varying conditions or design parameters may be conducted. Without such tools, the development of modern aircraft and spacecraft would not have been practicable. Physical simulations do not eliminate the need for exhaustive testing, but they are capable of studying a great variety of situations and of eliminating all but a few alternative designs. The savings in development time can be enormous.

Examples: Aircraft, Automobiles, and Space Vehicles. Few technical problems are as complicated as the design of high-speed aircraft. The aerodynamic forces are quite nonlinear and change drastically in going between subsonic and supersonic regimes. The stresses on airplane structures can be extremely high, resulting in flexure of wings and control surfaces. There are flow interference effects between the wings and tail structure that depend sharply on altitude, speed, and flight attitude. Simulation permits all of these forces and effects to be realistically represented in six-degree-of-freedom models (three position and three rotation coordinates).

The basic motions of an automobile are, of course, far simpler than those of an aircraft. However, modern automobiles possess features that call on very sophisticated dynamic analysis. The control dynamics of antilock brakes are complex and critical, as are those of traction control devices. The action of airbag deployment devices is even more critical and sensitive. Being intimately associated with passenger safety, these devices must be reliable under all expected conditions. Here again, simulation is an essential tool.

Without modern simulation, there would be no space program as we know it. The task of building a spacecraft and booster assembly that can execute several burns to put the spacecraft into orbit, that can survive launch, deploy solar panels, and antennae, control its attitude for reasons of illumination, observation, or communication, and perform a series of experiments in space would simply be impossible without a variety of simulations. The international space station program achieved remarkable sustainability as each mission was simulated and rehearsed to near perfection.

Hardware-in-the-Loop Simulation

This is a form of physical simulation in which actual system hardware is coupled with a computer-driven simulation. An example of such a simulation is a missile homing guidance facility. For realistic experiments of homing dynamics, such a facility is equipped with microwave absorbing materials, movable radiation sources, and actual seeker hardware. This constitutes a dynamic "hardware-in-the-loop" simulation, which realistically represents a complex environment.

Another example of a hardware-in-the-loop simulation is a computer-driven motion table used in the development testing of inertial components and platforms. The table is caused to subject the components to movement and vibration representing the motion of its intended platform, and is instrumented to measure the accuracy of the resulting instrument output. Figure 9.6 shows a developmental inertial platform mounted on a

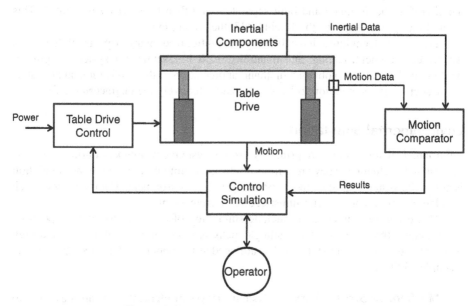

Figure 9.6. Hardware-in-the-loop simulation.

motion table, with a motor drive controlled by an operator and the feedback from the platform. A motion analyzer compares the table motion with the inertial platform outputs.

Engineering Simulation

At the component and subcomponent level, there are engineering tools that are extensions of mathematical models, described in the previous section. These are primarily used by design specialists, but their capabilities and limitations need to be understood by systems engineers in order to understand their proper applications.

Electronic circuit design is no longer done by cut-and-try methods using breadboards. Simulators can be used to design the required functionality, test it, and modify it until the desired performance is obtained. Tools exist that can automatically document and produce a hardware version of the circuit.

Similarly, the structural analysis of complex structures such as buildings and bridges can be done with the aid of simulation tools. This type of simulation can accommodate the great number of complicated interactions among the mechanical elements that make up the structure, which are impractical to accomplish by analysis and testing.

Development of the Boeing 777 Aircraft

As noted previously, virtually all of the structural design of the Boeing 777 was done using computer-based modeling and simulation. One of the aircraft's chief reasons for success was the great accuracy of interface data that allowed the various portions of the aircraft to be designed and built separately and then to be easily integrated. This technology set the stage for the Boeing 797, the Dreamliner.

The above techniques have literally revolutionized many aspects of hardware design, development, testing, and manufacture. It is essential for systems engineers working in these areas to obtain a firsthand appreciation of the application and capability of engineering simulation to be able to lead effectively the engineering effort.

Environmental Simulation

Environmental simulations are primarily used in system development during engineering test and evaluation. They are a form of physical simulation in which the simulation is not of the system but of elements of the system's environment. The majority of such simulations are dynamic, deterministic, and discrete events.

This category is intended to include simulation of (usually hazardous) operating environments that are difficult or unduly expensive to provide for validating the design of systems or system elements, or that are needed to support system operation. Some examples follow.

Mechanical Stress Testing. System or system elements that are designed to survive harsh environments during their operating life, such as missiles, aircraft systems, spacecraft, and so on, need to be subjected to stresses simulating such conditions. This is customarily done with mechanical shake tables, vibrators, and shock testing.

Crash Testing. To meet safety standards, automobile manufacturers subject their products to crash tests, where the automobile body is sacrificed to obtain data on the extent to which its structural features lessen the injury suffered by the occupant. This is done by use of simulated human occupants, equipped with extensive instrumentation that measures the severity of the blow resulting from the impact. The entire test and test analysis are usually computer driven.

Wind Tunnel Testing. In the development of air vehicles, an indispensable tool is an aerodynamic wind tunnel. Even though modern computer programs can model the forces of fluid flow on flying bodies, the complexity of the behavior, especially near the velocity of sound, and interactions between different body surfaces often require extensive testing in facilities that produce controlled airflow conditions impinging on models of aerodynamic bodies or components. In such facilities, the aerodynamic model is mounted on a fixture that measures forces along all components and is computer controlled to vary the model angle of attack, control surface deflection, and other parameters, and to record all data for subsequent analysis.

As noted in the discussion of scale models, analogous simulations are used in the development of the hulls and steering controls of surface vessels and submersibles, using water tunnels and tow tanks.

Virtual Reality Simulation

The power of modern computers has made it practical to generate a three-dimensional visual environment of a viewer that can respond to the observer's actual or simulated position and viewing direction in real time. This is accomplished by having all the coordinates of the environment in the database, recomputing the way it would appear to the viewer from his or her instantaneous position and angle of sight, and projecting it on a screen or other display device usually mounted in the viewer's headset. Some examples of the applications of virtual reality simulations are briefly described next.

Spatial Simulations. A spatial virtual reality simulation is often useful when it is important to visualize the interior of enclosed spaces and the connecting exits and entries of those spaces. Computer programs exist that permit the rapid design of these spaces and the interior furnishings. A virtual reality feature makes it possible for an observer to "walk" through the spaces in any direction. This type of model can be useful for the preliminary designs of houses, buildings, control centers, storage spaces, parts of ships, and even factory layouts. An auxiliary feature of this type of computer model is the ability to print out depictions in either two- or three-dimensional forms, including labels and dimensions.

Spatial virtual simulations require the input to the computer of a detailed three-dimensional description of the space and its contents. Also, the viewing position is input into the simulation either from sensors in the observer's headset or directed with a

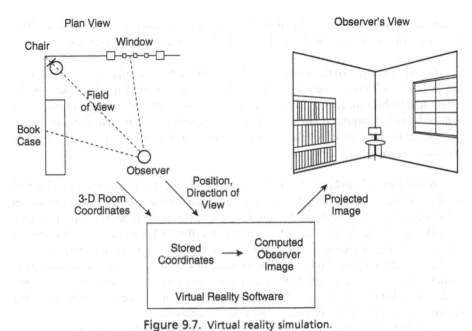

Figure 9.7. Virtual reality simulation.

joystick, mouse, or other input device. The virtual image is computed in real time and projected either to the observer's headset or on a display screen. Figure 9.7 illustrates the relationship between the coordinates of two sides of a room with a bookcase on one wall, a window on the other, and a chair in the corner, and a computer-generated image of how an observer facing the corner would see it.

Video Games. Commercial video games present the player with a dynamic scenario with moving figures and scenery that responds to the player's commands. In many games, the display is fashioned in such a way that the player has the feeling of being inside the scene of the action rather than of being a spectator.

Battlefield Simulation. A soldier on a battlefield usually has an extremely restricted vision of the surroundings, enemy positions, other forces, and so on. Military departments are actively seeking ways to extend the soldier's view and knowledge by integrating the local picture with situation information received from other sources through communication links. Virtual reality techniques are expected to be one of the key methods of achieving these objectives of situational awareness.

Development of System Simulations

As may be inferred from this section, the several major simulations that must be constructed to support the development of a complex system are complex in their own right. System effectiveness simulations have to not only simulate the system functionality but also to simulate realistically the system environment. Furthermore, they have to be designed so that their critical components may be varied to explore the performance of alternative configurations.

In Chapter 5, modeling and simulation were stated to be an element of the systems engineering management plan. In major new programs, the use of various simulations may well account for a substantial portion of the total cost of the system development. Further, the decisions on the proper balance between simulation fidelity and complexity require a thorough understanding of the critical issues in system design, technical and program risks, and the necessary timing for key decisions. In the absence of careful analysis and planning, the fidelity of simulations is likely to be overspecified, in an effort to prevent omissions of key parameters. The result of overambitious fidelity is the extension of project schedules and exceedance of cost goals. For these reasons, the planning and management of the system simulation effort should be an integral part of systems engineering and should be reflected in management planning.

Often the most effective way to keep a large simulation software development within bounds is to use iterative prototyping, as described in Chapter 11. In this instance, the simulated system architecture is organized as a central structure that performs the basic functions, which is coupled to a set of separable software modules representing the principal system operational modes. This permits the simulation to be brought to limited operation quickly, with the secondary functions added, as time and effort are available.

Simulation Verification and Validation

Because simulations serve an essential and critical function in the decision making during system development, it is necessary that their results represent valid conclusions regarding the predicted behavior of the system and its key elements. To meet this criterion, it must be determined that they accurately represent the developer's conceptual description and specification (verification) and are accurate representations of the real world, to the extent required for their intended use (validation).

The verification and validation of key simulations must, therefore, be an integral part of the total system development effort, again under the direction of systems engineering. In the case of new system effectiveness simulations, which are usually complex, it is advisable to examine their results for an existing (predecessor) system whose effectiveness has been previously analyzed. Another useful comparison is with the operation of an older version of the simulation, if one exists.

Every simulation that significantly contributes to a system development should also be documented to the extent necessary to describe its objectives, performance specifications, architecture, concept of operation, and user modes. A maintenance manual and user guide should also be provided.

The above actions are sometimes neglected to meet schedules and in competition with other activities. However, while simulations are not usually project deliverables, they should be treated with equal management attention because of their critical role in the success of the development.

Even though a simulation has been verified and validated, it is important to remember that it is necessarily only a model, that is, a simplification and approximation to reality. Thus, there is no such thing as an *absolutely validated* simulation. In particular, it should only be used for the prescribed application for which it has been tested. It is the responsibility of systems engineering to circumscribe the range of valid applicability of a given simulation and to avoid unwarranted reliance on the accuracy of its results.

Despite these cautions, simulations are absolutely indispensable tools in the development of complex systems.

9.5 TRADE-OFF ANALYSIS

Performing a trade-off is what we do whenever we make a decision, large or small. When we speak, we subconsciously select words that fit together to express what we mean, instinctively rejecting alternative combinations of words that might have served the purpose, but not as well. At a more purposeful level, we use trade-offs to decide what to wear to a picnic or what flight to take on a business trip. Thus, all decision processes involve choices among alternative courses of action. We make a decision by comparing the alternatives against one another and by choosing the one that provides the most desirable outcome.

In the process of developing a system, hundreds of important systems engineering decisions have to be made, many of them with serious impacts on the potential success of the development. Those cases in which decisions have to be approved by management or by the customer must be formally presented, supported by evidence attesting to the thoroughness and objectivity of the recommended course of action. In other cases, the decision only has to be convincing to the systems engineering team. Thus, the trade-off process needs to be tailored to its ultimate use. To differentiate a formal trade-off study intended to result in a recommendation to higher management from an informal decision aid, the former will be referred to as "trade-off analysis" or a "trade study," while the latter will be referred to as simply a "trade-off." The general principles are similar in both cases, but the implementation is likely to be considerably different, especially with regard to documentation.

Basic Trade-Off Principles

The steps in a trade-off process can be compared to those characterizing the systems engineering methodology, as used in the systems concept definition phase for selecting the preferred system concept to meet an operational objective. The basic steps in the trade-off process at any level of formality are the following (corresponding steps in the systems engineering methodology are shown in parentheses).

Defining the Objective (Requirements Analysis). The trade-off process must start by defining the objectives for the trade study itself. This is carried out by identifying the requirements that the solution (i.e., the result of the decision) must fulfill.

The requirements are best expressed in terms of measures of effectiveness (MOE), as quantitatively as practicable, to characterize the merits of a candidate solution.

Identification of Alternatives (Concept Exploration). To provide a set of alternative candidates, an effort must be made to identify as many potential courses of action as will include all promising candidate alternatives. Any that fail to comply with an essential requirement should be rejected.

Comparing the Alternatives (Concept Definition). To determine the relative merits of the alternatives, the candidate solutions should be compared with one another with respect to each of their MOEs. The relative order of merit is judged by the cumulative rating of all the MOEs, including a satisfactory balance among the different MOEs.

Sensitivity Analysis (Concept Validation). The results of the process should be validated by examining their sensitivity to the assumptions. MOE prioritization and candidate ratings are varied within limits reflecting the accuracy of the data. Candidates rated low in only one or two MOEs should be reexamined to determine whether this result could be changed by a relatively straightforward modification. Unless a single candidate is clearly superior, and the result is stable to such variations, further study should be conducted.

Formal Trade-Off Analysis and Trade Studies

As noted above, when trade-offs are conducted to derive and support a recommendation to management, they must be performed and presented in a formal and thoroughly documented manner. As distinguished from informal decision processes, trade-off studies in systems engineering should have the following characteristics:

1. They are organized as defined processes. They are carefully planned in advance, and their objective, scope, and method of approach are established before they are begun.
2. They consider all key system requirements. System cost, reliability, maintainability, logistics support, growth potential, and so on, should be included. Cost is frequently handled separately from other criteria. The result should demonstrate thoroughness.
3. They are exhaustive. Instead of considering only the obvious alternatives in making a systems engineering decision, a search is made to identify all options deserving consideration to ensure that a promising one is not inadvertently overlooked. The result should demonstrate objectivity.
4. They are semiquantitative. While many factors in the comparison of alternatives may be only approximately quantifiable, systems engineering trade-offs seek to quantify all possible factors to the extent practicable. In particular, the various MOEs are prioritized relative to one another in order that the weighting of the

various factors achieves the best balance from the standpoint of the system objectives. All assumptions must be clearly stated.

5. They are thoroughly documented. The results of systems engineering trade-off analyses must be well documented to allow review and to provide an audit trail should an issue need reconsideration. The rationale behind all weighting and scoring should be clearly stated. The results should demonstrate logical reasoning.

A formal trade study leading to an important decision should include the steps described in the following paragraphs. Although presented linearly, many overlap and several can be, and should be, coupled together in an iterative subprocess.

Step 1: Definition of the Objectives. To introduce the trade study, the objectives must be clearly defined. These should include the principal requirements and should identify the mandatory ones that all candidates must meet. The issues that will be involved in selecting the preferred solution should also be included. The objectives should be commensurate with the phase of system development. The operational context and the relationships to other trade studies should be identified at this time. Trade studies conducted early in the system development cycle are typically conducted at the system level and higher. Detailed component-level trade studies are conducted later, during engineering and implementation phases.

Step 2: Identification of Viable Alternatives. As stated previously, before embarking on a comparative evaluation, an effort should be made to define several candidates to ensure that a potentially valuable one is not overlooked. A useful strategy for finding candidate alternatives is to consider those that maximize a particularly important characteristic. Such a strategy is illustrated in the section on concept selection in Chapter 8, in which it is suggested to consider candidates based on the following:

- the predecessor system as a baseline,
- technological advances,
- innovative concepts, and
- candidates suggested by interested parties.

In selecting alternatives, no candidate should be included that does not meet the mandatory requirements, unless it can be modified to qualify. However, keep the set of mandatory requirements small. Sometimes, an alternative concept that does not quite meet a mandatory requirement but is superior in other categories, or results in significant cost savings, is rejected because it does not reach a certain threshold. Ensure that all mandatory requirements truly are mandatory—and not simply someone's guess or wish.

The factors to consider in developing the set of alternatives are the following:

- There is never a single possible solution. Complex problems can be solved in a variety of ways and by a variety of implementations. In our experience, we have never encountered a problem with one and only one solution.

- Finding the optimal solution is rarely worth the effort. In simple terms, systems engineering can be thought of as the art and science of finding the "good enough" solution. Finding the mathematical optimum is expensive and many times near impossible.

- Understand the discriminators among alternatives. Although the selection criteria are not chosen at this step (this is the subject of the next step), the systems engineer should have an understanding of what discriminates alternatives. Some discriminators are obvious and exist regardless of the type of system you are developing: cost, technical risk, reliability, safety, and quality. Even of some of these cannot be quantified, yet a basic notion of how alternatives discriminate within these basic categories will enable the culling of alternatives to a reasonable quantity.

- Remain open to additional solutions surfacing during the trade study. This step is not forgotten once an initial set of alternatives has been identified. Many times, even near the end of the formal trade study, additional options may emerge that hold promise. Typically, a new option arises that combines the best features of two or more original alternatives. Many times, identifying these alternatives is not possible, or at least difficult, early in the process.

Staged Process. This step tends to occur in discrete stages. Initially, a large number and variety of alternatives should be considered. Brainstorming is one effective method of capturing a variety of alternatives, without evaluating their merits. Challenge participants to think "out of the box" to ensure that no option is overlooked. And while some ridiculous ideas are offered, this tends to stimulate thinking on other, plausible options. In our experience, 40–50 alternatives can be identified initially. This set is not our final set of alternatives, of course. It needs to be reduced.

As long as there are more than three to five potential alternatives, it is suggested that the staged approach be continued, culling the set down to a manageable set. The process of reducing alternatives generally follows a rank-ordering process, rather than quantitative weighing and scoring, to weed out less desirable candidates. Options can be dismissed due to a variety of reasons: cost, technological feasibility, safety, manufacturability, operational risk, and so on. This process may also uncover criteria that are not useful differentiators. Follow-on stages would focus on a few candidates that include likely candidates. These would be subjected to a much more thorough analysis as described below.

Remember to document the choices and reasoning behind the decision. Include the specifications for the alternatives to make the trade-off as quantitative as possible. The result of this multistage process is a reasonable set of alternatives that can be evaluated formally and comprehensively.

Step 3: Definition of Selection Criteria. The basis of differentiating between alternative solutions is a set of selection criteria to be chosen from and referenced to the requirements that define the solution. Each criterion must be an essential attribute of the product, expressed as a MOE, related to one or more of its requirements. It is desirable that it be quantifiable so that its value for each alternative may be derived objectively. Cost is almost always a key criterion. Reliability and maintainability are also usually important characteristics, but they must be quantified. In the case of large systems, size, weight, and power requirements can be important criteria. In software products, ease of use and supportability are usually important differentiators.

Characteristics that are possessed by all candidates to a comparable degree do not serve to distinguish among them and hence should not be used, because their inclusion only tends to obscure the significant discriminators. Also, two closely interdependent characteristics do not contribute more discrimination than can be obtained by one of them with appropriate weighting. The number of criteria used in a particular formal trade study can vary widely but usually ranges between 6 and 10. Fewer criteria may not appear convincing of a thorough study. More criteria tend to make the process unwieldy without adding value.

Step 4: Assignment of Weighting Factors to Selection Criteria.

In a given set of criteria, not all of them are equally important in determining the overall value of an alternative. Such differences in importance are taken into account by assigning each criterion a "weighting factor" that magnifies the contribution of the most critical criteria, that is, those to which the total value is the most sensitive, in comparison to the less critical. This procedure often turns out to be troublesome to carry out because many, if not most, of the criteria are incommensurable, such as cost versus risk, or accuracy versus weight. Also, judgments of relative criticality tend to be subjective and often depend on the particular scenario used for the comparison.

Several alternative weighting schemes are available. All of them should engage domain experts to help with the decisions. Perhaps the simplest is to assign weights from 1 to n (with n having the greatest contribution). Although subjective, the criteria are measured relative to each other (as opposed to an absolute measure). A disadvantage with using the typical 1 to n scheme is that people tend to group around the median, in this case, $(1 + n)/2$. For example, using a 1–5 scale may really be using a 1–3 scale since many will simply not use 1 and 5 often. Other times, people tend to rate all criteria high, either a 4 or 5—resulting in the equivalent of using 1–2.

Adding some objectivity requires a trade-off decision in and of itself when assigning weights. For instance, we could still use the 1–5 scale, but use a maximum number of weighting points; that is, the sum of all of the weights must not exceed a maximum value. A good starting maximum sum might be to take the sum of all average weights,

$$\text{MaxSum} = \frac{(\text{MaxWeight} - \text{MinWeight})}{2} n,$$

where

    MaxSum is the total number of weighting points to be allocated;
    MaxWeight is the greatest weight allowed;
    MinWeight is the least weight allowed; and
    n is the number of criteria.

Thus, this scheme holds the average weight as a constant. If the engineer (or stakeholders, depending on who is weighting the criteria) wants to weight a criterion higher, then she must reduce the weight of another criterion. Keep in mind, however, with any subjective weighting scheme (any scheme that uses "1 to n"), you are making assumptions about the relative importance. A "5" is five times as relevant as a "1." These numbers are used in the calculations to compare alternatives. Make sure the scheme is appropriate.

If more mathematical accuracy is desired, the weights could be constrained to sum to 1.0. Thus, each weighting would be a number between 0 and 1.0. This scheme has some mathematical advantages that will be described later in this chapter. One logical advantage is that weightings are not constrained to integers. If one alternative is 50% more important than another, this scheme can represent that relationship; integers cannot. When using spreadsheets for the calculations, be sure not to allow too many significant figures! The credibility of the engineering judgment would fall quickly.

To summarize, deciding on a weighting scheme is important. Careful thinking about the types of relative importance of alternatives is required. Otherwise, the engineer can inadvertently bias the results without knowing.

Step 5: Assignment of Value Ratings for Alternatives. This step can be confusing to many people. You may ask, why can we not simply measure the criteria values for each alternative at this point and use those values in our comparison? Of course, we could, but it becomes hard to compare the alternatives without integrating the criteria in some manner. Each criterion may use different units; so how does the systems engineer integrate multiple criteria together to gain an understanding of an overall value assessment for each alternative? We cannot combine measures of area (square foot) with velocity (foot per second), for example. And what if a criterion is impossible to measure? Does that mean subjective criteria are simply not used? In fact, subjective criteria are used in system development frequently (though usually in combination with objective criteria). Thus, we need a method to combine criterion together without trying to integrate units that are different. Basically, we need an additional step beyond measuring criterion values for each alternative. We need to assign an effectiveness value.

There are several methods of assigning a value for each criterion to each alternative. Each has its own set of advantages and attributes. And the method ultimately used may not be a choice for the systems engineer, depending on what data can be collected. Three basic options are available: (1) the subjective value method, (2) the step function method, and (3) the utility function method.

The first method relies on the systems engineer's subjective assessment of the alternative relative to each criterion. The latter two methods use actual measurements and translate the measurement to a value. For example, if volume is a criterion with cubic feet as the unit, then each alternative would be measured directly—what is the volume that each alternative fills, in cubic feet? Combinations of the three methods are also frequently used.

TABLE 9.3. Weighted Sum Integration of Selection Criteria

	For each alternative ...		
Selection criteria	Weights	Value	Score = weight × value
1	w_1	v_1	w_1v_1
2	w_2	v_2	w_2v_2
3	w_3	v_3	w_3v_3
4	w_4	v_4	w_4v_4

Subjective Value Method. When this method is chosen, the procedure begins with a judgment of the relative utility of each criterion on a scale analogous to student grading, say 1–5. Thus, 1 = poor, 2 = fair, 3 = satisfactory, 4 = good, and 5 = superior. (A candidate that fails a criterion may be given a zero, or even a negative score if the scores are to be summed, to ensure that the candidate will be rejected despite high scores on other criteria.) This is the effectiveness *value* for each criterion/alternative pair. The *score* assigned to the contribution of a given criterion to a specific candidate is the product of the weight assigned to the criterion and the assigned effectiveness value of the candidate in meeting the criterion.

Table 9.3 depicts a generic example that could be constructed for each alternative, for four selection criteria (they are not described, just numbered one through four).

In this method, the value v_i would be an integer between 1 and 5 (using our subjective effectiveness rating above), and would be assigned by the systems engineer.

Actual Measurement Method. If a more objective effectiveness rating is desired (more than "poor/fair/satisfactory/good/superior"), and alternatives could be measured for each criterion, then a simple mathematical step function could be constructed that translates an actual measurement into an effectiveness value. The systems engineer still needs to define this function and what value will be assigned to what range of measurements. Using our example of volume as a criterion, we could define a step function that assigns an effectiveness value to certain levels of volume. Assuming lesser volume is better effectiveness,

Volume (ft^3)	Value
0–2.0	5
2.01–3.0	4
3.01–4.0	3
4.01–5.0	2
>5.0	1

TABLE 9.4. Weighted Sum of Actual Measurement

		For each alternative ...		
Selection criteria	Weights	Measurement	Value	Score = weight × value
1	w_1	m_1	v_1	$w_1 v_1$
2	w_2	m_2	v_2	$w_2 v_2$
3	w_3	m_3	v_3	$w_3 v_3$
4	w_4	m_4	v_4	$w_4 v_4$

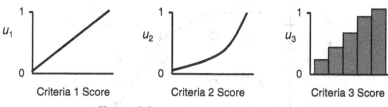

Figure 9.8. Candidate utility functions.

If an alternative fills $3.47\,\text{ft}^3$ of volume, it would be given an effectiveness value of 3. Keep this concept in mind as we will use something similar with our next method.

Table 9.4 illustrates this method. In this case, the alternative is actually measured for each criterion; the result is m_i. The step function is then used to translate the measurement to an effectiveness value, v_i. The final score for that criterion is the product of the measurement and value, $m_i v_i$. Once the measurements are converted to values, the actual measurements, m_i, are no longer used.

Utility Function Method. A refinement of the second approach is to develop a utility function for each criterion, which relates its measurable performance to a number between zero and one. Each criterion is measured, just as in the second method. But instead of allocating subjective values, a utility function is used to map each measurement to a value between zero and one.

Advantages to this method over the second are mathematical. As in using a utility function for weights (i.e., summing the weights to one), using utility functions places all criteria on an equal basis—the effectiveness of each criterion is constrained to a number between zero and one. Furthermore, if utility functions are used, mathematical properties of utility functions can be utilized. These are described in the next section.

Figure 9.8 illustrates some examples of utility functions. A utility function can be either continuous or discrete, linear or nonlinear.

If utility functions are used, calculating a total score for each criterion is similar to the second method. The score is simply the product of the weight and the utility. Table 9.5 depicts these relationships.

TABLE 9.5. Weighted Sum of Utility Scores

		For each alternative ...		
Selection criteria	Weights	Measurement	Utility	Score = weight × utility
1	w_1	m_1	u_1	$w_1 u_1$
2	w_2	m_2	u_2	$w_2 u_2$
3	w_3	m_3	u_3	$w_3 u_3$
4	w_4	m_4	u_4	$w_4 u_4$

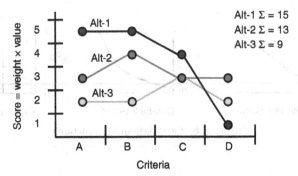

Figure 9.9. Criteria profile.

Step 6: Calculating Comparative Scores. The conventional method for combining the scores for the several alternatives is to calculate the sum of the weighted scores for each criterion to produce a total score. The candidate with the greatest summed value is judged to be the best candidate given the selection criteria and weightings, provided the score of the next highest alternative is statistically lower:

$$\text{Alternative total score} = w_1 v_1 + w_2 v_2 + w_3 v_3 + w_4 v_4.$$

This process is simple to implement, but lumping together the scores of the individual criteria tends to obscure factors that may be more important than initially supposed. For example, a candidate may receive a very low score on an essential MOE and high scores on several others. This lack of balance should not be obscured. It is strongly recommended that in addition to presenting the total scores, a graph of the criteria profile for each candidate be also included. Figure 9.9 presents a notional example of a criteria profile for three alternatives.

Deciding which alternative among the three is best is difficult since Alt-1 scores very low on criterion D but very high on criteria A, B, and C. Is this significant? If only the weighted sums are used, then Alt-1 would be the best candidate (with a sum of $5 + 5 + 4 + 1 = 15$). In its purest form, Alt-1 is selected due to its greatest weighted sum, but as always, numbers do not tell the whole story; we need analysis.

Step 7: Analyzing the Results. Because of the necessary reliance on qualitative judgments and the incommensurable nature of many of the criteria, the results of a trade study should be subjected to critical scrutiny. This process is especially important when the two or three top scores are close together and do not produce a decisive winner.

An essential step in analyzing the results is to examine the individual candidate profiles (scores for each criterion). Candidates that score poorly on one or more criteria may be less desirable than those with satisfactory scores in all categories. Cost is another factor that needs to be considered separately.

The conventional method of summing the individual scores is simple to use but has the unfortunate characteristic of underemphasizing low scores. A technique that does not suffer from this defect is to derive the composite score for a candidate by calculating the product (or geometric mean), rather than the sum, of the individual scores for the several criteria. If a candidate scores a zero on any criterion, the product function will also be zero, rejecting the alternative. An equivalent variant with the same property is to sum the logarithms of the individual scores.

A conventional approach to testing the robustness of trade study results is called "sensitivity analysis." Sensitivity analysis tests the invariance of the results to small changes in the individual weighting factors and scores. Because of uncertainties in the assignment of weighting and scores, substantial variations (20–30%) should be considered. A preferred approach is to sequentially set each criterion equal to zero and to recalculate the study. When such variations do not change the initial top choice, the procedure builds confidence in the result of the analysis.

An additional sensitivity test is to consider if there are important criteria that have not been included in the evaluation. Examples may be risk, growth potential, availability of support services, maturity of the product or of its supplier, ease of use, and so on. One of the alternatives may be considerably more attractive in regard to several of such additional issues.

Trade-Off Analysis Report. The results of a formal trade study represent an important milestone in the development of a system or other important operation and will contribute to decisions that will determine the future course. As such, they have to be communicated to all principal participants, who may include customers, managers, technical leaders, and others closely associated with the subject at issue. Such communication takes two forms: presentations and written reports.

Both oral and written reports must contain sufficient material to fully explain the method used and the rationale leading to the conclusions. They should include

- a statement of the issue and requirements on the solution;
- a discussion of assumptions and relationships to other components and subsystems;
- a setting of mission or operational considerations;
- a listing of relevant and critical system or subsystem requirements;

- a description of each alternative selected and the key features that led to its selection;
- an explanation of how the evaluation criteria were selected and the rationale for their prioritization (weighting);
- a rationale for assigning specific scores to each alternative for each criterion;
- a summary of the resulting comparison;
- a description of the sensitivity analysis and its results;
- the final conclusion of the analysis and an evaluation of its validity;
- recommendation for adoption of the study results or further analysis; and
- references to technical, quantitative material.

The presentation has the objective of presenting valuable information to program decision makers in order to make informed decisions. It requires a careful balance between sufficient substance to be clear and too much detail to be confusing. To this end, it should consist mainly of graphical displays, for which the subject is well suited, with a minimum of word charts. On the other hand, it is essential that the rationale for selection weighting and scoring is clear, logical, and persuasive. A copy of the comparison spreadsheet may be useful as a handout.

The purpose of the written trade study report is not only to provide a historical record of the basis for program decisions but also, more importantly, to provide a reference for reviewing the subject if problems arise later in the program. It represents the documented record of the analysis and its results. Its scope affords the opportunity for a detailed account of the steps of the study. For example, it may contain drawings, functional diagrams, performance analysis results, experimental data, and other materials that support the trade-off study.

Trade-Off Analysis Example

An example of a trade-off matrix is illustrated in Table 9.6, for the case of selecting a software code analysis tool. The table compares the ratings of five candidate commercial software tools with respect to six evaluation criteria:

- speed of operation, measured in minutes per run;
- accuracy in terms of errors per 10 runs;
- versatility in terms of number of applications addressed;
- reliability, measured by program crashes per 100 runs;
- user interface, in terms of ease of operation and clarity of display; and
- user support, measured by response time for help and repair.

Scoring. On a scale of 0–5, the maximum weight of 5 was assigned to accuracy—for obvious reasons. The next highest, 4, was assigned to speed, versatility, and reliability, all of which have a direct impact on the utility of the tool. User interface and

TABLE 9.6. Trade-Off Matrix Example

Criteria	Weight	Videx Score	Videx Weighted score	PeopleSoft Score	PeopleSoft Weighted score	CodeView Score	CodeView Weighted score	HPA Score	HPA Weighted score	Zenco Score	Zenco Weighted score
Speed	4	5	20	5	20	3	12	3	12	5	20
Accuracy	5	2	10	4	20	3	15	4	20	2	10
Versatility	4	5	20	5	20	3	12	5	20	5	20
Reliability	4	3	12	2	8	3	12	5	20	4	16
User interface	3	5	15	5	15	3	9	5	15	5	15
User support	3	2	6	1	3	3	9	4	12	5	15
Weighted sum			83		86		69		99		96
Cost			750		520		420		600		910
Weighted sum/cost			0.11		0.17		0.16		0.17		0.11

support were assigned a medium weight of 3 because, while they are important, they are not quite as critical as the others to the successful use of the tool.

Cost was considered separately to enable the consideration of cost/effectiveness as a separate evaluation factor.

The *subjective value method* was used to determine raw scores. The raw scores for each of the candidates were assigned on a scale of 5 = superior, 4 = good, 3 = satisfactory, 2 = weak, 1 = poor, and 0 = unacceptable. The row below the criteria lists the summed total of the weighted scores. The cost for each candidate tool and the ratio of the total score to the cost are listed in the last two rows.

Analysis. Comparing the summed scores in Table 9.6 shows that HPA and Zenco score significantly higher than the others. It is worth noting, however, that CodeView scored "satisfactory" on all criteria and is the least expensive by a substantial margin. Videx, CodeView, and HPA are essentially equal in cost/effectiveness.

Sensitivity analysis by varying criteria weightings does not resolve the difference between HPA and Zenco. However, examining the profiles of the candidates' raw scores highlights the weak performance of Zenco with respect to accuracy. This, coupled with its very high price, would disqualify this candidate. The profile test also highlights the weak reliability and poor user support of PeopleSoft, and the weak accuracy and high price of Videx. In contrast, HPA scores satisfactory or above in all categories and superior in half of them.

The above detailed analysis should result in a recommendation to select HPA as the best tool, with an option of accepting CodeView if cost is a determining factor.

Limitations of Numeric Comparisons

Any decision support method provides information to decision makers; it does not make the decision for them. Stated another way, trade-off analysis is a valuable aid to decision making rather than an infallible formula for success. It serves to organize a set of inputs in a systematical and logical manner, but is wholly dependent on the quality and sufficiency of the inputs.

The above trade-off example illustrates the need for a careful examination of all of the significant characteristics of a trade-off before making a final decision. It is clear that the total candidate scores in themselves mask important information (e.g., the serious weaknesses in some of the candidates). It is also clear that conventional sensitivity analysis does not necessarily suffice to resolve ties or to test the validity of the highest-scoring candidate. The example shows that the decision among alternatives should not be reduced to merely a mathematical exercise.

Furthermore, when, as is very often the case, the relative weightings of MOE are based on qualitative judgments rather than on objective measurements, there are serious implications produced by the automated algorithms that compute the results. One problem is that such methods tend to produce the impression of credibility well beyond the reliability of the inputs. Another is that the results are usually presented to more significant figures than are warranted by the input data. Only in the case of existing products whose characteristics are accurately known are the inputs truly quantitative.

For these reasons, it is absolutely necessary to avoid blindly trusting the numbers. A third limitation is that the trade-off studies often fail to include the assumptions that went into the calculations. To alleviate the above problems, it is important to accompany the analysis with a written rationale for the assignment of weighting factors, rounding off the answer to the relevant number of significant figures, and performing a sanity check on the results.

Decision Making

As was stated in the introduction to this section, all important systems engineering decisions should follow the basic principles of the decision-making process. When a decision does not require a report to management, the basic data gathering and reasoning should still be thorough. Thus, all decisions, formal and informal, should be conducted in a systematic manner, use the key requirements to derive the decision criteria, define relevant alternatives, and attempt to compare the candidates' utility as objectively as practicable. In all important decisions, the opinions of colleagues should be sought to obtain the advantage of collective judgment to resolve complex issues.

9.6 REVIEW OF PROBABILITY

The next section discusses the various evaluation methods that are available to the systems engineer when making decisions among a set of alternatives. All of the evaluation methods involve some level of mathematics, especially probability. Therefore, it is necessary to present a quick review of basic probability theory before describing the methods.

Even in the classical period of history, people noticed that some events could not be predicted with certainty. Initial attempts at representing uncertainty were subjective and nonquantitative. It was not until the late Middle Ages before some quantitative methods were developed. Once mathematics had matured, probability theory could be grounded in mathematical principles. It was not long before probability was applied beyond games of chance and equipossible outcomes (where it started). Before long, probability was applied to the physical sciences (e.g., thermodynamics and quantum mechanics), social sciences (e.g., actuarial tables and surveying), and industrial applications (e.g., equipment failures).

Although modern probability theory is grounded in mathematics, there still exists different perspectives on what probability is and how best it should be used:

- *Classical.* Probability is the ratio of favorable cases to the total equipossible cases.
- *Frequentist.* Probability is the limiting value as the number of trials becomes infinite of the frequency of occurrence of a random event that is well-defined.
- *Subjectivist.* Probability is an ideal rational agent's degree of belief about an uncertain event. This perspective is also known as Bayesian.

Probability Basics. At its core, probability is a method of expressing someone's belief or direct knowledge about the likelihood of an event occurring, or having occurred. It is expressed as a number between zero and one, inclusive. We use the term probability to always refer to uncertainty—that is, information about events that either have yet to occur or have occurred, but our knowledge of their occurrence is incomplete. In other words, probability refers only to situations that contain uncertainty.

As a common example, we can estimate the probability of rain for a certain area within a specified time frame. Typically referred to as "chance," we commonly hear, "The chance of rain today for your area is 70%." What does that mean? It actually may have different meanings than is commonly interpreted, unless a precise description is given. However, after the day is over, and it indeed rained for a period of time that day, we cannot say that the probability of rain yesterday was 100%. We do not use probability to refer to known events.

Probability has been described by certain axioms and properties. Some basic properties are provided below:

1. The probability of an event, A, occurring is given as a real number between zero and one.

$$P(A) \in [0, 1]$$

2. The probability of an event, A, NOT occurring is represented by several symbols including ~A, ¬A, and A' (among others), and is expressed as

$$P(\sim A) = 1 - P(A).$$

3. The probability of the domain of events occurring (i.e., all possible events) is always

$$P(D) = 1.0.$$

4. The probability of the union of two events, A and B, is given by the equation

$$P(A \cup B) = P(A) + P(B) - P(A \cap B)$$

$$P(A \cup B) = P(A) + P(B), \text{ if } A \text{ and } B \text{ are independent.}$$

This concept is depicted in Figure 9.10.

5. The probability of an event, A, occurring given that another event, B, has occurred is expressed as P(A|B) and is given by the equation

$$P(A \mid B) = \frac{P(A \cap B)}{P(B)}.$$

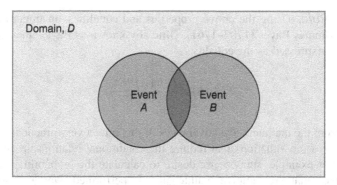

Figure 9.10. Union of two events.

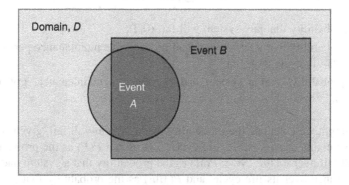

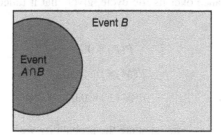

Figure 9.11. Conditional events.

This concept is depicted in Figure 9.11. In essence, the domain is reduced to the event B, and the probability of the event A is only relevant to the domain of B.

6. The probability of the intersection of two events, A and B, is given by the equation

$$P(A \cap B) = P(A \mid B)P(B)$$

$P(A \cap B) = P(A)P(B)$, if A and B are independent.

Bayes' Rule. Using the above properties and equalities, an important rule was derived by Thomas Bayes (1702–1761). Officially known as Bayes' theorem, the rule is commonly expressed as the equality

$$P(A \mid B) = \frac{P(B \mid A)P(A)}{P(B)}.$$

Apart from the mathematical advantages of this rule, a very practical usage of this equality stems from situations that require the conditional relationship among events to reverse. For example, suppose we desire to calculate the probability that a system will fail given that preventative maintenance is performed over a period of time. Unfortunately, we may not have measured data to directly calculate this probability. Suppose that we only have the following probabilities:

- the probability that any system will fail (0.2),
- the probability that a system has had preventative maintenance performed on it over its life cycle (0.4), and
- the probability that a system had preventative maintenance, given it failed (0.02).

How might we calculate the probability that a system will fail, given we perform preventative maintenance over its life cycle? Let us call $P(F)$ as the probability that a system will fail over its life cycle, $P(M)$ as the probability that a system had preventative maintenance over its life cycle, and $P(M \mid F)$ as the probability that a system had preventative maintenance over its life cycle, given that it failed at some point. This is represented as

$$P(F) = 0.2;$$

$$P(M) = 0.3; \text{ and}$$

$$P(M \mid F) = 0.02.$$

We can use Bayes' rule to calculate the probability we seek:

$$P(F \mid M) = \frac{P(M \mid F)P(F)}{P(M)};$$

$$P(F \mid M) = \frac{(0.02)(0.2)}{0.3}; \text{ and}$$

$$P(F \mid M) = 0.013.$$

The probability that our system will fail, given we perform preventative maintenance throughout its life cycle, is very low, 0.013, or almost 20 times lower than the probability of any system failing.

Bayes' rule is a powerful tool for calculating conditional probabilities. But it does have its limitations. Bayes' rule assumes that we have a priori knowledge in order to apply it. In most cases, in engineering and science, we either do have a priori knowledge of the domain or can collect data to estimate it. In our example, the a priori knowledge was the probability that any system would fail, $P(F)$. If we did not have this knowledge, then we could not apply Bayes' rule.

We could collect statistical data on historical system failures to obtain an estimate of $P(F)$. We could also test systems to collect these data. But if the system is new, with new technologies, or new procedures, we may not have sufficient historical data. And applying Bayes' rule would not be possible.

Now that we have reviewed the basics of probability, we are able to survey and discuss a sample of evaluation methods used in systems engineering today.

9.7 EVALUATION METHODS

In the section above, we described a systematic method for performing trade-off analyses. We used a rather simple scheme for evaluating a set of alternatives against a set of weighted selection criteria. In fact, we used a method that is part of a larger mathematical method, known as multiattribute utility theory (MAUT). Other methods exist that allow systems engineers to evaluate a set of alternatives. Some use a form of MAUT incorporating more complex mathematics to increase accuracy or objectivity, while others take an entirely different approach. This section introduces the reader to five types of methods, commonly used in decision support, starting with a discussion of MAUT. Others exist as well, to include linear programming, integer programming, design of experiments, influence diagrams, and Bayesian networks, to name just a few.

This section is simply an introduction of several, selected mathematical methods. References at the end of this chapter provide sources of more detail on any of these methods.

MAUT

This form of mathematics (which falls under operations research) is used quite extensively in all types of engineering, due to its simplicity. It can easily be implemented via a spreadsheet.

As described above, the basic concept involves identifying a set of evaluation criteria with which to select among a set of alternative candidates. We would like to combine the effectiveness values for these criteria into a single metric. However, these criteria do not have similar meanings that allow their integration. For example, suppose we had three selection criteria: reliability, volume, and weight. How do we evaluate the three together? Moreover, we typically need to trade off one attribute for another. So, how much reliability is worth x volume and y weight? In addition, criteria typically have different units. Reliability has no units as it is a probability; volume may use cubic meter and weight may use kilogram. How do we combine these three criteria into a single measure?

MAUT's answer to this dilemma is to use the concept of utility and utility functions. A utility function, $U(m_i)$, translates the selection criterion, m_i, to a unitless measure of utility. This function may be subjective or objective, depending on the data that are available. Typically, utility is measured using a scalar between zero and one, but any range of values will do.

Combining weighted utilities can be accomplished in a number of ways. Three were mentioned above: weighted sum, weighted product, and sum of the logarithms of the weighted utility. Typically, the weighted sum is used, at least as a start. During sensitivity analysis, other methods of combining terms are attempted.

Analytical Hierarchy Process (AHP)

A widely used tool to support decisions in general, and trade studies in particular, is based on the AHP. AHP may be applied using an Excel spreadsheet, or a commercial tool, such as Expert Choice. The latter produces a variety of analyses as well as graphs and charts that can be used to illustrate the findings in the trade study report.

The AHP is based on pairwise comparisons to derive both weighting factors and comparative scores. In deriving criterion-weighting factors, each criterion is compared with every other, and the results are entered into a computation that derives the relative factors. For informal trade-offs, the values obtained by simple prioritization are usually within 10% of those derived by AHP, so the use of the tool is hardly warranted in such cases. On the other hand, for a formal trade study, graphs and charts produced through the use of AHP may lend an appearance of credibility to the presentation.

Weighting factors are calculated using eigenvectors and matrix algebra. Thus, the method has a mathematical basis to it, although the pairwise comparisons are usually subjective, adding uncertainty to the process. The result is a weighting factor distribution among the criteria, summing to one. Figure 9.12 shows the results using the AHP of an example decision to select a new car. Three criteria were used: style, reliability, and fuel economy. After a pairwise comparison among these three criteria, AHP calculated the weights, which sum to one.

Once weighting factors are calculated, a second set of pairwise comparisons is performed. These comparisons are among the alternatives, for each criterion. Two results are provided during this stage of the method. First, the alternatives are evaluated within each criterion individually. Each alternative is provided with a criterion score

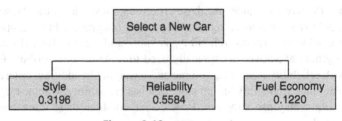

Figure 9.12. AHP example.

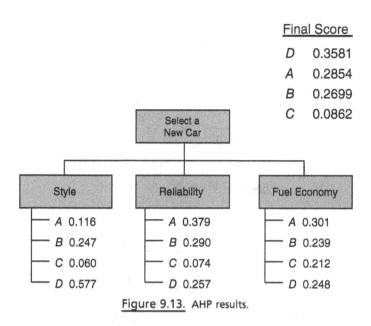

Figure 9.13. AHP results.

between zero and one, with the sum equal to one. Second, the method produces a final score for each alternative across all criteria, between zero and one, with the sum equal to one. Figure 9.13 displays both sets of results—each alternative car (lettered A through D) is given a score for each criterion, and then the scores are combined into a single, final score.

Sensitivity analysis is still needed to check results and to make any changes necessary to arrive at a preferred alternative.

Decision Trees

Decisions were developed to assist decision makers in identifying alternative decision paths and in evaluating and comparing different courses of action. The concept utilizes probability theory to determine the value or utility of alternative decision paths.

As the name suggests, a tree is used to formulate a problem. Typically, two symbols are used—one for decisions and one for events that could occur and are out of the decision maker's control. Figure 9.14 depicts a simple decision tree in which two decisions and two events are included. The decisions are depicted by rectangles and are lettered A and B; the events are depicted by circles and are designated E_1 and E_2. In this example, each decision has two possible choices. Events also have more than one outcome, with probabilities associated with each. Finally, the value of each decision path is shown to the right. A value can be anything that represents the quantitative outcome of a decision path. This includes money, production, sales, profit, wildlife saved, and so on.

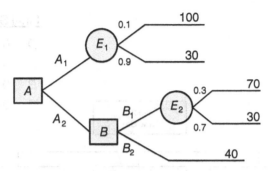

Figure 9.14. Decision tree example.

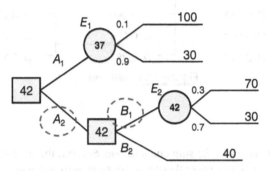

Figure 9.15. Decision path.

In this example, an engineer is faced with an initial decision, A. She has two choices, A_1 and A_2. If she chooses A_2, then an event will occur that provides a value to her of either 100 or 30, with a probability of 0.1 or 0.9, respectively. If she chooses A_2, she is immediately faced with a second decision, B, which also has two choices, B_1 or B_2. Choosing B_2 will result in a value of 40. Choosing B_1 will result in an event, E_2, with two possible outcomes. These outcomes result in values of 70 and 30, with probabilities of 0.3 and 0.7, respectively. Which decision path is the "best?"

The answer to the last question is dependent on the objective(s) of the trade-off study. If the study objective is to maximize the expected value of the decision path, then we can solve the tree using a defined method (which we will not go through in detail here). Basically, an analyst or engineer would start at the values (to the right) and work left. First, calculate the expected value for each event. Then at each decision point, choose the greatest expected value. In our example, calculating the events yield an expected value of 37 for E_1 and 42 for E_2. Thus, decision B is between choosing B_1 and gaining a value of 42, over B_2, with a value of 40. Decision A is now between two expected values: A_1 yields a value of 37, while A_2 yields an expected value of 42. Thus, choosing A_2 yields the greatest expected value.

The decision tree solution is depicted in Figure 9.15.

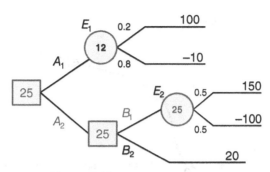

Figure 9.16. Decision tree solved.

Of course, the objective may not be to maximize expected value. It may be to minimize expected loss, or to minimize the maximum loss, or even to maximize value. If the objective was the last of these three, maximum value, then choosing A_1 would be preferred, since only A_1 yields a possibility of achieving a value of 100. Choosing A_2 yields a maximum possible value of only 70. Thus, the objective of the trade-off study determines how to solve the tree.

An alternative method of using decision trees is to add a utility assessment. Basically, instead of using values, we use utilities. The reason we may want to substitute utilities for actual values is to incorporate risk into the equation. Suppose, for example, that we have the decision tree shown in Figure 9.16, already solved to maximize the expected value. However, the customer is extremely risk adverse. In other words, the customer would forego larger profits than lose large amounts of value (in this case, the value could be profits).

We can develop a utility curve that provides a mathematical representation of the customer's risk tolerance. Figure 9.17 provides such a curve. The utility curve reveals the customer is conservative—large profits are great, but large losses are catastrophic. Small gains are good, and small losses are acceptable.

By substituting utilities for value (in this case, profit), we get a new decision tree and a new solution. The conservative nature of the customer, reflected by the utility curve, reveals a conservative decision path: A_2–B_2, which yield a utility of 5, which is a profit of 20. Figure 9.18 provides the new decision tree.

Decision trees are powerful tools for decision makers to make trade-off decisions. They have the advantage of combining decisions that are interdependent. Although the methods we have discussed can also represent this case, the mathematics becomes more complicated. Their disadvantage includes the fact that a priori knowledge of the event probabilities is required. Methods can be combined—each decision in a decision tree can be represented as a formal trade-off study in itself.

Cost–Benefit Analysis (CBA)

If time and resources permit, a more detailed type of trade-off study can be performed than what is described above. These types of studies are often mandated by policy and

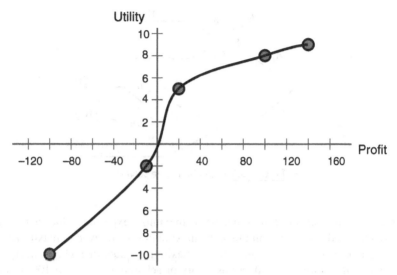

Figure 9.17. Utility function.

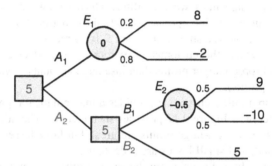

Figure 9.18. Decision tree solved with a utility function.

are known as an AoA. In many of these situations, the straightforward trade-off study methodology of the last section is not sufficient. Detailed analysis using models, and high-fidelity simulations, are typically required to measure the alternative systems' effectiveness. In these cases, a CBA is warranted.

The basic concept of the CBA is to measure the effectiveness and estimate the cost of each alternative. These two metrics are then combined in such a way as to shed light on their cost-effectiveness, or put another way, their effectiveness per unit cost. More often than not, the effectiveness of an alternative is a multidimensional metric, and cost is typically divided into its major components: development, procurement, and operations (which include maintenance). In some cases (such as with nuclear reactors), retirement and disposal costs are included.

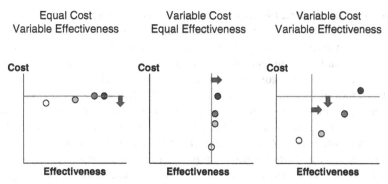

Figure 9.19. Example of cost-effectiveness integration.

Combining cost and effectiveness results is crucial in offering decision makers the information they need to make informed decisions. Three basic types of cost-effectiveness analyses exist, each offering advantages. Figure 9.19 illustrates the three types for a single-dimensional analysis.

Equal Cost–Variable Effectiveness. This type constrains the alternatives to a single cost level or a maximum cost threshold. If all of the alternatives are constrained to similar or the same costs, then the results offer an observable difference in effectiveness—enabling a simple ranking of alternatives. In essence, cost is taken out of the equation in comparing alternative systems.

The disadvantages of this CBA include the difficulty in constraining the alternatives to the same, or a maximum cost. Examples include selecting a system within a cost range, such as selecting a new car or purchasing equipment. Of course, one could argue that decisions such as these do not need detailed analysis—a straightforward trade-off study would be sufficient! More detailed examples include a new strike weapon system for the military. A maximum cost level is typically included in a new system's requirements, including its key performance parameter (KPP). All alternatives are required to be less than the cost threshold. Only effectiveness of these system alternatives varies.

Variable Cost–Equal Effectiveness. This type constrains the alternatives to a single effectiveness level or a minimum effectiveness threshold. If all of the alternatives are constrained to similar or the same effectiveness levels, then the results offer an observable difference in cost, enabling a simple ranking of alternatives. In essence, effectiveness is taken out of the equation in comparing alternative systems.

The disadvantages of this CBA include the difficulty in constraining the alternatives to the same or minimum effectiveness level. Examples include selecting a power plant to provide a selected amount of energy. In this case, the energy level, or amount of electricity, would be the minimum effectiveness threshold. Options would then be judged largely on cost.

Variable Cost–Variable Effectiveness. This type constrains the alternatives to both a maximum cost level and a minimum effectiveness level. However, beyond the limits, the alternatives can be any combination of cost and effectiveness. In some cases, no limits are established, and the alternatives are "free" to be at any cost and effectiveness levels. This is rare for government CBAs, but there can be advantages to this form of analysis. Out-of-the-box alternatives can be explored when cost and effectiveness constraints are removed. In some cases, a possible alternative that provides effectiveness that is just under the minimum (say, 5% under the threshold) may cost 50% less than any other alternative. Would not a decision maker at least want to be informed of that possibility? By and large, however, minimum and maximum levels are established to keep the number of alternatives manageable, with the exceptional case being handled separately.

The disadvantages of this CBA type include the risk that no alternative is clearly "better" than the rest. Each alternative offers effectiveness that is commensurate with its cost. Of course, this is not necessarily bad; the decision maker then must decide which alternative he wants. In these cases, calculating the effectiveness per unit cost is an additional measure that can shed light on the decision.

Most systems fall into this category: a new vehicle design, a new spaceship or satellite, a new software system, a new energy system, and so on.

Of course, the examples and notional Figure 9.19 all address single-dimensional applications. Multidimensional costs and effectiveness increase the complexity but still fall into one of the three types of CBA. Two general methods for handling multidimensional CBA are (1) combining effectiveness and cost into a single metric, typically by employing MAUT, then applying one of the three methods described; or (2) using an effectiveness and cost profile vector, with mathematical constraints on the vector as opposed to a single scalar threshold.

Quality Function Deployment (QFD)

QFD originated in Japan during the 1960s as a quality improvement program. Dr. Yoji Akao pioneered the modern version of QFD in 1972 with his article in the journal *Standardization and Quality Control*, followed by a book describing the process in 1978. Ford Motor Company brought the process to America by adopting it in the 1980s. By the 1990s, some agencies within the U.S. government had adopted the process as well.

At the heart of the process is the QFD matrix, known as the house of quality. Figure 9.20 depicts the general form of this tool, which consists of six elements. More complex forms of the QFD house of quality are also available but are not presented here. The basic use of QFD is in the design process—keeping design engineers, manufacturers, and marketers focused on customer requirements and priorities. It has also been used in decision making.

QFD is an excellent tool for developing design objectives that satisfy key customer priorities. It has also been used in trade studies as a method for developing selection criteria and weightings. The output of the house of quality process and analy-

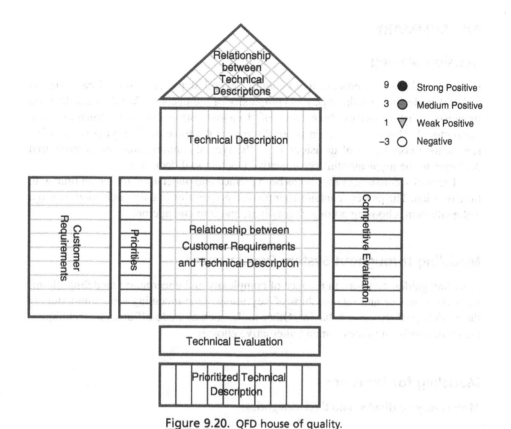

Figure 9.20. QFD house of quality.

sis is a technical evaluation of alternative subsystems and the relative importance and technical difficulty of developing and manufacturing each component in the technical description. This output is at the bottom of the figure. This evaluation is accomplished by comparing prioritized customer requirements with technical component options and by determining the characteristics of their relationships. Generally, a subset of relationship types, or strengths, is determined. In the figure, four distinct relationships are given: strong, medium, weak, and negative. Additionally, each technical component is compared against other components using the same relationship scale (represented by the triangle at the top or roof of the house). The mathematics (which are not described here but are based on matrix algebra) are then used to determine the technical evaluation.

QFD is typically used in conjunction with trade studies—either to generate inputs to a formal trade study or to conduct the trade studies as part of a design development effort.

9.8 SUMMARY

Decision Making

Decision making is a process that contains several steps. How formal each step is undertaken depends on the type and complexity of the decision. We define a decision framework that examines three types of decisions: structured, semistructured, and unstructured. This categorization is not discrete as the three distinct types suggest but represents a continuum of decisions from the typical/common/understood structured decisions to the atypical/intuitive/subjective unstructured decisions.

The decision-making process has been defined and understood for a long time with little revision. The process contains four phases: preparation and research, model design and evaluation, choosing among alternatives, and implementation.

Modeling throughout System Development

Modeling guides decisions in the face of complexity and uncertainty; modeling illuminates the behavior and relationships of key issues. One modeling tool, simulation, is the modeling of dynamic behavior. Other tools, such as trade-off analysis techniques, model the decision process among alternative choices.

Modeling for Decisions

Models may be divided into three categories.

1. *Schematic Models* use diagrams to represent system elements or processes. An architect's sketches, such as floor layouts, are examples of schematic models. System block diagrams model system organizations. They are often arranged in a treelike structure to represent hierarchical organizations, or they use simple rectangular boxes to represent physical or other elements.

   System context diagrams show all external entities that interact with the system, where the system is represented as a "black box" (not showing internal structure). The diagram describes the system's interactions with its environment.

   FFBDs model functional interactions, where functional elements are represented by rectangles, and arrows represent interactions and flow of information, material, or energy between elements. The names of the elements begin with a verb, denoting action. Examples and extensions of FFBDs include system life cycle models, IDEF0 diagrams, and F^2D^2.

   FFPDs are similar—they form a hierarchical description of a complex process. They also interrelate process design with requirements and specifications.

   The diagrams defined by UML and Systems Modeling Language (SysML) are examples of schematic models (see Chapter 8).

2. *Mathematical Models* use mathematical notation to represent relationships. They are important aids to system development and can be useful both for design and systems engineering. They also perform sanity checks on results of complex analyses and simulations.
3. *Physical Models* are physical representations of systems or system elements. They are extensively used in system design and testing, and include test models, mock-ups, and prototypes.

Simulation

System simulations deal with the dynamic behavior of systems and system elements and are used in every phase of system development. Management of simulation effort is a systems engineering responsibility.

Computer "war games" are an example of operational simulations, which involve a simulated adversarial system operated by two teams of players. They are used to assess the operational effectiveness of tactics and system variants.

System effectiveness simulations assess alternative system architectures and are used during conceptual development to make comparative evaluations. The design of effectiveness simulations is itself a complex systems engineering task. Developing complex simulations such as these must seek a balance between fidelity and cost since such simulations can be systems in their own right. Scope must be controlled to obtain effective and timely results.

Physical or physics-based simulations are used in the design of high-performance vehicles and other dynamic systems, and they can save enormous amounts of development time and cost.

Hardware-in-the-loop simulations include hardware components coupled to computer-driven mechanisms. They are a form of physical simulation, modeling dynamic operational environments.

Environmental simulations subject systems and system elements to stressful conditions . They generate synthetic system environments that test systems' conformance to operational requirements.

Finally, computer-based engineering tools greatly facilitate circuit design, structural analysis, and other engineering functions.

Trade-Off Analysis

Trade-off processes are involved consciously or subconsciously in every decision we make (personally as well as professionally). An important issue with respect to trade studies is the stimulation of alternatives. Trades ultimately select the "best" course of action from two or more alternatives. Major decisions (which are typical within systems engineering) require formal trade-off analysis.

A trade-off, formal or informal, consists of the following steps:

1. Define the objective.
2. Identify qualified alternative candidates.

3. Define selection criteria in the form of MOE.
4. Assign weights to selection criteria in terms of their importance to the decision.
5. Identify or develop a value rating for each criterion.
6. Calculate or collect comparative scores for each alternative's criterion; combine the evaluations for each alternative.
7. Analyze the basis and robustness of the results.

Revise findings if necessary and reject any alternatives that fail to meet an essential requirement. For example, delete MOEs that do not discriminate significantly among alternatives. Limit the value of assignments to the least accurate quantity and examine the total "profile" of scores of the individual candidates.

Trade-off studies and analyses are aids to decision making—they are not infallible formulae for success. Numerical results produce an exaggerated impression of accuracy and credibility. Finally, if the apparent winner is not decisively superior, further analysis is necessary.

Review of Probability

At its core, probability is a method of expressing someone's belief or direct knowledge about the likelihood of an event occurring or having occurred. It is expressed as a number between zero and one, inclusive. We use the term probability to always refer to uncertainty—that is, information about events that either have yet to occur or have occurred, but our knowledge of their occurrence is incomplete.

Evaluation Methods

As systems engineering is confronted with complex decisions about uncertain outcomes, it has a collection of tools and techniques that can be useful support aids. We present five such tools:

1. *MAUT* uses a utility function to translate a selection criterion to a unitless utility value, which can then be combined with other utility functions to derive a total value score for each alternative.
2. *AHP* is a mathematically based technique that uses pairwise comparisons of criteria and alternatives to general weightings and combines utility scores for alternatives.
3. *Decision Trees* are graphical networks that represent decision choices. Each choice can be assigned a value and an uncertainty measure (in terms of probabilities) to determine expected values of alternative decision paths.
4. *CBA* is a method typically used with modeling and simulation to calculate the effectiveness or a system alternative per unit cost.

5. *QFD* defines a matrix (the house of quality) that incorporates relationships between customers' needs, system specifications, system components, and component importance to overall design. The matrix can be solved to generate quantitative evaluations of system alternatives.

PROBLEMS

9.1 Suppose you needed to make a decision regarding which engine type to use in a new automobile. Using the process in Figure 9.1, describe the five steps in deciding on a new engine type for an advanced automobile.

9.2 Identify the stakeholders for the following decisions:
   (a) the design of a traffic light at a new intersection,
   (b) the design of a new weather satellite,
   (c) the choice of a communications subsystem on a new mid-ocean buoy designed to measure ocean temperature at various depths,
   (d) the choice of a security subsystem for a new power plant, and
   (e) the design of a new enterprise management system for a major company.

9.3 Give two examples of each decision type: structured, semistructured, and unstructured.

9.4 Write an essay describing the purpose of each type of model: schematic, mathematical, and physical. What are their advantages?

9.5 Develop a context diagram for a new border security system. This system would be intended to protect the land border between two countries.

9.6 In an essay, compare and contrast the three types of functional diagrams: functional block diagram, functional flow diagram, and IDEF0. A table that lists the characteristics of each of the three would be a good start to this problem.

9.7 Describe three examples of problems or systems where gaming would be useful in their development and ultimate design.

9.8 Perform a trade study on choosing a new car. Identify four alternatives, between three and five criteria, and collect the necessary information required.

9.9 To illustrate some important issues in conducting trade studies, consider the following simplified example. The trade study involved six alternative system concepts. Five MOEs were used, each weighted equally. For simplicity's sake, I have titled the MOEs *A*, *B*, *C*, *D*, and *E*. After assigning values to each MOE of the six alternatives, the results were the following:

Note that two stood out well above the rest, both receiving the same total number of points:

Weighted MOE	A	B	C	D	E	Total
Concept I	1	1	5	4	2	13
Concept II	3	3	2	5	4	17
Concept III	4	1	5	5	5	20
Concept IV	2	2	3	5	1	13
Concept V	4	4	4	4	4	20
Concept VI	1	1	1	3	3	9

On the basis of the above rating profiles,

(a) Would you conclude that concept III to be superior, equal, or inferior to concept V?

Explain your answer.

(b) If you were not entirely satisfied with this result, what further information would you try to obtain?

(c) Discuss potential opportunities for further study that might lead to a clearer recommendation between concepts III and V.

9.10 Supposed that you are looking to purchase a new vacuum cleaner, and you have decided to conduct a trade study to assist you in your decision. Conduct product research and narrow down your choices to five products.

Please conduct the following steps:

(a) Identify exactly *four* selection criteria, not including purchase price or operating cost.

(b) Assign weights to each criterion, explaining in one sentence your rationale.

(c) Construct a utility function for each criterion—describe it verbally or graphically.

(d) Research the actual values for your criteria for each alternative.

(e) Perform the analysis, calculating a weighted sum for each alternative.

(f) Calculate the effectiveness/unit cost for each alternative using purchase price for cost.

(g) Describe your choice for purchase, along with any rationale.

FURTHER READING

R. Clemen and T. Reilly. *Making Hard Decisions with DecisionTools Suite*. Duxbury Press, 2010.

Defense Acquisition University. *Systems Engineering Fundamentals*. DAU Press, 2001, Chapter 12.

G. M. Marakas. *Decision Support Systems*. Prentice Hall, 2001.

C. Ragsdale. *Spreadsheet Modeling and Decision Analysis: A Practical Introduction to Management Science*. South-Western College Publishing, 2007.

A. P. Sage. *Decision Support Systems Engineering*. John Wiley & Sons, Inc., 1991.

H. Simon. *Administrative Behavior*, Third Edition. New York: The Free Press, 1976.

R. H. Sprague and H. J. Watson. *Decision Support Systems: Putting Theory into Practice*. Prentice Hall, 1993.

E. Turban, R. Sharda, and D. Delan. *Decision Support Systems and Intelligence Systems*, Ninth Edition. Prentice Hall, 2010.

PART III

ENGINEERING DEVELOPMENT STAGE

Part III is concerned with the implementation of the system concept into hardware and software components, their integration into a total system, and the validation of the systems operational capability through a process of developmental and operational testing. Systems engineering plays a decisive part in these activities in the form of analysis, oversight, and problem solving.

A critical application of systems engineering is to identify and reduce potential difficulties inherent in the use of unproven components based on new technology, highly stressed system elements, and other sources of risk. This subject is discussed in detail in Chapter 10, which describes typical sources of potential risk, the use of prototype development, and the process of validation testing and analysis. The identification, prioritization, and reduction of program risks is a vital contribution of systems engineering.

Chapter 11 introduces the special and unique features of software systems engineering and highlights differences between hardware and software development. Common life cycle models are introduced for software-intensive systems, and the primary steps for developing software functionality are discussed.

Systems Engineering Principles and Practice, Second Edition. Alexander Kossiakoff, William N. Sweet, Samuel J. Seymour, and Steven M. Biemer
© 2011 by John Wiley & Sons, Inc. Published 2011 by John Wiley & Sons, Inc.

The engineering design phase is concerned with the implementation of the system architectural units into engineered components that are producible, reliable, maintainable, and can be integrated into a system meeting performance requirements. The systems engineering responsibilities are to oversee and guide this process, to supervise the configuration management function, and to resolve problems that inevitably arise in this process. Chapter 12, Engineering Design, deals with these issues.

The engineered system components are integrated into a fully operational system and are evaluated in the integration and evaluation phase of the life cycle. Thorough systems engineering planning is necessary to organize and execute this process efficiently, with the best practical combination of realism and economy of time and resources. Chapter 13 describes the elements of the successful accomplishment of the integration and evaluation processes, which qualifies the system for production and operational use.

10

ADVANCED DEVELOPMENT

10.1 REDUCING PROGRAM RISKS

The advanced development phase is that part of the system development cycle in which the great majority of the uncertainties inherent in the selected system concept are resolved through analysis, simulation, development, and prototyping. The principal purpose of the advanced development phase is to reduce the potential risks in the development of a new complex system to a level where the functional design of all previously unproven subsystems and components has been validated. At its conclusion, the risks of discovering serious problems must be sufficiently low that full-scale engineering may be begun with confidence. This phase's primary objectives are to develop, where necessary, and validate a sound technical approach to the system design and to demonstrate it to those who must authorize the full-scale development of the system.

The general methodology of accomplishing risk reduction is discussed in Chapter 5 in the section on risk management. The components of risk management are described as risk assessment, in which risks are identified and their magnitude assessed, and risk mitigation, in which the potential damage to the development is eliminated or reduced.

Systems Engineering Principles and Practice, Second Edition. Alexander Kossiakoff, William N. Sweet, Samuel J. Seymour, and Steven M. Biemer
© 2011 by John Wiley & Sons, Inc. Published 2011 by John Wiley & Sons, Inc.

This chapter is concerned with typical sources of risks encountered in the early phases of developments of complex systems and the methods for their mitigation.

To accomplish the above objectives, the degree of definition of the system design and its description must be advanced from a system functional design to a physical system configuration consisting of proven components coupled with a design specification, to serve as the basis for the full-scale engineering of the system. In most new complex systems, this calls for mature designs of the subsystems and components. All ambiguities in the initial system requirements must be eliminated, and often, some of the more optimistic design goals of the original concept of the system must be significantly curtailed.

It should be noted that all new system developments do not have to go through a formal advanced development phase. If all major subsystems are directly derivable from proven predecessor or otherwise mature subsystems, and their characteristics can be reliably predicted, then the system development can proceed on to the engineering design phase. Such is the case with most new model automobiles, in which the great majority of components are directly related to those of previous models. In that case, such critical items as the airbag system or pollution control may be individually built and tested in parallel with the engineering of the new model.

Place of the Advanced Development Phase in the System Life Cycle

The advanced development phase marks the transition of the system development from the concept development to the engineering development stage. As seen in Figure 10.1, it follows the concept definition phase, from which it derives the inputs of system functional specifications and a defined system concept. Its outputs to the engineering design phase are system design specifications and a validated development model. It thus converts the requirements of *what* the system is to do and a conceptual approach of its configuration into a specification of generally *how* the required functions are to

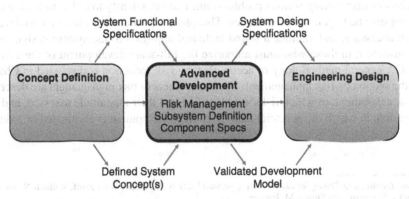

Figure 10.1. Advanced development phase in system life cycle.

be implemented in hardware and software. Other required outputs, not shown in the figure, include an updated work breakdown structure (WBS), a revised systems engineering management plan (SEMP) or its equivalent, and related planning documents. Additionally, the system architecture is updated to reflect changes to date.

As noted above, this phase is especially critical in the development of complex systems that involve extensive use of advanced technology and/or novel unproven concepts. It may require several years of intensive development effort before the new features are sufficiently mature and well demonstrated to warrant initiating full-scale engineering. In addition, it may also be necessary to develop new manufacturing processes to support the proposed new technology. In such cases, the advanced development phase is frequently contracted separately from the follow-on engineering.

At the other end of the spectrum, those systems that do not involve major technological advances over similar previous systems, and hence require only a minor amount of development, may not have a separately defined and managed advanced development phase. Instead, the corresponding work may be included in the front end of the engineering design phase. However, the tasks embodied in the translation of the system functional requirements into a system implementation concept and system-level design specifications must still be accomplished prior to undertaking detailed engineering.

Design Materialization Status

Table 10.1 depicts the system materialization status during the advanced development phase. It is seen that the principal change in the system status is designated as "validation"—validation of the soundness of the selected concept, validation of its partitioning into components, and validation of the functional allocation to the component and subcomponent levels. The focus of development in this phase is thus the definition of how the components will be built to implement their assigned functions. The manner in which these tasks are accomplished is the subject of this chapter.

Systems Engineering Method in Advanced Development

The organization of this chapter is arranged according to the four steps of the systems engineering method (see Chapter 4) followed by a brief section that discusses risk reduction, a methodology used throughout system development, but is especially important in this phase. The principal activities during this phase in each of the four steps in the systems engineering method, as applied to those subsystems and components requiring development, are briefly summarized below and are illustrated in Figure 10.2.

Requirements Analysis. Typical activities include

- analyzing the system functional specifications with regard to both their derivation from operational and performance requirements and the validity of their translation into subsystem and component functional requirements and
- identifying components requiring development.

TABLE 10.1. Status of System Materialization at the Advanced Development Phase

Phase	Concept development			Engineering development		
Level	Needs analysis	Concept exploration	Concept definition	Advanced development	Engineering design	Integration and evaluation
System	Define system capabilities and effectiveness	Identify, explore, and synthesize concepts	Define selected concept with specifications	Validate concept		Test and evaluate
Subsystem		Define requirements and ensure feasibility	Define functional and physical architecture	Validate subsystems		Integrate and test
Component			Allocate functions to components	Define specifications	Design and test	Integrate and test
Subcomponent		Visualize		Allocate functions to subcomponents	Design	
Part					Make or buy	

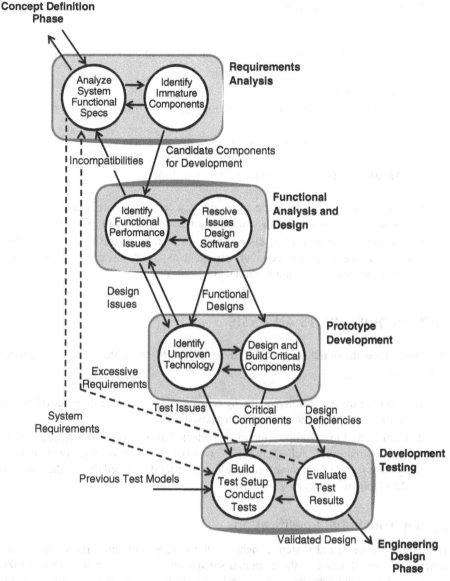

Figure 10.2. Advanced development phase flow diagram.

Functional Analysis and Design. Typical activities include

- analyzing the allocation of functions to components and subcomponents, and identifying analogous functional elements in other systems; and
- performing analyses and simulations to resolve outstanding performance issues.

Prototype Development. Typical activities include

- identifying issues of physical implementation involving unproven technology and determining the level of analysis, development, and test required to reduce risks to acceptable values;
- designing critical software programs;
- designing, developing, and building prototypes of critical components and subsystems; and
- correcting deficiencies fed back from test and evaluation.

Development Testing. Typical activities include

- creating test plans and criteria for evaluating critical elements, and developing, purchasing, and reserving special test equipment and facilities; and
- conducting tests of critical components, evaluating results, and feeding back design deficiencies or excessively stringent requirements as necessary for correction, leading to a mature, validated system design.

10.2   REQUIREMENTS ANALYSIS

As stated above, the initial effort in the advanced development phase is mainly devoted to two areas:

1. reexamining the validity of the system functional specifications developed in or following the concept definition phase and
2. identifying those components of the selected system concept that are not sufficiently mature for full-scale engineering (i.e., have not been proven in existing systems), and which therefore should be further developed during the advanced development phase.

System Functional Specifications

In defining the preferred system concept in the concept definition phase, the system functions were allocated to the principal subsystems, and these were further broken down into functional elements. These functional design concepts were then embodied in the system specifications document prepared as an input to the advanced development phase.

The analysis of these specifications should take into account the circumstances under which the concept definition phase took place. If, as is frequently the case, it was performed in the space of a few months and with limited funding, and especially if it was done in a competitive environment, then the results should be viewed as preliminary and subject to modification, and must be analyzed very thoroughly. Prior design decisions must be viewed with some skepticism until they are examined and demonstrated to be well founded. This does not mean that the selected technical approaches should necessarily be changed, only that they should not be accepted without understanding their derivation.

Requirements Derivation

The key to understanding the significance and sensitivity of system functional specifications is to trace them back to their derivation from the system performance requirements. Such an understanding is essential to making the design decisions required for the physical implementation of the functions in hardware and software.

The system life cycle support scenario should be revisited to identify functions necessary to sustain the different circumstances to which the system will be exposed during its preoperational as well as its operational life. In addition, the requirements for compatibility; reliability, maintainability, availability (RMA); and environmental susceptibility should be examined, as well as those for operational performance. At this time, specifications concerning human–system interface issues and safety are incorporated into subsystem and component specifications.

As stated previously, some requirements are frequently unstated, and others are immeasurable. For example, affordability and system growth potential are frequently not explicitly addressed. User interface requirements are often qualitative and are not susceptible to measurement. The relation of each of the above issues to the functional design needs to be understood and documented.

Relation to Operational Requirements

If some system specifications cannot be readily met, it is necessary to gain an even deeper understanding of their validity by tracing them back one step further, namely, to their relationship to the execution of the system's mission, that is, to the system operational requirements. This relationship is often lost in the early phases of system definition and needs to be recaptured to provide the systems engineer with an informed rationale for dealing with problems that invariably arise during development.

One of the means for gaining such understanding, as well as for obtaining an appreciation of operational factors beyond those formally stated, is to develop contacts with prospective system users. Such contacts are not always available, but when they are, they can prove to be extremely valuable. Organizations that specialize in operational analyses and those that conduct system field evaluations are also valuable sources in many system areas. Involving the user as a team member during development should be considered where appropriate.

Relation to Predecessor Systems

If the new system has a predecessor that fulfills a similar function, as is usually the case, it is important to fully understand the areas of similarity and difference, and how and why the new requirements differ from the old. This includes the understanding of the perceived deficiencies of the predecessor system and how the new system is intended to eliminate them.

The degree of benefit to be gained from this comparison, of course, depends on the accessibility of key persons and records from the predecessor system development. However, at a minimum, the comparison should provide added confidence in the chosen approach or suggest alternatives to be explored. Where key participants in the development are accessible, their advice with regard to potential problems and lessons learned can be invaluable.

Identification of Components Requiring Development

The principal purpose of the advanced development phase has been stated as ensuring that all components of the system have been demonstrated to be ready for full-scale engineering. This means that component design is sound and capable of being implemented without significant risk of functional or physical deficiencies that would require different approaches to satisfy the requirements.

The above statement implies that all system components must be brought up to a level of maturity where all significant design issues have been resolved. The process that raises the level of maturity is called "development," and therefore the advanced development phase consists largely of development effort focused on those system components that have not previously been brought to the necessary level of proven performance. This, in turn, means that all components that are determined to be insufficiently mature for full-scale engineering would be further developed and their design validated. Those components that are deemed to be sufficiently mature that they do not require development still need to be validated through analysis or test prior to their acceptance for engineering.

Assessment of Component Maturity. The determination of whether or not a given component is sufficiently proven for full-scale engineering can only be made by comparing the component with analogous components that have been successfully engineered and produced. If no proven analogous component is similar to the new one, the comparison may often be made in two parts, functional and physical, by asking the following questions:

1. Are there proven components that have very similar functionality and performance characteristics? Where significant differences exist, are they within the demonstrated performance boundaries of this type of component?

2. Are there existing components whose physical construction uses similar materials and architectures? Are the projected stresses, tolerances, safety, and lifetime characteristics within the demonstrated limits of similar existing components?

If both of these questions are answered in the affirmative, a case may be made that development is not necessary. However, an additional critical question is whether or not the functional interactions and physical interfaces of the components with their operational environment are understood well enough not to require development and experimentation. The answer to this question depends on whether the differences between the proposed component design and those previously proven are reliably predictable from known engineering relationships, or whether the relationship is too remote or complicated to be predicted with assurance. A common example of the latter case has to do with human–machine interfaces, which are seldom well enough understood to obviate the need for experimental verification.

Risk Analysis. After identifying those system elements that require further development, the next step is to determine the appropriate nature and extent of such development. This is where systems engineering knowledge and judgment are especially important because these decisions involve a careful balance between the cost of a thorough development effort on one hand and the risks inherent in insufficient development and consequent residual uncertainties on the other. Reference to the application of risk assessment to system development is contained in the paragraphs below, and this methodology is enlarged upon in a separate section at the end of this chapter.

Development Planning. It is clear from the above discussion that the planning of the advanced development phase should be based on a component-by-component assessment of the maturity of the proposed system design to define (1) the specific character of the unproven design features and (2) the type of analysis, development, and test activities required to resolve the residual issues. In most new systems, the uncertainties are concentrated in a limited number of critical areas, so that the development effort can be focused on those components deficient in design maturity.

Risk Reduction Budget. The result of the above analysis of risks and definition of appropriate risk reduction efforts should be incorporated into a detailed development plan to guide the analysis, development, and testing effort of the advanced development phase. In doing so, an essential step is to revise carefully the relative allocation of effort to the individual components or subsystems that are planned for development. Do the relative allocations correspond to an appropriate balance from the standpoint of a potential gain to investment ratio? Is each allocation adequate to acquire the needed data? If, as is often the case, the available resources do not cover all the proposed effort, it is usually better to replace some of the most risky components with more conservative choices than to fail to validate their use in the system. Thus, the risk reduction/ development plan should contain a risk mitigation budget broken down into the significant individual development efforts.

Example: Unproven Components. Table 10.2 illustrates the above considerations by listing several representative examples of hypothetical unproven components that use new functional or physical design approaches or new production methods. The first column indicates the relative maturity of the functional, physical, and production characteristics of the design approach. The second column is a bar graph representing the maturity of these three characteristics (names abbreviated) by the relative heights of the three vertical bars. The third column shows the type of development that is usually appropriate to resolving the resulting issues of each of the new designs. The

TABLE 10.2. Development of New Components

Design approach	Maturity	Development	Validation

New function
  Proven physical medium
  and production method
	(Func low, Phys mid, Prod high; scale 0–1)	Design, build, and test rapid prototype	Functional performance

New implementation
  Proven function and
  production method
	(Func high, Phys low, Prod high; scale 0–1)	Design, build, and test rapid development model	Engineering design

New production method
  Proven function and
  implementation
	(Func high, Phys high, Prod low; scale 0–1)	Perform critical experiments on the production method	Production method

Extended function Proven component	(Func mid, Phys mid, Prod high; scale 0–1)	Design and run functional simulation	Functional performance

fourth column lists the particular characteristic to be validated. These examples are, of course, very much simplified compared to the factors that must be considered for an actual complex system, but they indicate the component-by-component analysis and planning that is associated with the advanced development process. The table illustrates that components in a new system may have a variety of different types of unproven features, each requiring a development approach tailored to its specific character. The decisions as to the choice of development strategy are the primary responsibility of systems engineering. The subsequent three sections describe the application of each of the remaining steps of the systems engineering method to the resolution of the above design issues.

Example: Natural Gas-Powered Automobile. The development of an automobile that uses natural gas as a fuel in place of gasoline offers an example of some of the principles discussed above. This development has the dual objective of conforming to future strict auto pollution standards while at the same time preserving all the desirable characteristics of conventional modern automobiles, including affordability. Thus, it seeks to minimize the required changes in standard auto design by limiting them to the fuel system and its immediate interfaces. Other changes to the body, engine, and other components are kept to a minimum.

The changes to the fuel subsystem, however, are considerable and also impact the design of the rear section of the body. Storing a sufficient amount of natural gas to obtain the desired travel distance between refueling, and also keeping the volume small enough to have adequate trunk space, requires gas storage pressures higher than those used in conventional storage cylinders. To minimize weight, fiber-wrapped composites are used in place of steel. To maximize safety, the container design consists of a cluster of cylinders, anchored to the frame so as to withstand severe rear-end impacts.

This example falls in the third category in Table 10.2. The physical construction of the fuel container is a major departure from conventional containers in its physical design and materials. Furthermore, the determination of its safety from explosion in case of a collision is not derivable from engineering data but must be established by experiment. The fuel control and refill provisions will also be new designs. Thus, a substantial development effort will have to be undertaken to validate the design and probably will involve comparative tests of several design variations.

The components that interface directly with the fuel subsystem, such as the engine and the rear body structure, especially the trunk and suspension, will also need to be tested in conjunction with the fuel container. Components not associated with this system element will not require development but must be examined to ensure that significant interactions are not overlooked.

The above example illustrates a common case of a new system that differs from its predecessor in a major way, but one that is restricted to a few components.

10.3 FUNCTIONAL ANALYSIS AND DESIGN

Because of the rapid advance of modern technology, a new system that is to be developed to replace an obsolescent current system will inevitably have performance requirements well beyond those of its predecessor. Moreover, in order for the new system to have a long, useful operational life in the face of further projected increases in the capability of competitive or opposing systems, the requirements will specify that its performance more than meets current needs. While the concept definition phase should have eliminated excessively risky approaches, these requirements will necessitate the application of advanced development and therefore development of some advanced system elements.

The increase in system performance frequently requires a significant increase in component complexity, as in many of today's automated computer-based systems. The means for achieving such projected extensions are often not reliably predictable by analytical or simulation methods and have to be determined experimentally. System elements involving dynamic behavior with feedback may be analyzed through simulation but usually require the construction and testing of experimental models to establish a firm basis for engineering.

A common instance where system functions may require development is where the user needs and the environment are not well understood, as is often the case with decision support and other complex automated systems. In such instances, the only sound approach (especially if user interfaces are concerned) is to build prototype components corresponding to the critical system elements and to test their suitability by experimentation.

In summary, three types of components that frequently require development are

1. components required to have extended functional performance beyond previously demonstrated limits,
2. components required to perform highly complex functions, and
3. components whose interactions with their environment are imperfectly understood.

Each of these is described in greater detail in the succeeding paragraphs.

Extended Functional Performance

The identification of system elements (components or subsystems) whose required performance may exceed demonstrated limits can be illustrated by reference to the set of functional system building blocks discussed in Chapter 3. Table 3.2 lists 23 basic functional elements grouped into four classes: signal, data, material, and energy. Each functional element has a number of key characteristics that define its functional capability. Most of these characteristics have limits established by the physical properties of their implementing technologies and often by the basic interdependence between functions (e.g., accuracy vs. speed). A functional requirement for a new system that poses demands on a system element beyond its previously demonstrated limits signals the potential need for either a component development effort or a reallocation of the requirement.

To illustrate this type of comparison, Table 10.3 lists the functional elements along with some of the characteristics that most often turn out to be critical in new systems. The table represents the application of the systems engineering approach to the analysis of system functional requirements and the identification of development objectives.

In using system building blocks to identify functional elements requiring development, the first step is to relate each system element to its functionally equivalent generic element and then to compare the required performance with that of corresponding physical components whose capabilities have been demonstrated as a part of existing systems.

Given an approximate correspondence, the next step is to see whether the differences between the required and existing elements can be compared quantitatively by established engineering relations so as to make a convincing case that the new element can be engineered with confidence, on the basis of proven performance and straightforward engineering practice. When such a case cannot be made, it is necessary either to reduce the specified performance requirement to a level where it can be so adapted or to plan a development and test program to obtain the necessary engineering data.

The process of identifying elements requiring development is often part of the process of "risk identification" or "risk assessment." Risk assessment considers the likely effect of a given decision, in this case, the choice of a particular technical

TABLE 10.3. Selected Critical Characteristics of System Functional Elements

Functional elements	Critical characteristics
Input signal	Fidelity and speed
Transmit signal	High-power, complex waveform
Transduce signal	Gain, beam pattern, and multielement
Receive signal	Sensitivity and dynamic range
Process signal	Capacity, accuracy, and speed
Output signal	Resolution and versatility
Input data	Fidelity and speed
Process data	Versatility and speed
Control data	User adaptability and versatility
Control processing	Architecture, logic, and complexity
Store data	Capacity and access speed
Output data	Versatility
Display data	Resolution
Support material	Strength and versatility
Store material	Capacity and input/output capability
React material	Capacity and controls
Form material	Capacity, accuracy, and speed
Join material	Capacity, accuracy, and speed
Control position	Capacity, accuracy, and speed
Generate thrust	Power, efficiency, and safety
Generate torque	Power, efficiency, and control
Generate electricity	Power, efficiency, and control
Control temperature	Capacity and range
Control motion	Capacity, accuracy, and response time

approach, on the success or failure of the overall objective. Thus, the utilization of unproven system components involves a degree of risk depending on the likelihood that the system will fail to meet its design goals. If the risk is considerable, as when the element is both unproven and critical to the overall system operation, then the element must be developed to a point where its performance may be demonstrated and validated (i.e., low risk). The subject of risk management is discussed in Chapter 5 and is encountered in all phases of the system life cycle.

Highly Complex Components

Consideration of the functional building blocks as system architectural components is also useful in identifying highly complex functions. Equally important is to identify complex interfaces and interactions because elements of even moderate complexity may interact with one another in complicated ways. Interfaces are especially important because complexities internal to elements are likely to be detected and resolved during design, while problems resulting from interface complexities may not reveal themselves until integration testing, at which time changes required to make them operate properly are likely to be very costly in time and effort. The existence of excessively complicated interfaces is a sign of inappropriate system partitioning and is the particular responsibility of the systems engineer to discover and to resolve. This concern is particularly important when several organizations are involved in the system development.

Specialized Software. Certain customized software components are inherently complex and hence are sources of program risk, and should be treated accordingly. Three types of software in particular are especially difficult to analyze without prototyping. These are (1) real-time software, (2) distributed processing, and (3) graphical user interface software. In real-time systems, the control of timing can be especially complicated, as when system interrupts occur at unpredictable times and with different priorities for servicing. In distributed software systems, the designer gives up a large degree of control over the location of system data and processing among networked data processors and memories. This makes the course of system operation exceedingly difficult to analyze. In graphical user interfaces, the requirements are often incomplete and subject to change. Further, the very flexibility that makes such systems useful is itself an invitation to complexity. Thus, the above special software modes, which have made computer systems so powerful and ubiquitous in today's information systems, inherently create complexities that must be resolved by highly disciplined design, extensive experimentation, and rigorous verification, including formal design reviews, code "walk-throughs," and integration tests. Chapter 11 is devoted to the subject of software engineering and its special challenges.

Dynamic System Elements. Another form of complexity that usually requires development and testing is inherent in closed-loop dynamic systems such as those that are used for automated controls (e.g., autopilots). While these lend themselves to digital or analog simulation, they often involve coupling and secondary effects (e.g., flexure of the mounting of an inertial component) that cannot be readily separated from their physical implementation. Thus, the great majority of such system elements must be built and tested to ensure that problems of overall system stability are well in hand.

Ill-Defined System Environments

Poorly defined system environments and imprecise external interface requirements are also design issues that must be carefully examined and clarified. For example, a radar system designed to detect targets in the presence of clutter due to weather or surface returns is impossible to characterize in a well-defined fashion due to the great diversity of possible operational and environmental conditions and the limited understanding of the physics of radar scattering by clutter and of anomalous radar propagation. Similarly, space environments are difficult to understand and characterize due to the limited data available from past missions. The expense of placing systems into the space environment means testing and operational data are not as prevalent as atmospheric data.

The operation of user-interactive systems involves the human–machine interface, which is also inherently difficult to define. The parts of the system that display information to the user and that accept and respond to user inputs are often relatively uncomplicated physically but are very intricate logically. This complexity operates at several levels, sometimes beginning with the top-level objective of the system, as in the conceptual design of a medical information system where the needs of the physicians, nurses, clerical staff, and others that interact with the system tend to be not only ill-defined but also highly variable and subject to argument. At lower levels, the form of the display, the format of information access (menu, commands, speech, etc.), portability, and means of data entry may all constitute system design issues that are not likely to be settled without an extensive testing of alternatives.

The design of automobile air bags represents another type of component with a complex environmental interface that has required extensive development. In this case, the conditions for actuating the air bag had to be explored very thoroughly to establish a range between excessively frequent (and traumatic) false alarms and assured response to real collisions. The shape, size, and speed of inflation and subsequent deflation of the air bag had to provide maximum safety for the individual with minimum chance for injury by the force of inflation of the bag. This example is representative of system–environment interactions that can only be accurately defined experimentally. It also illustrates a system component whose operational and functional performance cannot be separated from its physical implementation.

Functional Design

Beyond identifying system elements requiring further development, the functional design and integration of the total system and all its functional elements must be completed during this phase. This is a necessary step to developing the system design specifications, which are a prerequisite to the start of the engineering design phase.

Functional and Physical Interfaces. Prior to initiating full-scale engineering, it is especially important to ensure that the overall system functional partitioning is sound and will not require significant alteration in the engineering design phase. Before a major commitment is made to the detailed design of individual components, the proposed functional allocations to subsystems and components and their interactions must be carefully examined to ensure that a maximum degree of functional independence and minimum interface complexity has been achieved. This is necessary so that each component can be designed, built, tested, and assembled with other components without significant fitting or adjustment, not to mention adaptation. This examination must take into account the availability of test points at the interfaces for fault isolation and maintenance, environmental provisions, opportunity for future growth with minimum change to associated components, and all the other systems engineering characteristics of a good product. The system functional and physical architectures are emphasized in this phase because the design should be sufficiently advanced to make such judgments meaningful but is not as yet so committed as to make modifications unduly time-consuming and expensive.

Software Interfaces. It was noted above that many new software components are too complex to be validated only through analysis and need therefore to be designed and tested in this phase. Further, many hardware elements are controlled by or interface with software. Hence, as a general rule, it can be assumed that many, if not most, software system elements will have to be first designed and subsequently implemented in this phase of the system development.

Use of Simulations

While many of the above problem areas require resolution by prototyping actual hardware and software, a number of others can be effectively explored by simulation. Some examples are the following:

- *Dynamic Elements.* Except for very high frequency dynamic effects, most system dynamics can be simulated with adequate fidelity. The six-degree-of-freedom dynamics of an aircraft or missile can be explored in great detail.
- *Human–Machine Interfaces.* User interfaces are control elements of most complex systems. Their proper design requires the active participation of potential users in the design of this system element. Such participation can best be obtained by providing a simulation of the interface early in the development and by enhancing it as experience accumulates.
- *Operational Scenarios.* Operational systems are usually exposed to a variety of scenarios that impact the system in different ways. A simulation with variable input conditions is valuable in modeling these different effects well before system prototypes or field tests can be conducted.

Example: Aircraft Design. Illustrating a use of simulation, assume, as in the example in Chapter 7, that an aircraft company is considering the development of a new medium-range commercial airplane. The two basic options being considered are to power it with either turbo-prop or jet engines. While the gross characteristics of these options are known, the overall performance of the aircraft with various types and numbers of engines is not sufficiently well-known to make a choice. It is clearly not practical to build a prototype aircraft to obtain the necessary data. However, in this case, simulation is a practical and appropriate method for this purpose because extensive engineering data on aircraft performance under various conditions are available.

Since the primary issue at this stage is the type and number of engines, it is only necessary to have a first-order, two-dimensional (i.e., vertical and longitudinal) model of the aerodynamic and flight dynamics of the airplane. The performance of various engines can be represented by expressions of thrust as a function of fuel flow, speed, altitude, and so on, known from their measured performance data. From this simple model, basic performance in terms of such variables as take-off distance, climb rate, and maximum cruise speed can be determined for various design parameters such as gross weight, number of engines, and payload. Assuming that this process led to a recommended configuration, extension of this simple simulation to higher orders of detail could provide the necessary data for advanced analysis. Thus, such simulations can save cost and can build on the experience gained at each stage of effort.

To validate or amplify the results of the above type of analysis of a prototype engine, it could be operated in an engine test facility where the airflow and atmospheric conditions are varied over the range of predicted flight conditions. The measured engine thrust and fuel consumption can then be factored into the overall performance analysis. A still more realistic test would be to mount a prototype engine in a special pod under the wing of a "mother" aircraft, which would fly at various speeds and altitudes. In this case, the mother aircraft itself can be thought of as a development facility.

10.4 PROTOTYPE DEVELOPMENT AS A RISK MITIGATION TECHNIQUE

In the previous chapters, we have discussed the principles and techniques to identify, manage, and ultimately mitigate risks. Significant problem areas have been identified at this point, and individual strategies are in full implementation by the advanced development stage. However, in the development of a new complex system, the decisions as to which components and subsystems require further development and testing prior to full-scale engineering, and issues regarding their physical implementation, are frequently more difficult and critical than those regarding their functional design and performance. One of the reasons is that many physical characteristics (e.g., fatigue cracking) do not easily lend themselves to analysis or simulation, but rather require the component to be designed, built, and tested to reveal potential problems. The paragraphs below describe general approaches to identifying and resolving problems in areas that do not lend themselves to mitigation through these methods.

During early risk management activities, the systems engineering approach to the identification of potential problem areas is to take a skeptical attitude, especially to design proposals unsupported by relevant precedent or hard engineering data. The systems engineer asks:

1. What things could go wrong?
2. How will they first manifest themselves?
3. What could then be done to make them right?

Potential Problem Areas

In looking for potential problems, it is essential to examine the entire system life cycle—engineering, production, storage, operational use, and operational maintenance. Special attention must be devoted to manufacturing processes, the "ilities" (RMA), logistic support, and the operational environment. The approach is that of risk assessment: what risks may be involved at each phase and where are the unknowns such as areas in which prior experience is scanty? For each potential risk, the likelihood and impact of a failure in that area must be determined.

As in the case of functional characteristics, the most likely areas where proposed component implementation may be significantly different from previous experience can be classified in four categories:

1. components requiring unusually stringent physical performance, such as reliability, endurance, safety, or extremely tight manufacturing tolerances;
2. components utilizing new materials or new manufacturing methods;
3. components subjected to extreme or ill-defined environmental conditions; and
4. component applications involving unusual or complex interfaces.

Examples of each of these categories are discussed below.

Unusually High Performance. Most new systems are designed to provide performance well in excess of that of their predecessors. When such systems are at the same time more complex, and also demand greater reliability and operating life, it is almost always necessary to verify the validity of the design approach experimentally.

Radars used in air traffic control systems are examples of complex devices requiring extremely high reliability. These radars are frequently unmanned and must operate without interruption for weeks between maintenance periods. The combination of performance, complexity, and reliability requires special attention to detailed design and extensive validation testing. All key components of these radars require development and testing prior to full-scale engineering.

Modern aircraft are another example of systems required to perform under high stress with very high reliability. Many aircraft have operating lifetimes of 30–40 years, with only a limited renewal of the more highly stressed structural and power components. The development and testing of aircraft components is notably extensive.

The components used in manned space flight must be designed with special consideration for safety as well as reliability. The launch and reentry environment places enormous stresses on all parts of the space vehicle and on the crew. Special procedures are employed to conceive of all possible accidents that might occur and to ensure that causes of such eventualities are eliminated or otherwise dealt with, for example, by extensive design redundancy.

More familiar systems do not have quite such dramatic requirements, but many require remarkable performance. The engines of some of today's automobiles do not require maintenance until 50,000–100,000 miles. Such reliable performance has required years of development and testing to achieve.

Special Materials and Processes. Advances in technology and new processes and manufacturing techniques continue to produce new materials with remarkable properties. In many instances, it is these new materials and processes that have made possible the advances in component performance discussed in the previous paragraphs.

Table 10.4 lists some examples of the many special materials developed in recent years that have made a major impact on the performance of the components in which they are used. In each new application, however, these components have undergone extensive testing to validate their intended function and freedom from unwanted side effects. Titanium has proven extremely effective in many applications but has been found to be more difficult to machine than the steel or aluminum that it replaced. Sintered metals can be formed easily into complex forms but do not have the strength

TABLE 10.4. Some Examples of Special Materials

Material	Characteristics	Typical applications
Titanium	High strength-to-weight ratio, corrosion resistant	Lightweight structures
Tungsten	Temperature resistant, hard to work	Power sources
Sintered metal	Easy to mold	Complex shapes
Glues	High strength	Composite structures
Gallium arsenate	Temperature resistant	Reliable microelectronics
Glass fibers	Optical transmission	Fiber optic cable
Ceramic components	Strength, temperature resistant	Pressure vessels
Plastics	Ease of forming, low weight and low cost	Containers

of the conventionally formed metal. Some of the new adhesives are remarkably strong but do not retain their strength at elevated temperatures. These examples show that the use of a special material in the critical elements of a component needs to be carefully examined and, in most cases, tested in a realistic environment before acceptance.

The same considerations apply to the use of new processes in the manufacture of a component. The introduction of extensive automation of production processes has generally increased precision and reproducibility and has decreased production costs. But it has also introduced greater complexity, with its risks of unexpected shutdowns, and has usually required years of development and testing of the new equipment.

Unfortunately, it is very difficult to appraise the time and cost of introducing a new manufacturing process in advance of its development and full-scale testing. For this reason, a new system that counts on the availability of projected new production processes must ensure that adequate time and resources are invested in process development and engineering, or it must have a fallback plan that does not rely on the availability of the process.

Extreme Environmental Conditions. The proper operation of every system component depends on its ability to satisfactorily operate within its environment, including such transport, storage, and other conditions as it may encounter during its life cycle. This includes the usual factors of shock, vibration, extreme temperatures, and humidity, and, in special instances, radiation, vacuum, corrosive fluids, and other potentially damaging environments.

The susceptibility of components to unfavorable environments can often be inferred from their basic constitution. For example, cathode ray tube components (e.g., displays) tend to be inherently fragile. Some thermomechanical components, such as jet engines, operate at very high internal temperatures in very cold external environments (7 miles above the surface of the earth), placing great stress on their internal parts. The endurance of such components as the turbines in aircraft engines is always a potential problem.

Military equipment has to be designed to operate over a large temperature range and to withstand rough handling in the field. The recent trend in using standard commercial components (e.g., computers) in military systems, and the relaxation of military specifications (milspecs) to save cost, has created potential problems that require special attention. Fortunately, such commercial equipment is usually inherently reliable and designed to be rugged enough to withstand shipping and handling by inexperienced operators. However, each component needs to be carefully examined to ensure that it will in fact survive in the projected environment. These circumstances place an even greater responsibility on systems engineering than when milspecs were rigidly enforced.

Component Interfaces. Perhaps the most neglected aspect of system design is component interfaces. Since these are seldom identified as critical elements, and since they fall between the domains of individual design specialists, often only systems engineering feels responsible for their adequacy. And the press of more urgent problems frequently crowds out the necessary effort to ensure proper interface management. Aggravating this problem is the fact that physical interfaces require detailed design, and frequently construction, of both components to ensure their compatibility—a costly process.

To overcome the above obstacles, special measures are required, such as establishment of interface control groups, interface documentation and standards, interface design reviews, and other similar means, for revealing deficiencies in time to avoid later mismatches. Such measures also provide a sound basis for the continuation of this activity in the engineering design phase.

Component Design

The previous sections described a number of criteria that may be used to identify components that require development effort to bring their design to a level of maturity sufficient to qualify them for full-scale engineering. Such development effort involves some combination of analysis, simulation, design, and testing according to the specific nature of the proposed design approach and its departure from proven practice.

The extent of development required may, naturally, vary widely. At one extreme, the design may be taken only to the stage where its adequacy can be verified by inspection and analysis. This may be done for components whose departure from their predecessors is mainly related to size and fit rather than to performance or producibility. At the other extreme, components for which the validation of new materials, or the verification of stringent production tolerances (or other characteristics of the production article), are required may need to be designed, constructed, and extensively tested. Here again, the decisions involve systems engineering trade-offs between program risk, technical performance, cost, and schedule.

Concurrent Engineering. It is evident from the above that such issues as RMA, safety, and producibility must be very seriously considered at this stage in the program rather than deferred until the engineering design phase. Failure to do so runs a high risk of major design modifications in the subsequent phase, with their likely impact on other components and on the system as a whole. This is an area where many system developments encounter serious difficulties and resultant overruns in cost and schedule.

To minimize the risks inherent in such circumstances, it has been recognized that specialty engineers who are particularly versed in production, maintenance, logistics, safety, and other end-item considerations should be brought into the advanced development process to inject their experience into decisions on design and early validation. This practice is referred to as "concurrent engineering" and is part of the function of integrated product teams (IPTs), which are used in the acquisition of defense systems. The phrase concurrent engineering should not be confused with the term "concurrency," which is often applied to the practice of carrying out two phases of the system life cycle, such as advanced development and engineering design, concurrently (i.e., at the same time) rather than sequentially. The effective integration of specialty engineers into the development process is not easy and must be orchestrated by systems engineers.

The problem in making concurrent engineering effective is that design specialists, as the name implies, have a deep understanding of their own disciplines but typically have only a limited knowledge of other disciplines, and hence lack a common vocabulary (and frequently interest) for communicating with specialists in other disciplines. Systems engineers, who by definition should have such a common knowledge, vocabulary, and interest, must serve as coordinators, interpreters, and, where necessary, as mentors. It is essential that the specialty engineers be led to acquire a sufficient level of understanding of the specific design requirements to render their opinions relevant and meaningful. It is equally essential that the component design specialists become sufficiently knowledgeable in the issues and methods involved in designing components that will result in reliable, producible, and otherwise excellent products. Without such mutual understanding, the concurrent engineering process can be wholly ineffectual. It is noteworthy that such mutual learning builds up the effectiveness of those involved with each successive system development, and hence the proficiency of the engineering organization as a whole.

Software Components. Software components should be addressed similarly. Each component is assessed for complexity, and a risk strategy is developed and implemented. Particularly complex components, especially those controlling system hardware elements, may necessitate the design and test of many system software components in prototype form during this phase of system development. This generally constitutes an effort of major proportions and is of critical importance to the system effort as a whole.

To support software design, it is necessary to have an assortment of support tools (computer-aided software engineering [CASE]), as well as a set of development and documentation standards. The existence of such facilities and established quality practices are the best guarantee for successful software system development. The Software Engineering Institute (SEI), operated by Carnegie Mellon University, is the current source of standards and evaluation criteria to rate the degree of software engineering maturity of an organization. As noted previously, Chapter 11 is entirely devoted to the special systems engineering problems associated with software-embedded and software-intensive systems.

Design Testing

The process of component design is iterative, just as we have seen the system development process to be. This means that testing must be an integral part of design rather than just a step at the end to make sure it came out properly. This is especially true in the design of components with new functionality or those utilizing unproven implementation approaches. The appropriate process in such cases is "build a little, test a little," providing design feedback at every step of the way. This may not sound very orderly but is often the fastest and most economical procedure. The objective is to validate the large majority of design elements at lower levels, where the results are more easily determined in less complex test configurations and errors corrected at the earliest time.

As stated earlier, the degree of completion to which the design of a given component is carried during this phase is very much a function of what is required to ensure a sound basis for its subsequent engineering. Thus, if a component's design issues are largely functional, they may be resolved by comparative simulation to establish which will best fulfill the required functional needs of the system. However, if the design issues relate to physical characteristics, then the component usually needs to be designed and built in prototype form, which can then be tested in a physical environment simulating operational conditions. The design of such tests and of the corresponding test equipment will be discussed in the next section.

Rapid Prototyping

This is a term describing the process of expedited design and building of a test model of a component, a subsystem, and sometimes the total system to enable it to be tested at an early stage in a realistic environment. This process is employed most often when the user requirements cannot be sufficiently well defined without experimenting with an operating model of the system. This is particularly true of decision support systems, dynamic control systems, and those operating in unusual environments. Rapid prototyping can be thought of as a case of carrying development to a full-scale demonstration stage prior to committing the design to production engineering.

When engaging in rapid prototyping, the term "rapid" means that adherence to strict quality standards, normally a full part of system development, is suspended. The goal is to produce a prototype that features selected functionality of the system for demonstration as quickly as possible. The article that is produced is not intended to survive—once requirements are developed and validated using the prototype, the article itself should be discarded. At times, the prototype article is used as a basis for another iteration of rapid prototyping. The risk in this process is that eventually, the pressure to use the prototype article as the foundation for the production article becomes too great. Unfortunately, because the prototype was developed without the strict quality standards, it is not appropriate for production.

Examples abound where rapid prototyping was engaged and a prototype article was developed without quality controls (e.g., development standards, documentation, and testing). Unfortunately, the customer deems the article sufficient and requires the developer to provide the article for production (after all, the customer paid for the prototype—he owns it!). Once production starts, the flaws in this process quickly become evident, and the system fails its development and operational testing. In the end, development and production cause slippages in schedule and overruns in cost.

Rapid prototyping was pioneered in software development and will be discussed further in the next chapter.

Development Facilities

A development facility or environmental test facility, as referred to here, is a physical site dedicated to simulating a particular environmental condition of a system or a part thereof in a realistic and quantitative manner. It is usually a fixed installation capable of use on a variety of physical and virtual models (or actual system components with embedded software) representing different systems or components. It can be used for either development or validation testing, depending on the maturity of the system/ component subjected to the environment. Such facilities contain a set of instrumentation to control the simulated environment and to measure its effects on the system. They may be used in conjunction with a system simulation and usually have computing equipment to analyze and display the outputs.

A development facility usually represents a substantial investment; it is often enclosed in a dedicated building and/or requires a significant amount of real estate. A wind tunnel is an example of a facility used to obtain aerodynamic data. It contains a very substantial amount of equipment test chambers, air compressors, precise force measuring devices, and data reduction computers and plotters. Often the cost to build and operate a wind tunnel is so high that support is shared by a number of commercial and government users. When a wind tunnel is used to obtain data on a number of candidate aerodynamic bodies or control surfaces, it can be thought of as a development tool; when it is used to supply a source of high-speed airflow to check out a full-scale airplane control surface, it serves as a validation test facility.

Automobile manufacturers use test tracks to help design and test new model cars and to prove-in the final prototypes before production begins. Test tracks can simulate various wear conditions under accelerated aging, for example, by driving heavily loaded cars at high speed or over rough pavement. Other development facilities use electromagnetic radiation to test various electronic devices, for example, to measure antenna patterns, to test receiver sensitivity, to check for radio frequency (RF) interference, and so on.

Most development facilities use some form of models and simulations when conducting tests. It is common for some part of the system under test to be the actual article while other parts are simulated. An RF anechoic chamber that tests a tracking device in the presence of various RF interference signals is an example. In this case, the flight of the vehicle can also be simulated by a computer, which solves the equations of motion using appropriate aerodynamic and dynamic models of the system.

The engineering design of hardware components that are subjected to external stresses, high temperatures, and vacuum conditions in space requires the extensive use of stress testing, environmental chambers, and other special test facilities. The same facilities are also used in the development of these components. Thus, shake and shock facilities, vacuum chambers, hot and cold chambers, and many other engineering test facilities are as necessary in the development as in the engineering phases. The main difference is that development testing usually requires the acquisition of more performance data and more extensive analyses of the results.

10.5 DEVELOPMENT TESTING

The determination that all of the design issues identified during the advanced development phase have been satisfactorily resolved requires a systematic program of analysis, simulation, and test of not only the particular components and subsystem directly involved but also of their interfaces and interactions with other parts of the system. It also requires explicit consideration of the operational environment and its effect on system performance.

Development testing should not be confused with what is traditionally referred to as "developmental testing" and "operational testing." Developmental testing typically involves the engineered system within a series of test environments, under controlled scenarios. This type of testing is conducted by the developer. Operational testing is also on the engineered system, but involves the customer, and is conducted under more realistic operational conditions, including environments and scenarios. "Development testing," on the other hand, is on subsystems and components and is conducted by the developer.

A well-planned development test program generally requires the following steps:

1. development of a test plan, test procedures, and test analysis plan;
2. development or acquisition of test equipment and special test facilities;
3. conduct of demonstration and validation tests, including software validation;
4. analysis and evaluation of test results; and
5. correction of design deficiencies.

These steps are discussed briefly below.

Test and Test Analysis Plans

An essential but sometimes insufficiently emphasized step in the advanced development process is the development of a well-designed test plan for determining whether or not the system design is sufficiently mature to proceed to the engineering design phase.

Test Planning Methodology. The overall testing approach must be designed to uncover potential design deficiencies and acquire sufficient test data to identify sources of these deficiencies and provide a sound basis for their elimination. This is very different from an approach that presupposes success and performs a minimal test with scanty data acquisition. Whereas the latter costs less initially, its inadequacies often cause design faults to be overlooked, which later result in program interruptions and delays, and in a far greater ultimate cost. The following steps provide a useful checklist:

1. Determine the objectives of the test program. The primary purpose, of course, is to test the subsystems and system against a selected set of operational and performance requirements. However, other objectives might be introduced as well: (1) increasing customer confidence in particular aspects of the system, (2) uncovering potential design flaws in high-risk areas, (3) demonstrating the selected capability publicly, and (4) demonstrating interfaces with selected external entities.

2. Review the operational and top-level requirements. Determine what features and parameters must be evaluated. Key performance parameters identified early in the development process must be included in this set. However, testing every requirement usually is not possible.

3. Determine the conditions under which these items will be tested. Consider upper and lower limits and tolerances.

4. Review the process leading to the selection of components requiring development and of the design issues involved in the selection.

5. Review development test results and the degree of resolution of design issues.

6. Identify all interfaces and interactions between the selected components and other parts of the system as well as the environment.

7. On the basis of the above factors, define the appropriate test configurations that will provide the proper system context for testing the components in question.

8. Identify the test inputs necessary to stimulate the components and the outputs that measure system response.

9. Define requirements for test equipment and facilities to support the above measurements.

10. Determine the costs and manpower requirements to conduct the tests.

11. Develop test schedules for preparation, conduct, and analysis of the tests.

12. Prepare detailed test plans.

The importance of any one task and the effort required to execute it will depend on the particular system element under test, the resources available to conduct the tests, and the associated risk. In any case, the systems engineer must be familiar with each of these items and must be prepared to make decisions that may have a major impact on the success of the overall development program. It is evident that the above tasks involve a close collaboration between systems engineers and test engineers.

Test Prioritization. The test planning process is often conducted under considerable stress because of time and cost constraints. These restrictions call for a strict prioritization of the test schedule and test equipment to allocate the available time and resources in the most efficient manner. Such prioritization should be a particular responsibility of systems engineering because it requires a careful balancing of a wide range of risks based on a comparative judgment of possible outcomes in terms of performance, schedule, and cost.

The above considerations are especially pertinent to defining test configurations. The ideal configuration would place all components in the context of the total system in its operating environment. However, such a configuration would require a prototype of the entire system and of its full environment, which is usually too costly in terms of resources. The minimum context would be an individual component with simple simulations of all its interfacing elements. A more practical middle ground is incorporating the component under test in a prototype subsystem, within a simulation of the remainder of the system and the relevant part of the operating environment. The choice of a specific test configuration in each case requires a complex balancing of risks, costs, and contingency plans requiring the highest level of systems engineering judgment.

Test Analysis Planning. The planning of how the test results are to be analyzed is just as important as how the tests are to be conducted. The following steps should be taken:

1. Determine what data must be collected.
2. Consider the methods by which these data can be obtained—for example, special laboratory tests, simulations, subsystem tests, or full-scale system tests.
3. Define how all data will be processed, analyzed, and presented.

Detailed analysis plans are especially important where a test is measuring the dynamic performance of a system, thus producing a data stream that must be analyzed in terms of dynamic system inputs. In such cases, where a large volume of data is produced, the analysis must be performed with the aid of a computer program that is either designed for the purpose or is a customized version of an existing program. The analysis plan must, therefore, specify exactly what analysis software will be needed and when.

The test analysis plan should also specify that the test configuration has the necessary test points and auxiliary sensors that will yield measurements of the accuracy needed for the analysis. It also must contain the test scenarios that will drive the system during the tests. Whereas the details of the test analysis plan are usually written by test engineers and analysts, the definition of the test and test analysis requirements is the task of systems engineering. The loop needs to be closed between the definition of test configuration, test scenarios, test analysis, and criteria for design adequacy. These relationships require the expertise of systems engineers who must ensure that the test produces the data needed for analysis.

Special consideration is needed when testing human–machine interactions and interfaces. The evaluation of such interactions usually does not lend itself to quantitative measurement and analysis, but must nevertheless be provided for in the test and analysis plan. This is an area where the active participation of specialists is essential. All the above plans should be defined during the early to middle phases of the advanced development phase to provide the time to develop or otherwise to acquire the necessary supporting equipment and analysis software before formal testing is scheduled to begin.

Test and Evaluation Master Plan (TEMP). In government projects, the development of a comprehensive test plan is a formal requirement. Designated the TEMP, the plan is to be prepared first as a part of concept definition and then expanded and detailed at each phase of the development. The TEMP is not so much a *test* plan as a *test management* plan. Thus, it does not spell out *how* the system is to be evaluated or the procedures to be used but is directed to *what* is planned to be done and *when*. The typical contents of a system TEMP are the following:

- *System Introduction*
  Mission description
  Operational environment
  Measures of effectiveness and suitability
  System description
  Critical technical parameters
- *Integrated Test Program Summary*
  Test program schedule
  Management
  Participating organizations
- *Developmental Test and Evaluation*
  Method of approach
  Configuration description
  Test objectives
  Events and scenarios
- *Operational Test and Evaluation*
  Purpose
  Configuration description
  Test objectives
  Events and scenarios
- *Test and Evaluation Resource Summary*
  Test articles
  Test sites
  Test instrumentation

Test environment and sites
Test support operations
Computer simulations and models
Special requirements

Special Test Equipment and Test Facilities

It has been noted in previous chapters that the simulation of the system operational environment for purposes of system test and evaluation can be a task of major proportions, sometimes approaching the magnitude of the system design and engineering effort itself. In the advanced development phase, this aspect of system development is not only very important but frequently is also very expensive. Thus, the judgment as to the degree of realism and precision that is required of such simulation is an important systems engineering function. This and related subjects are also discussed in Chapter 13.

The magnitude of the effort to provide suitable test equipment and facilities naturally depends on the nature of the system and on whether the developer has had prior experience with similar systems. Thus, the development of a new spacecraft requires a host of equipment and facilities ranging from vacuum chambers and shake and vibration facilities that simulate the space and launch environment, and space communication facilities to send commands and receive data from the spacecraft, to clean rooms that prevent contamination during the building and testing of the spacecraft. Some of these facilities were described in the subsection on development facilities. Having a full complement of such equipment and facilities enables an established spacecraft developer to limit the cost and time for developing the necessary support for a new development. However, even if the bulk of such equipment may be available from previous system developments, every new program inevitably calls for different equipment combinations and configurations. The rate of technological change creates both new demands and new opportunities, and this is no less true in the area of system testing than in the area of system design.

Creating the Test Environment. The design and construction of the test environment to validate a major component or subsystem requires equipment for the realistic generation of all the input functions and the measurement of the resulting outputs. It also requires the prediction and generation of a set of outputs representing what the system element should produce if it operates according to its requirements. The latter, in turn, requires the existence of mathematical or physical models designed to convert the test inputs into predicted system outputs for comparison with test results.

The above operations are represented by a functional flow diagram (Figure 10.3) that is an expansion of the test and evaluation block of Figure 8.2. The four functions on the left side of the figure show how the design of the test environment creates a predictive test model and a test scenario, which in turn activates a test stimulus generator. The test stimuli activate the system element (component or subsystem) under test and are also used by the mathematical or physical model of the system element to create

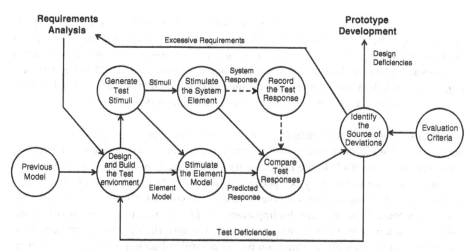

Figure 10.3. Test and evaluation process of a system element.

a corresponding set of predicted outputs for comparison with the actual test outputs. The functions on the right side of Figure 8.3 represent the analysis and evaluation of test results, as further described below in a subsequent subsection bearing that name.

Test Software. Test support and analysis software requires special attention in virtually all developments and has to be tailored very specifically to the system at hand. Establishing its objectives and detailed requirements is a major systems engineering task. Where user (human—machine) interfaces are also involved, the task becomes even more complex. Such support software is usually best developed by rapid prototyping, with strong inputs from the test engineers and analysts who will be responsible for installing and using it. For this reason, and because of the inherent difficulty in predicting software development time, it is important to begin this task as early as possible.

Test Equipment Validation. Like any system element, test equipment for system design validation itself requires test and validation to ensure that it is sufficiently accurate and reliable to serve as a measure of system performance. This process requires careful analysis and consideration because it often stresses the limits of equipment measurement capabilities. This task is often underestimated and is not allocated with sufficient time and effort.

Demonstration and Validation Testing

The actual conduct of tests to demonstrate and validate the system design is often the most critical period in the development of a new system. The primary effort during advanced development has been seen to be concerned with the resolution of identified design issues—in other words, eliminating the known unknowns or "unks." And, with luck, it will succeed in resolving the great majority of the initial uncertainties in the system design. But every new complex system inevitably also encounters unanticipated "unknown unknowns," or "unk-unks." Thus, it is also a major objective of the advanced development phase to discover such features before committing to full-scale engineering. To this end, the validation tests are designed to subject the system to a broad enough range of conditions to reveal hitherto undiscovered design deficiencies.

Dealing with Test Failures. It can be seen that the above process is at once necessary and at the same time poses program risks. When a test uncovers an unk-unk, it usually manifests itself in the failure of the system element to function as expected. In some cases, the failure may be spectacular and publicly visible, as in testing a new aircraft or guided missile. Because the failure is unexpected, there is a period of time before a proposed solution can be implemented. During this time, the impact of the failure on system development may be serious. Because the decision to proceed with the engineering design phase hinges on the successful validation of the system design, a hiatus in the program may be in prospect, and if no adequate solution is found relatively quickly, the entire program may be jeopardized.

It is when eventualities such as the above occur that systems engineers are most indispensable. They are the only members of the program staff who are equipped to bring together the breadth of knowledge and experience necessary to guide the effort to find solutions to unexpected system problems. Quite often, a deficiency found in the design of a given component cannot be overcome by a local fix but can be compensated for by a change in a related part of the system. In other cases, analysis may show the fault to be in the test equipment or procedure rather than in the system itself. In some instances, analysis can demonstrate that the particular system performance requirement that was at issue cannot be fully justified on the basis of operational need. In these and other cases, the expedited search and identification of the most desirable solution to the problem is led by systems engineering, as is the task of persuading program management, the customer, and other decision makers that the recommended solution is worthy of their confidence and support.

Testing and the System Life Cycle. It has been noted in previous chapters that a new system not only has to perform in its operational environment but also must be designed to survive conditions to which it will be exposed throughout its life, such as shipping, storage, installation, and maintenance. These conditions are often insufficiently addressed, especially in the early stages of system design, only to unexpectedly cause problems at a stage when their correction is extremely costly. For these reasons, it is essential that the design validation tests include an explicit imposition of all conditions that the system is expected to encounter.

Testing of Design Modifications. As noted above, the test programs must anticipate that unexpected results that reveal design deficiencies may occur. Accordingly, it must provide scheduled time and resources to validate design changes that correct such deficiencies. Too often, test schedules are made on the assumption of 100% success, with little or no provision for contingencies. The frequent occurrence of time and cost overruns in the development of new complex systems is in considerable part due to such unrealistic test planning.

Analysis and Evaluation of Test Results

The operations involved in evaluating test results are illustrated in the right half of Figure 10.3. The outputs from the component or subsystem under test are either recorded for subsequent analysis or compared in real time with the predicted values from the simulated element model. The results must then be analyzed to disclose all significant discrepancies, to identify their source, and to assess whether or not remedial measures are called for, as derived with reference to a set of evaluation criteria. These criteria should be developed prior to the test on the basis of careful interpretation of system requirements and understanding of the critical design features of the system element.

It should be noted that one of the first places to look for as a cause of a test discrepancy is a defect in the test equipment or procedure. This is largely because there is usually less time and effort available to validate the test setup than has gone into the design of the system element under test.

The successful use of test results to either confirm the design approach or to identify specific design deficiencies is wholly dependent on the acquisition of high-quality data and its correct interpretation in terms of system requirements. An essential factor in effective test analysis is a versatile and experienced analysis team composed of analysts, test engineers, and systems engineers. The function of the analysts is to apply analytical tools and techniques to convert the raw test results to a measurement of the performance of specific system elements. The test engineers contribute their intimate knowledge of the test conditions, sensors, and other test variables to the systems analysis. The systems engineers apply the above knowledge to the interpretation of the tests in terms of system performance as related to requirements.

Tracing deficiencies in performance to the stated system requirements is especially important when remedying the deficiencies may require significant redesign. In such cases, the requirements must be critically reviewed to determine whether or not they may be relaxed without significant loss in system effectiveness, in preference to expending the time and cost required to effect the system changes required to meet them fully. In view of the potential impact of any deficiencies uncovered in the test analysis process, it is essential that the analysis be accomplished quickly and its results used to influence further testing, as well as to initiate such further design investigations as may be called for.

Evaluation of User Interfaces

A special problem in the validation of system design is posed by the interface and interaction between the user/controller and the system. This is especially true in decision support systems where the system response is critically dependent on the rapid and accurate interpretation of complex information inputs by a human operator aided by displays driven by computer-based logic. The air traffic controller function is a prime example of such an interface. However, even in much less information-intensive systems, the trends toward increased automation have made user interfaces more interactive and hence more complex. Even the basic interface between a personal computer and the user, while becoming more intuitive and powerful, has nevertheless become correspondingly more complicated and challenging to the nonexpert user.

The test and evaluation of user interface controls and displays poses difficult problems because interfaces are inherently incapable of objective quantitative measurement, except in their most primitive features (e.g., display luminosity). Large variations in the experience, visual and logical skills, and personal likes and dislikes of individual users also color their reactions to a given situation. Moreover, it is essential that members of the design team do not serve as sole subjects for the assessment of user interfaces. Rather, to the maximum extent possible, operators of similar systems should be employed for this purpose.

Nonetheless, the importance of an effective user interface to the performance of most systems makes it essential to plan and conduct the most substantive evaluation of this systems feature as may be practicable. This is especially relevant because of the inherent difficulty of establishing user requirements at the outset of the development. Thus, there are bound to be surprises when users are first confronted with the task of operating the system.

User interfaces are areas where rapid prototyping can be particularly effective. Before the full system or even the full human–computer interface is designed, prototypes can be developed and demonstrated with potential users to solicit early feedback on preferences of information representation.

The evaluation of the user interface may be considered in four parts:

1. ease of learning to use the operational controls,
2. clarity of visual situational displays,
3. usefulness of information content to system operation, and
4. online user assistance.

Of these, the first and last are not explicitly parts of the basic system operation, but their effectiveness can play a decisive role in the user's performance. It is, therefore, important that sufficient attention be paid to user training and basic user help to ensure that these factors do not obscure the evaluation of the basic system design features.

Even more than most other design characteristics, user interfaces should be tested in anticipation of discovering and having to fix inadequacies. To this end, wherever practicable, users should be presented with design alternatives to choose from rather than having to register their level of satisfaction with a single design option. This may usually be accomplished in software rather than in hardware.

As in the case of other operational characteristics, such as reliability, producibility, and so on, the design related to the human–machine interface should have involved human factor experts as well as potential users. For the developer to obtain the participation of the latter, it may be necessary to obtain customer assistance. In these and other cases, customer participation in the development process can materially enhance the utility and acceptance of the final product.

The evaluation of the effectiveness of the user interface is not subject to quantitative engineering methods and extends the systems engineer into the field of human–machine interaction. The experts (usually psychologists) in certain aspects of such interactions are mostly specialists (e.g., in visual responses) and must be integrated into the evaluation process along with other specialists. It is a systems engineering responsibility to plan, lead, and interpret the tests and their analysis in terms of what system design changes might best make the user most effective. To do so, the systems engineer must learn enough of the basics of human–machine interactions to exercise the necessary technical leadership and system-level decision making.

Correction of Design Deficiencies

All of the previous discussions have centered on discovering potential deficiencies in the system design that may not have been eliminated in the development and test process. If the development has been generally successful, the deficiencies that remain will prove to be relatively few, but how to eliminate them may not always be obvious nor the effort required trivial. Further, there is almost always little time and few resources available at this point in the program to carry out a deliberate program of redesign and retest. Thus, as noted earlier, there must be a highly expedited and prioritized effort to quickly bring the system design to a point where full-scale engineering can begin with a relatively high expectation of success. The planning and leadership of such an effort is a particularly critical systems engineering responsibility.

10.6 RISK REDUCTION

As described in Chapter 5, a major fraction of risk reduction during the system life cycle should be accomplished during the advanced development phase. To reiterate, the principal purpose of the advanced development phase is to reduce the potential risks in the development of a new complex system to a level where the functional design of all previously unproven subsystems and components has been validated.

The typical sources of development risks are described in the sections on Functional Analysis and Design and Prototype Development. Most of them are seen to arise because of a lack of adequate knowledge about new technologies, devices, or processes that are intended to be key elements in the system design. Thus, the process of risk reduction in this phase amounts to the acquisition of additional knowledge through analysis, simulation, or implementation and testing.

We have advocated two primary methods to reduce risk within this phase: prototype development (both hardware and software) and development testing. While both methods could certainly be implemented earlier (and should be in many cases), it is not until the advanced development phase that sufficient information on the system architecture (both functional and physical) are available to properly implement prototyping and advanced testing.

Other risk reduction strategies are available to both the program manager and the systems engineer. From the program manager's perspective, several acquisition strategies are available to reduce risk, depending on the level of resources: (1) parallel development efforts developing alternative technologies or processes in case a primary technology or process fails to mature, (2) alternative integration strategies to emphasize alternative interface options, and (3) one of the incremental development strategies to engineer functional increments while technologies mature.

The systems engineer also has several strategies available beyond those of prototyping and testing: (1) increase use of modeling and simulation over physical prototyping to ensure an increased understanding of the environment and system processes and (2) interface development and testing before engineered components are available to reduce interface risks. Regardless of strategies ultimately employed to reduce risks, the program manager and systems engineer work hand in hand to ensure risk reduction occurs at the proper time.

How Much Development?

A key decision that must be made in planning the risk reduction effort is by what means and how far each risk area should be developed. If the development is too limited, the residual risk will remain high. If it is very extensive, the time and cost consumed in risk reduction may unnecessarily inflate the total system development cost. Striking the proper balance calls on the exercise of expert systems engineering judgment.

The decision as to how much development should be undertaken on a given component should be part of the risk management plan, as described in Chapter 5. The objective of the plan is to minimize the total cost of managing each significant risk area. This "risk cost" is the sum of the cost of such analysis, simulation, and design and testing that may be undertaken, that is, the "development cost," and the cost of mitigating the residual risk to the low level required to proceed to the engineering design phase, that is, the "mitigation cost." By varying the nature and amount of development, a judgment can be made as to the most favorable balance. Thus, for a critical, immature component, the balance may call for development up to the prototype stage, while for a noncritical or mature component, it would only call for analysis.

10.7   SUMMARY

Reducing Program Risks

Objectives of the advanced development phase are to resolve the majority of uncertainties (risks) through analysis and development and to validate the system design approach as a basis for full-scale engineering. The outputs of advanced development are a system design specification and a validated development model.

Advanced development is especially critical for systems containing extensive advanced development or unproven concepts that may involve several years of development effort.

Activities encompassed by advanced development are the following:

- *Requirements Analysis*—relating functional requirements to needs,
- *Functional Analysis and Design*—identifying performance issues,
- *Prototype Development*—building and testing prototypes of critical components, and
- *Test and Evaluation*—validating the maturity of critical components.

Requirements Analysis

Analysis of system functional specifications is required to relate them to their origin in operational requirements, especially those not readily met. Their differences from those of a predecessor system are also noted.

Functional Analysis and Design

Components that may require further development include those that

- implement a new function;
- are a new implementation of an existing function;
- use a new production method for an existing type of component;
- extend the function of a proven component; and
- involve complex functions, interfaces, and interactions.

Prototype Development as a Risk Mitigation Technique

Program risks requiring development may result from a number of conditions:

- unusually high performance requirements,
- new materials and processes,
- extreme environmental conditions,
- complex component interfaces, and
- new software elements.

Development Testing

Validation testing to confirm the resolution of risks requires the development of a formal test plan (TEMP). Furthermore, test equipment must be developed; validation tests must be conducted; and test results must be analyzed and evaluated. The results of this testing lead to the correction of design deficiencies. However, special test equipment and facilities often represent a major investment. Therefore, early experimental exploration of the interface design is essential.

Models of systems and components are used extensively in system development. Simulations are increasingly important in all stages of development and are essential in the analysis of dynamic systems and software that require development and a staff of analysts and operators.

Development facilities are installations simulating environmental conditions and are used for development tests and component evaluation. They represent a major investment and require a permanent operating staff.

Risk Reduction

Risk assessment is a basic systems engineering tool, which is used throughout development, but especially during advanced development. It involves identifying sources of risk, risk likelihood, and criticality.

PROBLEMS

10.1 The systems engineering method applies to the advanced development phase in a similar set of four steps, as it does to the preceding concept definition phase. For each step in the method, compare the activities in the two phases with one another, stating in your own words (a) how they are similar and (b) how they are different.

10.2 What specific activities in the advanced development phase sometimes cause it to be referred to as a "risk reduction" phase? Give an example of each activity considering a real or hypothetical system.

10.3 Why do so many new complex system developments incur large risks by choosing to apply immature technology? Give an example of where and how such choices paid off and one where they did not.

10.4 Table 10.2 illustrates four cases of developments involving different aspects of a system. Each is shown to require a different set of development activities to validate the result. Explain the rationale for each of the four development processes in terms of the given conditions.

10.5 In the development of a major upgrade to a terminal air traffic control system, what would you except to be three significant risks and what systems engineering approaches would you recommend to mitigate each of these risks? (Consider problems of failing to meet the schedule as well as safety problems.)

10.6 Components that are required to have extended functional performance well beyond previously demonstrated limits frequently need further development. Give an example of one such component in each of the four functional element categories (signal, data, material, and energy) as shown in Table 10.3. Give reasons for your choice of examples.

10.7 Graphical user interface software is generally difficult to design and test. Explain why this is true, giving at least three situations to illustrate your points. What types of development tests would you propose for each situation?

10.8 Closed-loop dynamic systems are often difficult to analyze and test. Special test facilities are often constructed for this purpose. Diagram such a test setup for evaluating an unmanned air vehicle (UAV) designed for remote surveillance using an optical sensor. Assume that the test equipment includes an actual optical sensor, while other system components are simulated. Indicate which elements in the simulation are part of the system under test and which elements represent external inputs. Label all blocks and input/output lines.

One systems engineering responsibility of advanced development is to understand how the system concept will accept, transform, consume, and produce each of the four functional elements of signals, data, materials, and energy. To illustrate this concept, for Problems 10.9–10.13, use a standard automated car wash found at most service stations in which a car enters an enclosed car wash via an automated conveyor belt and goes through several phases of activities before exiting the facility.

For each problem, construct a table with four columns labeled "Accept," "Transform," "Consume," and "Produce."

10.9 In the Accept column, describe what signals the system will accept from all external entities. In the Transform column, describe the transformation of these signals and what the system will transform these signals into. In the Consume column, describe what signals the system will consume and for what purpose. Note that the system will either transform or consume all of its input signals. In the Produce column, describe what signals the system will produce for output.

10.10 In the Accept column, describe what data the system will accept from all external entities. In the Transform column, describe the transformation of these data and what the system will transform these data into. In the Consume column, describe what data the system will consume and for what purpose. Note that the system will either transform or consume all of its input data. In the Produce column, describe what data the system will produce for output.

10.11 In the Accept column, describe what materials the system will accept from all external entities. In the Transform column, describe the transformation of these materials and what the system will transform these materials into. In the Consume column, describe what materials the system will consume and for what purpose. Note that the system will either transform or consume all of its input materials. In the Produce column, describe what materials the system will produce for output.

10.12 In the Accept column, describe what energy the system will accept from all external entities. In the Transform column, describe the transformation of these energies and what the system will transform these energies into. In the Consume column, describe what energies the system will consume and for what purpose. Note that the system will either transform or consume all of its input energy. In the Produce column, describe what energies the system will produce for output. Remember that energy may take several forms.

FURTHER READING

B. Blanchard and W. Fabrycky. *System Engineering and Analysis*, Fourth Edition. Prentice Hall, 2006, Chapter 5.

F. P. Brooks, Jr. *The Mythical Man Month—Essays on Software Engineering*. Addison-Wesley, 1995.

W. P. Chase. *Management of Systems Engineering*. John Wiley & Sons, Inc., 1974, Chapter 9.

P. DeGrace and L. H. Stahl. *Wicked Problems, Righteous Solutions*. Yourdon Press, Prentice Hall, 1990.

H. Eisner. *Computer-Aided Systems Engineering*. Prentice Hall, 1988, Chapter 13.

M. Maier and E. Rechtin. *The Art of Systems Architecting*. CRC Press, 2009.

J. N. Martin. *Systems Engineering Guidebook: A Process for Developing Systems and Products*. CRC Press, 1997, Chapter 10.

R. S. Pressman. *Software Engineering: A Practitioner's Approach*. McGraw Hill, 1982.

N. B. Reilly. *Successful Systems for Engineers and Managers*. Van Nostrand Reinhold, 1993, Chapter 13.

A. P. Sage. *Systems Engineering*. McGraw Hill, 1992, Chapter 6.

A. P. Sage and J. E. Armstrong, Jr. *Introduction to Systems Engineering*. John Wiley & Sons, Inc., 2000, Chapter 6.

S. M. Shinners. *A Guide for Systems Engineering and Management*. Lexington Books, 1989, Chapter 5.

R. Stevens, P. Brook, K. Jackson, and S. Arnold. *Systems Engineering, Coping with Complexity*. Prentice Hall, 1998, Chapter 11.

Systems Engineering Fundamentals. SEFGuide-12-00, Defense Acquisition University (DAU Press), 2001, Chapter 4.

11

SOFTWARE SYSTEMS ENGINEERING

Advancing information technology (IT) is the driving element to what many have called the "information revolution," changing the face of much of modern industry, commerce, finance, education, entertainment—in fact, the very way of life in developed countries. IT has accomplished this feat largely by automating tasks that had been performed by human beings, doing more complex operations than had been possible, and doing them faster and with great precision. Not only has this capability given rise to a whole range of new complex software-controlled systems but it has also been embedded in nearly every form of vehicle and appliance, and even in children's toys.

The previous chapters discussed the application of systems engineering principles and practice to all types of systems and system elements without regard to whether they were implemented in hardware or software. Software engineering, however, has advanced along a separate path than systems engineering. And only recently have the two paths begun to converge. Many principles, techniques, and tools are similar for both fields, and research has fostered the evolving merger.

The term *software systems engineering* was proposed by Dr. Winston Royce, father of the waterfall chart, early in the history of software engineering to represent the

Systems Engineering Principles and Practice, Second Edition. Alexander Kossiakoff, William N. Sweet, Samuel J. Seymour, and Steven M. Biemer

natural relationship between the two. However, the term was not adopted by the growing software community, and the term *software engineering* became the moniker for the field.

Within the first decade of the twenty-first century, the fact that the two fields have more in common has been recognized by both communities. And the "old" term was resurrected to represent the application of systems engineering principles and techniques to software development. Of course, the flow of ideas has gone in both directions, spawning new concepts in systems engineering as well—object-oriented systems engineering (OOSE) being one example. Today, the expanding role of software in modern complex systems is undeniable.

The two terms, software engineering and software systems engineering, are not synonymous, however. The former refers to the development and delivery of software products, stand-alone or embedded. The latter refers to the application of principles to the software engineering discipline.

Accordingly, this chapter will focus on software systems engineering—and how software engineering relates to systems. In other words, we take the perspective of using software to implement the requirements, functionality, and behaviors of a larger system. This excludes stand-alone commercial applications in our discussions, such as the ubiquitous office productivity products we all use today. While systems engineering principles could certainly be applied to the development of these types of products, we do not address these challenges.

Components of Software

We define software by its three primary components:

- *Instructions.* Referred to as a "computer program" or simply as "code," software includes the list of instructions that are executed by a variety of hardware platforms to provide useful features, functionality, and performance. These instructions vary in levels of detail, syntax, and language.
- *Data Structures.* Along with the set of instructions are the definitions of data structures that will store information for manipulation and transformation by the instructions.
- *Documentation.* Finally, software includes the necessary documents that describe the use and operation of the software.

Together, these three components are referred to as "software." A *software system* is software (as defined above) that also meets the definition of a system (see Chapter 1).

11.1 COPING WITH COMPLEXITY AND ABSTRACTION

One of the most fundamental differences between engineering software and engineering hardware is the abstract nature of software. Since modern systems are dependent on software for many critical functions, it is appropriate to focus on the unique challenges of engineering the software components of complex systems and to provide an overview of the fundamentals of software engineering of most interest to systems engineers.

In earlier chapters, we discussed the relationships between the systems engineer and design, or specialty engineers. Typically, the systems engineer acts in the role of a lead engineer responsible for the technical aspects of the system development. Concurrently, the systems engineer works with the program manager to ensure the proper programmatic aspects of system development. Together, the two work hand in hand, resulting in a successful program. Design engineers usually work for systems engineers (unofficially, if not directly reporting to them) in this split between responsibilities.

One perspective that can be taken with respect to software engineering is that the software engineer is simply another design engineer responsible for a portion of the system's functionality. As functions are allocated to software, the software engineer is called upon to implement those functions and behaviors in software code. In this role, the software engineer sits alongside his peers in the engineering departments, developing subsystems and components using programming code as his tool, rather than physical devices and parts. Figure 11.1 is an IEEE software systems engineering process chart that depicts this perspective using the traditional "Vee" diagram.

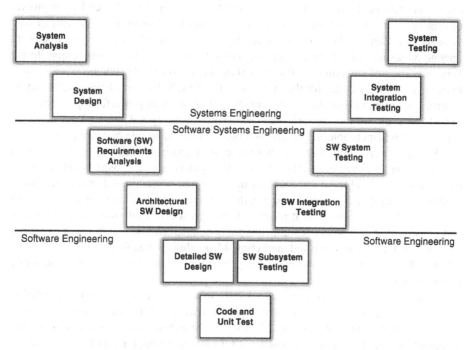

Figure 11.1. IEEE software systems engineering process.

Once a subsystem has been allocated for software development (or a combined software/hardware implementation), a subprocess of developing software requirements, architecture, and design commences. A combination of systems engineering and software engineering steps occurs before these software components are integrated into the overall system.

Unfortunately, this perspective tends to promote "independence" between the systems and software development teams. After design, hardware and software engineers begin their respective developments. However, the nature of software requires that software development strategies be devised early—during system design, depicted as the second major step in the Vee. If hardware and software are "split" during the design phase (i.e., functionality and subsystem components are allocated to hardware and software implementation) during or at the end of system design, then the differences in processes developing and implementing these components will cause the system development effort to become unbalanced in time.

Therefore, software development must be integrated earlier than what has been traditional—in the systems analysis phase. Although not shown in the figure, systems architecting is now a major portion of what this process constitutes as *systems analysis*. It is during this activity that software systems engineering is considered.

Role of Software in Systems

The development of software has coincided with the evolution of digital computing in the second half of the twentieth century, which in turn has been driven by the growth of semiconductor technology. Software is the control and processing element of data systems (see Chapter 3). It is the means by which a digital computer is directed to operate on sources of data to convert the data into useful information or action. In the very early days of computers, software was used to enable crude versions of computers to calculate artillery tables for the World War II effort. Software is being used today to control computers ranging from single chips to tremendously powerful supercomputers to perform an almost infinite variety of tasks. This versatility and potential power makes software an indispensable ingredient in modern systems, simple and complex.

While software and computer hardware are inextricably linked, the histories of their development have been very different. Computers, which consist largely of semiconductor chips, tend to be standardized in design and operation. All of the processing requirements of specific applications are, therefore, incorporated into the software. This division of function has made it possible to put great effort into increasing the speed and capability of computers while maintaining standardization and keeping computer costs low by mass production and marketing. Meanwhile, to handle increasing demands, software has grown in size and complexity, becoming a dominant part of the majority of complex systems.

A traditional view of the role of software in a computer system is represented in Figure 11.2. The figure shows the layering of software and its relationship to the user and to the machine on which it runs. The user can be either a human operator or another computer. The user is seen to interact with all layers through a variety of interfaces. The figure shows that the user interface is wrapped around all the software layers, as

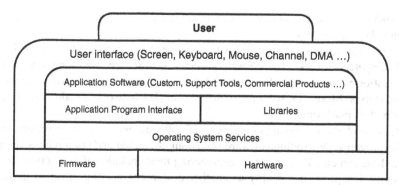

Figure 11.2. Software hierarchy.

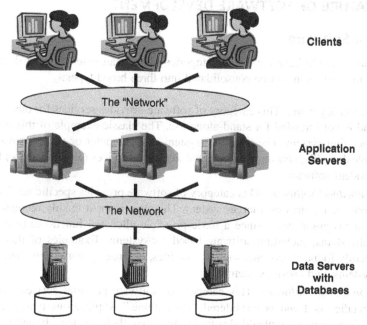

Figure 11.3. Notional three-tier architecture.

well as having some minimal interaction directly with the hardware. Software at the application layer is the essence of the computer system, and it is the application that is supported by the other layers.

Modern software systems are rarely found within single, stand-alone computers, such as that represented in this figure. Today, software is found across complex networks of routers, servers, and clients, all within a multitiered architecture of systems. Figure 11.3 depicts a simplified three-tier architecture utilizing thin clients over a series of networks. Within each component of the architecture, a similar hierarchy as depicted in Figure 11.2 is resident.

As one can imagine, the complexity of computer systems (which should not be called computer networks) has grown significantly. Software is no longer dedicated to single platforms, or even platform types, but must operate across heterogeneous hardware platforms. Moreover, software manages complex networks in addition to managing individual platforms.

Because of the increasing complexity of software and its ever-increasing role in complex systems, developing software is now an integrated and comprehensive part of system development. Thus, systems engineering must include software engineering as an integral discipline, not simply as another design engineering effort to implement functionality.

11.2 NATURE OF SOFTWARE DEVELOPMENT

Types of Software

While many people have presented categories of software over the past decades, we find that most of them can be consolidated into three broad types:

- *System Software.* This category of software provides services for other software and is not intended for stand-alone use. The classic example of this type is the operating system. The operating system of a computer or server provides multiple data, file, communications, and interface services (to name a few) for other resident software.
- *Embedded Software.* This category of software provides specific services, functions, or features of a larger system. This type is most readily recognized with systems engineering since a basic principle allocated functionality to specific subsystems, including software-based subsystems. Examples of this type are readily found in systems such as satellites, defense systems, homeland security systems, and energy systems.
- *Application Software.* This category of software provides services to solve a specific need and is considered "stand-alone." Applications typically interact with system and embedded software to utilize their services. Examples include the popular office productivity applications—word processors, spreadsheets, and presentation support.

Although these three categories cover the wide variety of software today, they do not provide any understanding of the multiple specialties that exist. Table 11.1 is presented to provide an additional categorization. The three major software categories are shown in the table for comparison. Four additional categories are presented: engineering/scientific, product line, Web based, and artificial intelligence. While all four fall under one or more of the three major categories, each type also addresses particular niches in the software community.

TABLE 11.1. Software Types

Software type	Short description	Examples
System	A system software provides services to other software.	Operating system, network manager
Embedded	An embedded software resides within a larger system and implements specific functions or features.	GUI, navigation software
Application	An application software is a stand-alone program that solves a specific need.	Business software, data processors, process controllers
Engineering/ scientific	An engineering/scientific software utilizes complex algorithms to solve advanced problems in science and engineering.	Simulations, computer-aided design
Product line	A product-line software is intended for wide use across a spectrum of users and environments.	Word processing, spreadsheets, multimedia
Web based	A Web-based software, sometimes called Web applications, is specifically designed for wide area network usage.	Internet browsers, Web site software
Artificial intelligence	An artificial intelligence software is distinguished by its use of nonnumerical algorithms to solve complex problems.	Robotics, expert systems, pattern recognition, games

Types of Software Systems

While software has become a major element in virtually all modern complex systems, the task of systems engineering a new system may be very different depending on the nature of functions performed by the software system elements. Despite the fact that there are no commonly accepted categories for different types of systems, it is useful to distinguish three types of software systems, which will be referred to as software-embedded systems, software-intensive systems, and computing-intensive systems. The term "software-dominated systems" will be used as inclusive of software systems in general.

The characteristics of the three categories of software-dominated systems and familiar examples are listed in Table 11.2 and are described more fully below.

Software-Embedded Systems. Software-embedded systems (also referred to as software-shaped systems, real-time systems, or sociotechnical systems) are hybrid combinations of hardware, software, and people. This category of systems is one in which the principal actions are performed by hardware but with software playing a major supporting role. Examples are vehicles, radar systems, computer-controlled manufacturing machinery, and so on. The function of software is usually that of performing critical control functions in support of the human operators and the active hardware components.

TABLE 11.2. Categories of Software-Dominated Systems

Characteristic	Software-embedded systems	Software-intensive systems	Data-intensive computing systems
Objective	Automate complex subsystems to perform faster and more accurately	Manipulate masses of information to support decisions or to acquire knowledge	Solve complex problems, model complex systems by computation and simulation
Functions	Algorithmic, logical	Transactional	Computational
Inputs	Sensor data, controls	Information, objects	Data numeric patterns
Processing	Real-time computation	Manipulation, GUI, networking	Non-real-time computation
Outputs	Actions, products	Information, objects	Information
Timing	Real time, continuous	Intermittent	Scheduled
Examples	Air traffic control Military weapons systems Aircraft navigation and control	Banking network Airline reservation system Web applications	Weather predictions Nuclear effect prediction Modeling and simulation
Hardware	Mini and micro processors	N-tier architectures	Supercomputers
Typical users	Operators	Managers	Scientists, analysts

Software-embedded systems usually run continuously, typically on embedded microprocessors (hence the designation), and the software must therefore operate in real time. In these systems, software is usually embodied in components designed in accordance with requirements flowed down from system and subsystem levels. The requirements may be specified for individual software components or for a group of components operating as a subsystem. In these systems, the role of software can range from control functions in household appliances to highly complex automation functions in military weapons systems.

Software-Intensive Systems. Software-intensive systems, which include all information systems, are composed largely of networks of computers and users, in which the software and computers perform virtually all of the system functionality, usually in support of human operators. Examples include automated information processing systems such as airline reservations systems, distributed merchandising systems, financial management systems, and so on. These software-intensive systems usually run intermittently in response to user inputs and do not have as stringent requirements on latency as real-time systems. On the other hand, the software is subject to system-level requirements directly linked to user needs. These systems can be very large and distributed over extended networks. The World Wide Web is an extreme example of a software-intensive system.

In software-intensive systems, software is key at all levels, including the system control itself. Hence, these must be systems engineered from the beginning. Most of them can be thought of as "transactional" systems (financial, airline reservation, command, and control). They are generally built around databases that contain domain information entities that must be accessed to produce the desired transaction.

Data-Intensive Computing Systems. A type of software system that is significantly different from the above software system categories includes large-scale computing resources dedicated to executing complex computational tasks. Examples are weather analysis and prediction centers, nuclear effects prediction systems, advanced information decryption systems, and other computationally intensive operations.

These data-intensive computing systems usually operate as facilities in which the computing is typically performed either on supercomputers or on assemblies of high-speed processors. In some cases, the processing is done by a group of parallel processors, with computer programs designed for parallel operation.

The development of data-intensive computing systems requires a systems approach like other systems. However, most of these are one of a kind and involve very specialized technical approaches. Accordingly, this chapter will be focused on the systems engineering problems associated with the much more common software-embedded and software-intensive systems.

Differences between Hardware and Software

It was noted at the beginning of this chapter that there are a number of fundamental differences between hardware and software that have profound effects on the systems engineering of software-dominated systems. Every systems engineer must have a clear appreciation of these differences and their import. The following paragraphs and Table 11.3 are devoted to describing software systems and how they differ significantly from hardware.

Structural Units. Most hardware components are made up of standard physical parts, such as gears, transistors, motors, and so on. The great majority are implementations of commonly occurring functional elements, such as "generate torque" or "process data" (see Chapter 3). In contrast, software structural units can be combined in countless different ways to form the instructions that define the functions to be performed by the software. There is not a finite set of commonly occurring functional building blocks, such as makeup hardware subsystems and components. The main exceptions are generic library functions (e.g., trigonometric) contained in some software programming environments and certain commercial software "components" mostly related to graphic user interface functions.

Interfaces. Because of its lack of well-defined physical components, software systems tend to have many more interfaces, with deeper and less visible interconnections than hardware systems. These features make it more difficult to achieve good system modularity and to control the effects of local changes.

TABLE 11.3. Differences between Hardware and Software

Attribute	Hardware	Software	Software engineering complications
Structural units	Physical parts, components	Objects, modules	Few common building blocks, rare component reuse
Interfaces	Visible at component boundaries	Less visible, deeply penetrating numerous	Difficult interface control, lack of modularity
Functionality	Limited by power, accuracy	No inherent limit (limited only by hardware)	Very complex programs, difficult to maintain
Size	Limited by space, weight	No inherent limits	Very large modules, difficult to manage
Changeability	Requires effort	Deceptively easy but risky	Difficult configuration management
Failure mode	Yields before failing	Fails abruptly	Greater impact of failures
Abstraction	Consists of physical elements	Textual and symbolic	Difficult to understand

Functionality. There are no inherent limits on the functionality of software as there are on hardware due to physical constraints. For this reason, the most critical, complex, and nonstandard operations in systems are usually allocated to software.

Size. While the size of hardware components is limited by volume, weight, and other constraints, there is no inherent limit to the size of a computer program, especially with modern memory technology. The large size of many software-based systems constitutes a major systems engineering challenge because they can embody an enormous amount of custom-built system complexity.

Changeability. Compared to the effort required to make a change in a hardware element, it is often falsely perceived to be easy to make changes in software, that is, "merely" by altering a few lines of code. The impacts of software changes are more difficult to predict or determine due to the complexity and interface problems cited above. A "simple" software change may require retesting of the entire system.

Failure Modes. Hardware is continuous in both structure and operation, while software is digital and discontinuous. Hardware usually yields before it fails and tends to fail in a limited area. Software tends to fail abruptly, frequently resulting in a system breakdown.

Abstraction. Hardware components are described by mechanical drawings, circuit diagrams, block diagrams, and other representations that are models of physical elements readily understood by engineers. Software is inherently abstract. Besides the actual code, architectural and modeling diagrams are highly abstract and each diagram restricted in its information context. Abstractions may be the single most fundamental difference between software and hardware.

The above differences, summarized in Table 11.3, profoundly affect the systems engineering of complex software-dominated systems. Not appreciating these differences and effectively accounting for them have contributed to a number of spectacular failures in major programs, such as an attempted modernization of the air traffic control system, the initial data acquisition system for the Hubble telescope, the Mars Lander spacecraft, and an airport baggage handling system.

For the majority of systems engineers who do not have experience in software engineering, it is essential that they acquire a grounding in the fundamentals of this discipline. The following sections are intended to provide a brief overview of software and the software development process.

11.3 SOFTWARE DEVELOPMENT LIFE CYCLE MODELS

As described in previous chapters, every development project passes through a series of phases as it evolves from its inception to its completion. The concept of a life cycle model is a valuable management tool for planning the activities, staffing, organization, resources, schedules, and other supporting activities required for a project's successful execution. It is also useful for establishing milestones and decision points to help keep the project on schedule and budget.

Chapter 4 described a system life cycle model appropriate for developing, producing, and fielding a typical, new large-scale complex system. It was seen to consist of a series of steps beginning with the establishment of a bona fide need for a new system and systematically progressing to devising a technical approach for meeting the need; engineering a hardware/software system embodying an effective, reliable, and affordable implementation of the system concept; validating its performance; and producing as many units as required for distribution to the users/customers.

The software elements in software-embedded systems perform critical functions, which are embodied in components or subcomponents. Therefore, their system life cycle is governed by the nature of the system and major subsystems and generally follows the steps characteristic of systems in general, as described in Chapters 4 and 6–10. A significant feature of the life cycle of software-embedded systems is the fact that there is no production for the software elements themselves, only of the processors on which the software runs. Also, there is cause for caution in that software elements are deceptively complex for their size and usually play critical roles in system operation. Hence, special measures for risk reduction in this area need to be considered.

Basic Development Phases. Just as the systems engineering method was seen to consist of four basic steps (Fig. 4.10),

1. requirements analysis,
2. functional definition,

3. physical definition, and
4. design validation,

so also the software development process can be resolved into four basic steps:

1. analysis;
2. design, including architectural, procedural, and so on;
3. coding and unit test, also called implementation; and
4. test, including integration and system test.

Although not strictly coincident with the systems engineering method, the general objectives of each of these steps correspond closely.

It should be noted that like the systems engineering method, different versions of the software process use variations in terminology in naming the steps or phases, and some split up one or more of the basic steps. For example, design may be divided into preliminary design and detailed design; unit test is sometimes combined with coding or made a separate step. System test is sometimes referred to as integration and test. It must be remembered that this stepwise formulation is a model of the process and hence is subject to variation and interpretation.

For the category of software-intensive systems, which have come to dominate communication, finance, commerce, entertainment, and other users of information, there are a variety of life cycle models in use. A few notable examples of these are discussed briefly in the following paragraphs. Detailed discussions of software life cycles may be found in the chapter references and in other sources.

As in the case of system life cycle models, the various software process models involve the same basic functions, differing mainly in the manner in which the steps are carried out, the sequencing of activities, and in some cases the form in which they are represented. Overall, software development generally falls into four categories:

1. *Linear.* Like formal system development life cycle models, the linear software development model category consists of a sequence of steps, typically with feedback, resulting in a software product. Linear development models work well in environments with well-understood and stable requirements, reasonable schedules and resources, and well-documented practices.
2. *Incremental.* Incremental models utilize the same basic steps as linear models but repeat the process in multiple iterations. In addition, not every step is performed to the same degree of detail within each iteration. These types of development models provide partial functionality at incremental points in time as the system is developed. They work well in environments with stable requirements where partial functionality is desired before the full system is developed.
3. *Evolutionary.* Evolutionary models are similar to the incremental concept but work well in environments where the final product's characteristics and attributes are not known at the beginning of the development process. Evolutionary models provide limited functionality in nonproduction forms (e.g., prototypes)

for experimentation, demonstration, and familiarization. Feedback is critical to evolutionary models as the system "evolves" to meet the needs of the users through these three procedures.

4. *Agile.* Agile development models deviate most from the four basic steps we have identified above. With linear, incremental, and evolutionary models, the four steps are manipulated into different sequences and are repeated in different ways. Within agile development environments, the four steps are combined in some manner and the delineations between them are lost. Agile methods are appropriate for environments where structure and definition are not available, and change is the constant throughout the process.

In addition to the four basic development model categories above, specialized development models have been proffered, practiced, and published. Two well-known examples are the component-based development model and the aspect-oriented development model. These special-purpose models have specific but limited applications warranting their use. We have chosen to omit these specialized models from our discussions.

Linear Development Models

The *waterfall model* is the classic software development life cycle, also called the "sequential" model (see Fig. 11.4). It consists of a sequence of steps, systematically

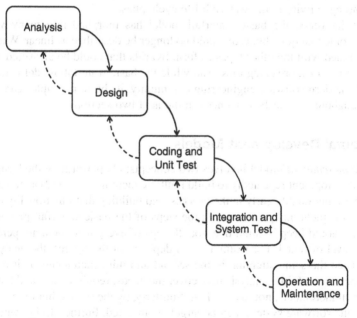

Figure 11.4. Classical waterfall software development cycle.

TABLE 11.4. Systems Engineering Life Cycle and the Waterfall Model

System phase	Objective	Waterfall phase
Needs analysis	Establish system need and feasibility	Analysis
Concept exploration	Derive necessary system	Analysis
Concept definition	Select a preferred system architecture	Design
Advanced development	Build and test risky system elements	Design (and prototype)
Engineering design	Engineer system components to meet performance requirements	Coding and unit test
Integration and evaluation	Integrate and validate system design	Integration and system test
Production	Production and distribution	None
Operations and support	Operation	Maintenance

proceeding from analysis to design, coding and unit test, and integration and system test. The waterfall model with feedback (see dashed arrows) depicts the adjustment of inputs from a preceding step to resolve unexpected problems before proceeding to the subsequent step. The waterfall model corresponds most closely to the conventional system life cycle. Table 11.4 lists the system life cycle phases, their objectives, and the corresponding activity in the waterfall life cycle phase.

Over the years, the basic waterfall model has morphed into many variations, including some that quite honestly could no longer be described as linear. Waterfall has been combined with the other types to form hybrids that could be classified as a combination of two or more categories. And while the basic waterfall model is rarely used in today's modern software engineering community, its basic principles can be recognized throughout, as will be evidenced in the next two sections.

Incremental Development Models

The *basic incremental* model involves two concepts: (1) performing the basic steps of software development repeatedly to build multiple increments and (2) achieving partial operational functionality early in the process, and building that functionality over time. Figure 11.5 depicts this process using the steps of the basic waterfall process model. The reader should keep in mind that not all steps of every increment are performed to the same level of detail. For example (and depicted in the figure), the analysis phase may not need the same attention in the second and third increments as it received in the first increment. Initial analysis may cover the needs, requirements, and feature definition for all increments, not just the first. Similarly, by the second iteration, the overall design of the software system may be largely completed. Further design would not be needed in the third iteration.

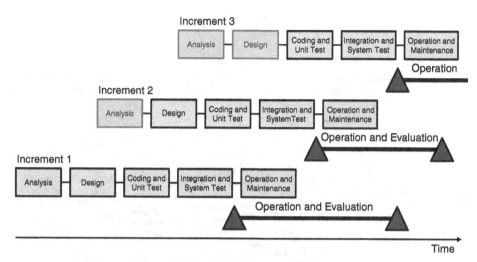

Figure 11.5. Software incremental model.

Another aspect of incremental development concerns the incremental releases, sometimes called "builds." As a new increment is released, older increments may be retired. In its purest form, once the last increment is released, all of the older increments are retired. Of course, situations arise when customers are fully satisfied with an increment—leading to multiple increments, and thus versions of the software—or future increments are cancelled. This is depicted in the figure by the triangles.

The *rapid application development* (RAD) model (sometimes called the "all-at-once" model) features an incremental development process with a very short cycle time. It is an iterative form of the waterfall model, depending on the use of previously developed or commercially available components. Its use is best suited to business application software of limited size that lends itself to relatively quick and low-risk development, and whose marketability depends on deployment ahead of an anticipated competitor.

Evolutionary Development Models

In situations where user needs and requirements are not well defined, and/or development complexity is sufficiently high to incur significant risk, an evolutionary approach may be best. The basic concept involves the development of an early software product, or prototype. The prototype is not intended for actual operations, sales, or deployment, but to assist in identifying and refining requirements, or in reducing development risks. If the purpose of the prototype is identifying and refining requirements, then typically, an experimental version of the system, or a representative portion that exhibits the characteristics of the user interface, is built early in the design phase of the development and operated by the intended user or a surrogate of the projected user. With the flexibility of software, such a prototype can often be designed and built relatively quickly and inexpensively. Attention to formal methods, documentation, and quality design need not be implemented, since the version is not intended for production.

Spiral

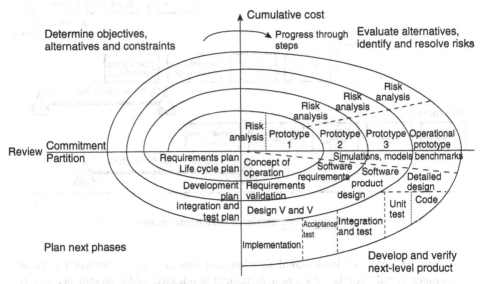

Figure 11.6. Spiral model.

In addition to refining requirements by building trial user interfaces, software prototyping is often used as a general risk reduction mechanism as in the advanced development phase. New design constructs can be prototyped early to refine the approach. Interfaces with other hardware and software can also be developed and tested early to reduce risk. As an example, consider an air traffic control system. It is often necessary to discover the real requirements of the system interfaces by testing preliminary models of the system in the field.

Perhaps the most common form of the evolutionary model is the *spiral model*. It is similar to that pictured in Figure 4.12 but is generally much less formal and with shorter cycles. Figure 11.6 depicts a version of the spiral development model. It differs in form by starting in the center and spiraling outward. The expanding spirals represent successive prototypes, which iteratively perfect the attainment of customer objectives by the system. Finally, the finishing steps are applied on the last spiral/prototype, resulting in a finished product.

With all evolutionary methods, it is important to plan for the disposition of the prototypes (or spirals) after they have been used. Examples abound where a spiral approach was adopted, and one or two prototypes were developed and tested using actual users or surrogates. However, after experiencing the prototype, the customer declared the product sufficient and requested immediate delivery. Unfortunately, without formal procedures and methods in place, nor general quality assurance followed in the prototype development, the "final product" was in no condition to be deployed

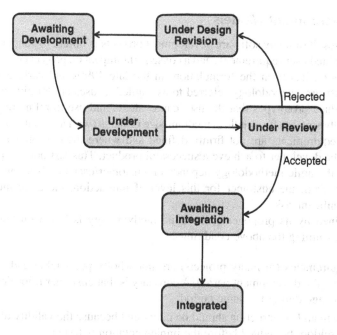

Figure 11.7. State transition diagram in the concurrent development model.

in the field (or sold to the market). Upon deployment, problems ensued quickly. Our recommendation is that prototypes should be discarded upon completion of their purpose—and the customer should be forewarned of the significant risks involved in deployment prototypes as operational systems.

The second model, which falls under the evolutionary category, is the *concurrent development model*. This approach eliminates the two concepts of sequence and increments, and develops all phases simultaneously. The model achieves this approach through the definition of software development states. Software modules are tagged with which state they belong. Formal state transition criteria are defined that enables software modules to transition from one state to the other. Development teams focus on specific activities within a single state. Figure 11.7 depicts an example state transition diagram (STD) associated with this type of model.

Software modules are initially assigned to the "awaiting development" state. This state could be thought of as a queue for the development teams. A module is not transitioned to the "under development" state until a team is assigned to its development. Once completed, the module is transitioned to the "under review" state, where a review team (or person) is assigned. Again, transition does not occur until a team is assigned to the module. This process is repeated. Since modules are developed simultaneously by different teams, modules can be in the same state. A push/pull system can be implemented to increase the efficiency of the associated teams.

Agile Development Models

A common result of many software development projects is failure to adapt to changing or poorly defined user requirements and a consequent impact on project cost. A response to this situation has been the formulation, in the late 1990s and early 2000s, of an adaptive software methodology referred to as "agile." It uses an iterative life cycle to quickly produce prototypes that the user can evaluate and use to refine requirements. It is especially suitable for small- to medium-size projects (with less than 30–50 people) where the requirements are not firmly defined and where the customer is willing to work with the developer to achieve a successful product. This last point is particularly important—the agile methodology depends on customer/user involvement. Without a commitment from the customer for this level of interaction, the agile methodology incurs a significant risk.

As defined by its proponents, the agile methodology is based on the following postulates, assuming the above conditions:

1. Requirements (in many projects) are not wholly predictable and will change during the development period. A corollary is that customer priorities are likely to change during the same period.
2. Design and construction should be integrated because the validity of the design can seldom be judged before the implementation is tested.
3. Analysis, design, construction, and testing are not predictable and cannot be planned with adequate levels of precision.

These methods rely heavily on the software development team to conduct simultaneous activities. Formal requirements analysis and design are not separate steps—they are incorporated in the coding and testing of software. This concept is not for the faint-of-heart customer—a great level of trust is required. Nevertheless, agile methods represent a leap in software development that can lead to highly robust software more quickly than traditional methods.

Agile methods include a number of recent process models:

- *Adaptive Software Development (ASD)* focuses on successive iterations of three activities: speculation, collaboration, and learning. The initial phase, speculation, focuses on the customer's needs and mission. The second phase, collaboration, utilizes the concept of synergistic talents working together to develop the software. The final phase, learning, provides feedback to the team, the customer, and the other stakeholders, and includes formal review and testing.
- *Extreme Programming* (XP) focuses on successive iterations of four activities: planning, design, coding, and testing. Requirements are identified through the use of user stories—informal user descriptions of features and functionality. These stories are organized and used through the iteration process, including as the basis for final testing.
- *Scrum* focuses on a short, 30-day iterative cycle—with strong teaming. This process yields several iterations in various maturities with which to learn, adapt,

and evolve. Within each cycle, a basic set of activities occurs: requirements, analysis, design, evolution, and delivery.

- *Feature-Driven Development* focuses on short iterations (typically about 2 weeks), each of which delivers tangible functionality (features) that the user values. Eventually, features are organized and grouped into modules that are then integrated in the system.
- *The Crystal* family of agile methods focuses on adaptation of a core set of agile methodologies to individual projects.

In all of the above approaches, quality and robustness are required attributes of products. Thus, the iterations are to be built on rather than thrown away (in contrast to the incremental and spiral methods). All projects that are based on uncertain requirements should consider the above principles in deciding on the methodology to be used.

In general, the software development life cycles follow the same pattern of progressive risk reduction and system "materialization" that has been described in Chapters 3 and 5–10. The remaining sections of this chapter follow a similar structure.

Software System Upgrades

Because of the rapid evolution of IT, the associated developments in data processors, peripherals, and networks, and the perceived ease of introducing software changes, there are relatively frequent cases where system software is subjected to significant modifications or "upgrades." In a large fraction of instances, the upgrades are planned and implemented by different individuals from those responsible for their development, with the resulting probability of inadvertent interface or performance deficiencies. Such cases call for participation of and control by systems engineering staff who can plan the upgrade design from a system point of view and can ensure an adequate requirements analysis, interface identifications, application of modular principles, and thorough testing at all levels.

When the system to be upgraded was designed before the general use of modern programming languages, there can be a severe problem of dealing with an obsolete language no longer supported by modern data processors. Such legacy software is generally not capable of being run on modern high-performance processors, and the programs, which total billions of lines of code, have to either be rewritten or translated into a modern language. The cost of the former is, in many cases, prohibitive, and the latter has not come into general practice. The result has been that many of these systems continue to use obsolete hardware and software and are maintained by a dwindling group of programmers still capable of dealing with the obsolete technology.

11.4 SOFTWARE CONCEPT DEVELOPMENT: ANALYSIS AND DESIGN

The analysis and design steps in the traditional software life cycle described in the previous sections generally correspond to the concept development stage that is embodied in Part II of this book. These are the activities that define the requirements and architecture of the software elements of the system. The line of demarcation between analysis and design may vary substantially among projects and practitioners, there being broad areas referred to as design analysis or design modeling. For this reason, the subsections below will focus more on approaches and problems that are of special interest to systems engineers than on issues of terminology.

Needs Analysis

The precondition for the development of any new system is that it is truly needed, that a feasible development approach is available, and that the system is worth the effort to develop and produce it. In the majority of software-intensive systems, the main role of software is to automate functions in legacy systems that have been performed by people or hardware, to do them at less cost, in less time, and more accurately. The issue of need becomes one of trading off the projected gains in performance and cost against the effort to develop and deploy the new system.

In new systems in which key operations performed by people or hardware are to be replaced with software, users are typically not unanimous regarding their needs, and the optimum degree of automation is seldom determinable without building and testing. Further, an extensive market analysis is usually necessary to gauge the acceptance of an automated system and the costs and training that this entails. Such an analysis also usually involves issues of market penetration, customer psychology, introduction trials, and corporate investment strategy.

Feasibility Analysis. The decision to proceed with system design has been seen to require the demonstration of technical feasibility. Within the realm of software, almost anything appears feasible. Modern microprocessors and memory chips can accommodate large software systems. There are no clear size, endurance, or accuracy limits such as there are on hardware components. Thus, technical feasibility tends to be taken for granted. This is a great advantage of software but also invites complexity and the assumption of challenging requirements. However, the resulting complexity may in itself prove too difficult and costly.

Software Requirements Analysis

The scope of the requirements analysis effort for a new system usually depends on whether the software is an element in a software-embedded system or if it embraces a total software-intensive system. In either case, however, the development of a concept of operations should play an important part.

Software-Embedded System Components. As noted previously, the software elements in software-embedded systems are usually at the component level, referred to as computer system configuration items (CSCIs). Their requirements are generated at the system and subsystem levels and are allocated to CSCIs, usually in a formal requirement specification document. The software team is expected to design and build a product to these specifications.

Too often, such specifications are generated by systems engineers with an inadequate knowledge of software capabilities and limitations. For example, a large dynamic range in combination with high precision may be prescribed, which may unduly stress the system computational speed. Other requirement mismatches may result from the communication gap that frequently exists between systems and software engineers and organizations. For such reasons, it is incumbent on the software development team to make a thorough analysis of requirements allocated to software and to question any that fail to have the characteristics described in Chapter 7. These reasons also constitute a good argument for including software engineers in the top-level requirements analysis process.

Software-Intensive System Requirements. As noted earlier, in a software-intensive system, software dominates every aspect and must be an issue at the highest level of system requirements analysis. Thus, the very formulation of the overall system requirements must be subject to analysis and participation by software systems engineers.

The basic problems in developing system requirements for software-intensive systems are fundamentally the same as for all complex systems. However, there are several aspects that are peculiar to requirements for systems that depend on the extensive software automation of critical control functions. One special aspect has been noted previously, namely, unreasonable performance expectations based on the extensibility of software. Another is the generally diverse customer base, with little understanding of what software automation is capable of doing, and hence is often not a good source of requirements.

The consequences of these and other factors that inhibit the derivation of a reliable set of requirements typically result in a considerable degree of uncertainty and fluidity in software-based system requirements. This is a major reason for the use of prototyping, RAD, or evolutionary development, all of which produce an early version of the system that can be subjected to experimentation by users to modify and firm up initial assumptions of desired system characteristics.

Several variations of developing software requirements exist today. Of course, many depend on the type of software development model being used; however, some generic features exist regardless of the model chosen. Figure 11.8 depicts a hierarchy of software requirements, starting with the user needs at the apex. These needs are decomposed into desired features, functional and performance requirements, and finally specifications. If the system in question is software embedded, the upper levels of the hierarchy are typically performed at the system level and requirements or specifications are allocated to software subsystems or components.

If the system in question is software intensive, the upper levels of the hierarchy are needed. In those cases, a separate process for developing and refining requirements may be needed. Several processes have been offered in the literature. A generic process is presented in Figure 11.9. Four steps, which can be further divided into separate steps, are critical to this effort:

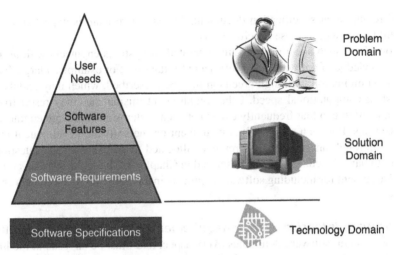

Figure 11.8. User needs, software requirements, and specifications.

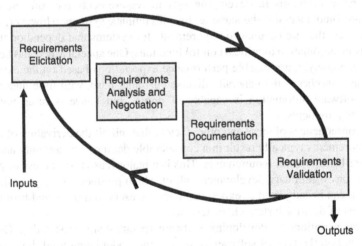

Figure 11.9. Software requirements generation process.

- *Requirements Elicitation.* This step seems straightforward but, in reality, can be challenging. Bridging the language barrier between users and developers is not simple. Although tools have been developed to facilitate this process (e.g., use cases, described below), users and developers simply do not speak the same language. Many elicitation methods exist—from direct interaction with stakeholders and users, involving interviews and surveys, to indirect methods, involving observation and data collection. Of course, prototyping can be of valuable use.

- *Requirements Analysis and Negotiation.* Chapter 7 described a series of methods to analyze and refine a set of requirements. These are applicable to software as much as they are to hardware. In general, these techniques involve checking four attributes of a requirements set: necessity, consistency, completeness, and feasibility. Once requirements have been refined, they need to be accepted—this is where negotiation begins. Requirements are discussed with stakeholders and are refined until agreement is reached. When possible, requirements are prioritized and problematic requirements are resolved. A more advanced analysis is then performed, examining the following attributes: business goal conformity, ambiguity, testability, technology requirements, and design implications.
- *Requirements Documentation.* Documentation is always the obvious step and can be omitted since everyone is expecting the requirements to be documented. We include it because of the criticality in articulating and distributing requirements to the entire development team.
- *Requirements Validation.* This step can be confusing because many engineers include "analysis" in this step, that is, the concept that each requirement is evaluated to be consistent, coherent, and unambiguous. However, we have already performed this type of analysis in our second step above. Validation in this context means a final examination of the requirements set in whole to determine whether the set will ultimately meet the needs of the users/customers/parent system. Several methods exist to enable requirement validation—prototyping, modeling, formal reviews, manual development, and inspection—even test case development can assist in the validation process.

Use Cases. As mentioned in Chapter 8, a popular tool available to requirements engineers is the *use case*. A use case has been best described as a story, describing how a set of actors interact with a system under a specific set of circumstances. Because the set of circumstances can be large, even infinite, the number of possible use cases for any system can also be large. It is the job of the requirements engineer, developers, users, and systems engineer to limit the number and variety of use cases to those that will influence the development of the system.

Use cases represent a powerful tool in bridging the language gap between users, or any stakeholder, and developers. All can understand sequences of events and activities that need to be performed. Although use cases were developed for describing software system behavior and features, they are regularly used in the systems world to describe any type of system, regardless of the functionality implemented by software.

Interface Requirements. Whichever the type of an essential tool of requirements analysis is the identification of all external interfaces of the system, and the association of each input and output with requirements on its handling within the system. This process not only provides a checklist of all relevant requirements but also a connection between internal functions required to produce external outcomes. In all software-dominated systems, this approach is especially valuable because of the numerous subtle interactions between the system and its environment, which may otherwise be missed in the analysis process.

System Architecture

It was seen in Chapter 8 that in complex systems, it is absolutely essential to partition them into relatively independent subsystems that may be designed, developed, produced, and tested as separate system building blocks, and similarly to subdivide the subsystems into relatively self-contained components. This approach handles system complexity by segregating groups of mutually interdependent elements and highlighting their interfaces. This step in the systems engineering method is referred to as functional definition or functional analysis and design (Fig. 4.10).

In hardware-based systems, the partitioning process not only reduces system complexity by subdividing it into manageable elements but also serves to collect elements together that correspond to engineering disciplines and industrial product lines (e.g., electronic, hydraulic, structural, and software). In software-intensive systems, the segregation by discipline is not applicable, while the inherent complexity of software makes it all the more necessary to partition the system into manageable elements. Software has numerous subdisciplines (algorithm design, databases, transactional software, etc.), which may, in certain cases, provide partitioning criteria. In systems that are distributed, the characteristics of the connective network can be used to derive the system architecture.

Software Building Blocks. The objective of the partitioning process is to achieve a high degree of "modularity." The principles that guide the definition and design of software components are intrinsically similar to those that govern hardware component design, but the essentially different nature of the implementation results in significant differences in the design process. One fundamental difference is in regard to commonly occurring building blocks such as those described in Chapter 3. There is a profusion of standard commercial software packages, especially for business and scientific applications (e.g., word processors, spread sheets, and math packages), but rarely for system components. Exceptions to this general situation are the commercial-off-the-shelf (COTS) software components heavily used in low-complexity information systems.

Another source of software building blocks is that of common objects (COs). These are somewhat the equivalents in software to standard hardware parts such as gears or transformers, or at higher levels to motors or memory chips. They are most often used in the graphical user interface (GUI) environment. The CO concept is represented by the Microsoft-developed distributed common object model (DCOM). A more vendor-independent implementation is the common object resource broker architecture (CORBA), which is a standard defined by the Object Management Group (OMG), an organization committed to vendor neutral software standards. However, these CO components comprise only a small fraction of system design. The result is that despite such efforts at "reuse," the great majority of new software products are very largely unique.

Modular Partitioning. Despite the lack of standard parts, software modules nevertheless can be well structured, with an ordered hierarchy of modular subdivisions and well-defined interfaces. The same principles of modularity to minimize the inter-

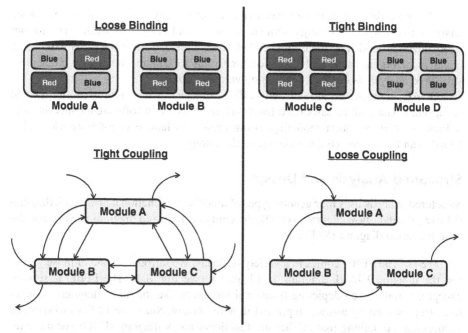

Figure 11.10. Principles of modular partitioning.

dependence of functional elements that apply to hardware components are applicable also to computer programs.

The principles of modular partitioning are illustrated in Figure 11.10. The upper patterns show the elements of "binding," also referred to as "cohesion," which measures the mutual relation of items within software modules (represented by boxes with the names of colors). It is desirable for binding to be "tight"—all closely related items should be grouped together in a single functional area. Conversely, unrelated and/or potentially incompatible items should be located in separate areas.

The lower two diagrams illustrate the elements of "coupling," which measures the interactions between the contents of different modules (boxes). With tight coupling as illustrated at the left, any change within a module will likely dictate changes in each of the other two modules. Conversely, with "loose" coupling, interactions between the modules are minimized. The ideal arrangement, usually only partially achievable, is illustrated in the right-hand diagram, where interactions between modules are kept simple and data flows are unidirectional. This subject is discussed further below as it relates to different design methodologies.

Architecture Modeling. As noted in Chapter 10, models are an indispensable tool of systems engineering for making complex structures and relationships understandable to analysts and designers. This is especially true in software-dominated systems where the abstract nature of the medium can make its form and function virtually incomprehensible.

The two main methodologies used to model software systems are called "structured analysis and design" and "object-oriented analysis and design (OOAD)." The former is organized around functional units called procedures and functions. It is based on a hierarchical organization and uses decomposition to handle complexity. Generally, structured analysis is considered a top-down methodology.

OOAD is organized around units called "objects," which represent entities and encapsulate data with its associated functions. Its roots are in software engineering and it focuses on information modeling, using classes to handle complexity. Generally, OOAD can be considered a bottom-up methodology.

Structured Analysis and Design

Structured analysis uses four general types of models: the functional flow block diagram (FFBD), the data flow diagram (DFD), the entity relationship diagram (ERD), and the state transition diagram (STD).

FFBD. The FFBD comes in a variety of forms. We introduced one of those varieties, the functional block diagram, in Chapter 8 (see Fig. 8.4). The FFBD is similar, except that rather than depicting functional interfaces like the block diagram, connections (represented by arrows) represent flow of control. Since the FFBD incorporates sequencing (something that neither the functional block diagram (FBD) nor the integrated definition 0 (IEDF0) formats do), logical breaking points are depicted by summing gates. These constructs enable the depiction of process-oriented concepts. Almost any process can be modeled using the FFBD. Figure 11.11 is an example of an FFBD.

As with all functional diagrams, each function within the hierarchy can be decomposed into subfunctions, and a corresponding diagram can be developed at each level. Functional diagrams are the standard method within structured analysis to depict a system's behavior and functionality.

DFD. This diagram consists primarily of a set of "bubbles" (circles or ellipses) representing functional units, connected by lines annotated with the names of data flowing between the units. Data stores are represented by a pair of parallel lines and external entities are shown as rectangles. Figure 11.12 shows a DFD for the checkout function of a small public library system.

A system is normally represented by DFDs at several levels, starting with a context diagram in which there is only one bubble, the system, surrounded by external entity rectangles (see Fig. 3.2). Successive levels break down each of the bubbles at the upper levels into subsidiary data flows. To systems engineers, a software DFD is similar to the functional flow diagram except for the absence of control flow.

ERD. The ERD model defines the relationships among data objects. In its basic form, the entities are shown as rectangles and are connected by lines representing the relationship between them (shown inside a diamond). In addition to this basic ERD notation, the model can be used to represent hierarchical relationships and types of associations among objects. These models are extensively used in database design.

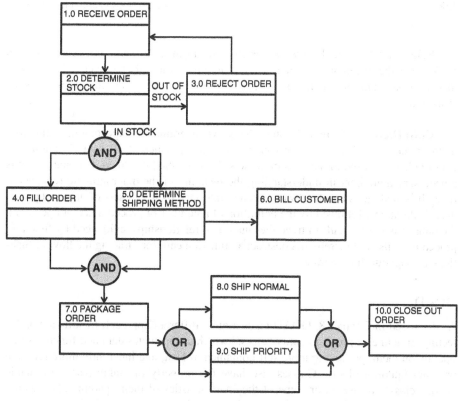

Figure 11.11. Functional flow block diagram example.

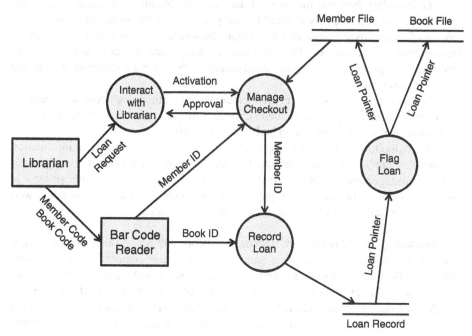

Figure 11.12. Data flow diagram: library checkout.

STD. An STD models how the system behaves in response to external events. An STD shows the different states that the system passes through, the events that cause it to transition from one state to another, and the actions taken to effect the state transition.

Data Dictionary. In addition to the above diagrams, an important modeling tool is an organized collection of the names and characteristics of all data, function, and control elements used in the system models. This is called the "data dictionary" and is a necessary ingredient in understanding the meaning of the diagrammatic representations. It is analogous to a hardware part and interface listing of sets of data and procedure declarations, followed by the definition of a number of procedures that operate on the data. It is not difficult to trace the functional relationships, evidenced by function/procedure calls, and thereby to construct a "function call tree" tracing the flow of functions throughout the program.

OOAD

As discussed in Chapter 8, OOAD takes a quite different approach to software architecting. It defines a program entity "class," which encapsulates data and functions that operate on them, producing more self-contained, robust, and inherently more reusable program building blocks. Classes also have the property of "inheritance" to enable "child" classes to use all or some of the characteristics of their "parent" class with a resultant reduction of redundancy. An object is defined as an instance of a class.

The boundary between the steps of analysis and design in object-oriented (OO) methodology is not precisely defined by the practitioners but generally is where the process of understanding and experimentation changes to one of synthesizing the architectural form of the system. This step also involves some experimentation, but its objective is to produce a complete specification of the software required to meet the system requirements.

The construction of the system architecture in OO methodology consists of arranging related classes into groups—called subsystems or packages—and of defining all of the relations/responsibilities within and among the groups.

OO methodology has been especially effective in many modern information systems that are largely transactional. In such programs as inventory management, financial management, airline reservation systems, and many others, the process is largely the manipulation of objects, physical or numerical. OO methods are not as well suited for primarily algorithmic and computational programs.

Modeling and Functional Decomposition. Object-oriented design (OOD) also has the advantage of using a precisely defined and comprehensive modeling language—the Unified Modeling Language (UML). This provides a powerful tool for all stages of program development. The characteristics of UML are described in Chapter 8.

A shortcoming of the OO methodology as commonly practiced is that it does not follow a basic systems engineering principle—that of managing complexity by partitioning the system into a hierarchy of loosely coupled subsystems and components.

This is accomplished by the systems engineering step of functional decomposition and allocation. By focusing on objects (things) rather than functions, OOD tends to build programs from the bottom-up rather than the top-down approach inherent in the systems engineering method.

OOD does have a structural element, the use case, which is basically a functional entity. As described above, use cases connect the system's external interfaces (actors) with internal objects. The application of use cases to design the upper levels of the system architecture and introducing objects at lower levels may facilitate the application of systems engineering principles to software system design. This approach is described in Rosenberg's book, *Use Case Driven Object Modeling with UML.*

Strengths of UML. The UML language combines the best ideas of the principal methodologists in the field of OOAD. It is the only standardized, well-supported, and widely used software modeling methodology. It therefore serves as a high-level form of communicating software architectural information within and among organizations and individuals engaged in a development program.

Moreover, UML has been applied successfully in software-intensive systems projects. Portions of UML are also used regularly in systems engineering to assist in communicating concepts and in bridging the language gaps between engineers and users (e.g., use case diagrams) and between software and hardware engineers (e.g., communications diagrams).

A major strength of UML is the existence of commercial tools that support the construction and use of its repertoire of diagrams. In the process, these tools store all the information contained in the diagrams, including names, messages, relationships, attributes, methods (functions), and so on, as well as additional descriptive information. The result is an organized database, which is automatically checked for completeness, consistency, and redundancy. In addition, many of the tools have the property of converting a set of diagrams into C++ or Java source code down to procedure headers. Many also provide a limited degree of reverse engineering—converting source code into one or several top-level UML diagrams. These capabilities can save a great deal of time in the design process.

Other Methodologies

The growing importance of software-dominated systems, and their inherent complexity and abstractness, has engendered a number of variants of structured and OO methodologies. Two of the more noteworthy ones are briefly discussed below.

Robustness Analysis. This is an extension of OO methodology that serves as a link between OO analysis (what) and design (how). It classifies objects into three types:

1. boundary objects, which link external objects (actors) with the system;
2. entity objects, which embody the principal objects that contain data and perform services (functions); and
3. control objects, which direct the interaction among boundary and entity objects.

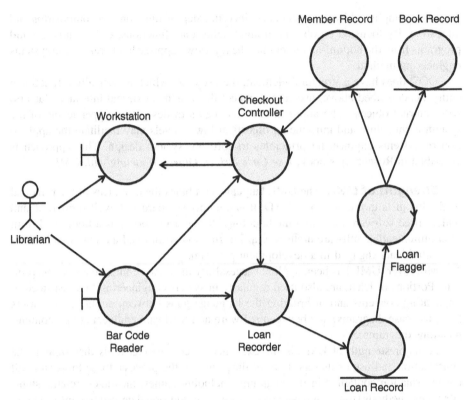

Figure 11.13. Robustness diagram: library checkout.

Robustness analysis creates a robustness diagram for each UML use case, in which the objects involved in the processing of the use case are classified as boundary and entity objects and are linked by control objects defined for the purpose. An example of a robustness diagram for the checkout use case for an automated library is shown in Figure 11.13. It is seen to resemble a functional flow diagram and to be easily understandable.

In the process of preliminary design, the robustness diagram is transformed into class, sequence, and other standard UML diagrams. Control objects may remain as controller types, or their functionality may be absorbed into methods of the other objects. To a systems engineer, robustness analysis serves as an excellent introduction to OOAD.

Function-Class Decomposition (FCD). This methodology, referred to as FCD, is a hybrid method that combines structured analysis with OO methodology. It is aimed at the top-down decomposition of complex systems into a hierarchy of functional sub-systems and components, while at the same time identifying objects associated with each unit.

As previously noted, conventional OO methodology tends to design a system from the bottom-up and has little guidance on how to group objects into packages. It is said to lead to a "flat" modular organization. The FCD method seeks to provide a top-down approach to system partitioning by using functional decomposition to define a hierarchical architecture into which objects are integrated. In so doing, it introduces the important systems engineering principle of functional decomposition and allocation into OO software system design.

FCD uses an iterative approach to partition successively lower levels of the system while at the same time also adding such objects as turn out to be needed for the lower-level functions. UML class diagrams are introduced after the first several levels are decomposed. The developers of the FCD method have demonstrated its successful use on a number of large system developments.

11.5 SOFTWARE ENGINEERING DEVELOPMENT: CODING AND UNIT TEST

The process of software engineering development consists of implementing the architectural design of system components, developed in the concept development stage, into an operational software that can control a processor to perform the desired system functions. The principal steps in this process and their systems engineering content are outlined below.

Program Structure

Software has been seen to be embodied in units called computer programs, each consisting of a set of instructions.

Program Building Blocks. A computer program may be considered to consist of several types of subdivisions or building blocks. In descending order of size, the subdivisions of a computer program and their common names are as follows:

1. A "module" or "package" constitutes a major subdivision of the overall program, performing one or more program activities. A medium to large program will typically consist of from several to tens or hundreds of modules.
2. In OO programs, a class is a unit composed of a set of "attributes" (data elements) combined with a set of associated "methods" or "services" (functions). An object is an instance of a class.
3. A function is a set of instructions that performs operations on data and controls the processing flow among related functions. A "utility" or "library function" is a commonly used transform (e.g., trigonometric function) that is supplied with an operating system.
4. A "control structure" is a set of instructions that controls the order in which they are executed. The four types of control structures are the following:

    (a) *Sequence:* a series of instructions;

    (b) *Conditional Branch:* **if** (condition) **then** (operation 1), **else** (operation 2);

    (c) *Loop:* **do while** (condition) or **do until** (condition); and

    (d) *Multiple Branch:* **case** (key 1): (operation 1) ... (key *n*) (operation *n*).

  5. An "instruction" is a "declarative" or "executable" order to the computer, composed of language key words, symbols, and names of data and functions.

  6. A language key word, symbol, or name of a data element or function.

Finally, a "data structure" is a definition of a composite combination of related data elements, such as a "record," "array," or "linked list."

As noted previously, software has no commonly occurring building blocks comparable to standard hardware parts and subcomponents such as pumps, motors, digital memory chips, cabinets, and a host of others that simplify designing and building production hardware. With few exceptions, software components are custom designed and built.

Program Design Language (PDL). A useful method for representing software designs produced by the conventional structured analysis and design methodology is PDL, sometimes called "structured English." This consists of high-level instructions formatted with control structures like an actual computer program, but consists of textual statements rather than programming language key words and phrases. PDL produces a program listing that can be readily understood by any software engineer and can be translated more or less directly into executable source codes.

OOD Representation. It was seen that OOD produces a set of diagrams and descriptive material, including defined objects that constitute intermediate program building blocks. Through the use of a UML support tool, the design information can be automatically converted into the architecture of the computer program.

Programming Languages

The choice of programming language is one of the major decisions in software design. It depends critically on the type of system—for example, whether software-embedded, software-intensive, or data-intensive computing, whether military or commercial, or whether real-time or interactive. While it is often constrained by the programming talents of the software designers, the nature of the application should have priority. A language may impact the maintainability, portability, readability, and a variety of other characteristics of a software product.

Except for very special applications, computer programs are written in a high-level language, where individual instructions typically perform a number of elementary computer operations. Table 11.5 lists a sample of past and current computer languages, their structural constituents, primary usage, and general description.

Fourth-Generation Language (4GL) and Special-Purpose Language. 4GLs are typically proprietary languages that provide higher-level methods to accomplish a

TABLE 11.5. Commonly Used Computer Languages

Language	Structural constituents	Primary usage	Description
Ada 95	• Objects • Functions • Tasks • Packages	• Military systems • Real-time systems	Designed expressly for embedded military systems, generally replaced C++
C	Functions	• Operating systems • Hardware interfaces • Real-time applications • General purpose	A powerful, general-purpose language with significant flexibility
C++	• Objects • Functions	• Simulations • Real-time applications • Hardware interfaces • General purpose	A powerful, general-purpose language that implements object-oriented constructs
COBOL	Subroutines	• Business and financial applications	A wordy language that is somewhat self-documenting, the primary language for legacy business systems
FORTRAN	• Subroutines • Functions	• Scientific • Data analysis • Simulation • General purpose	A long-standing general-purpose language used mainly for computation-intensive programs
Java	• Objects • Functions	• Internal applications • General purpose	Derived from C++, an interpretative language that is platform independent
Visual Basic	• Objects • Subroutines	• Graphical applications • User interfaces	A language that allows graphical manipulation of subprogram objects
Assembly language	• Subroutines • Macros	• Hardware control • Drivers	A language for primitive operations, enables complete machine control

problem solution in a specific domain. These 4GLs are usually coupled with a database system and are related to use of the structured query language (SQL). A key feature of 4GL tools is to bring the programming language environment as close to the natural language of the problem domain as possible and to provide interactive tools to create solutions. For example, the creation of a user input form on a workstation is carried out interactively with the programmer. The programmer enters the labels and identifies allowable entry values and any restrictions, and then the "screen" becomes part of the application. 4GLs can speed up the development time for specific applications but are generally not portable across products from different vendors.

There are many specialty areas where very efficient high-level languages have been developed. Such languages usually take on the jargon and constructs of the area they

TABLE 11.6. Some Special-Purpose Computer Languages

Language	Structural constituents	Primary usage	Description
Smalltalk and variants	Objects	• Database applications • Simulations	The original object-oriented language
LISP	Lists	• Artificial intelligence applications • Expert systems	A language based on operations of lists
Prolog	• Objects • Relationships	• Artificial intelligence applications • Expert systems	A powerful logic-based language with many variants
Perl	• Statements • Functions	• Data test manipulation • Report generation	A portable language with built-in text handling capabilities
HTML	• Tags • Identifiers • Test elements	Formatting and hyperlinking of documents	A document markup language with a unique but simple syntax
XML	• Tags • Identifiers • Strings/text	• Formatting • Field identification and linking	A textual data markup language with a unique complex syntax
PHP	• Tags • Identifiers • Strings/text • Commands	Server scripting	A document generation control language

are intended to serve. The intent of these special-purpose languages is to mimic the problem domain where possible, and to decrease development time while increasing reliability. In many cases, the special-purpose nature of such languages may limit performance for the sake of ease of use and development. When undertaking custom software development, the systems engineer should explore the availability and utility of languages in a required specialty area. Table 11.6 lists a number of special-purpose languages that have been developed for specific application domains, such as expert systems and Internet formatting.

Programming Support Tools

To support the effort of developing computer programs to implement software system design, a set of programming support tools and training in their effective use is essential. It is useful for the systems engineer and program manager to be knowledgeable about their uses and capabilities.

Editors. Editors provide programmers with the means to enter and change source code and documentation. Editors enhance the entry of programming data for specific languages. Some editors can be tailored to help enforce programming style guides.

Debuggers. Debuggers are programs that allow an application to be run in a controlled manner for testing and debugging purposes. There are two major types of debuggers: symbolic and numeric. The symbolic debugger allows the user to reference variable names and parameters in the language of the source code. A numeric debugger works at the assembly or machine code level. The computer instructions written in a programming language is called "source code." To convert the source code produced by the programmers into executable code, several additional tools are required.

Compilers. A compiler converts the source language into an intermediate format (often called object code) that is compatible for use by the hardware. In this process, the compiler detects syntax errors, omissions of data declarations, and many other programming errors, and identifies the offending statements.

A compiler is specific to the source language and usually to the data processor. Compilers for a given language may not be compatible with each other. It is important to know what standards govern the compiler that will be used and to be aware of any issues associated with code portability. Some compilers come with their own programming development environment that can increase programmer productivity and simplify the program documentation process.

Linkers and Loaders. A linker links several object code modules and libraries together to form a cohesive executable program. If there is a mixed language application (C and Java are common), the combination of a compiler and linker that works on multiple languages is required. Tools that help manage the linking of complex applications are essential in the management and control of software development. A loader converts linked object code into an executable module that will run in the designated environment. It is often combined with the linker.

Software Prototyping

The section on the software system life cycle described several models that used the prototyping approach, either once or recursively. The objective of software prototyping is the same as it is in hardware systems, where it is used to reduce risks by constructing and testing immature subsystems or components. In software systems, prototyping is generally used even more frequently for three reasons: (1) requirements are poorly defined; (2) the functionality is unproven; and (3) building the prototype does not require bending metal, only writing code.

Conventionally, a prototype is often taken to mean a test model that is to be discarded after being used. In practice, the system prototype often becomes the first step in an evolutionary development process. This strategy has the advantage of preserving the design features of the prototype after they have been improved as a result of user feedback, as well as building upon the initial programming effort. However, it requires that the prototype programs be engineered using a disciplined and well-planned and documented process. This places a limit on how fast the process can be. The choice of strategy must obviously be based on the particular requirements and circumstances of the project. Table 11.7 lists the typical characteristics of exploratory prototypes, which

TABLE 11.7. Characteristics of Prototypes

Aspect	Exploratory	Evolutionary
Objective	• Validate design • Explore requirements	• Demonstrate • Evaluate
Nature of product	• Algorithms • Concepts	• Engineered • Programed
Environment	Virtual	Operational
Configuration management	Informal	Formal
Testing	Partial	Rigorous
Ultimate use	Disposable	A foundation for further builds

are meant to be discarded, and of evolutionary prototypes, which are meant to be built upon.

The success of a prototyping effort is critically dependent on the realism and fidelity of the test environment. If the test setup is not sufficiently realistic and complete, the prototype tests are likely to be inadequate to validate the design approach and sometimes can be actually misleading. The design of the test should receive a comparable degree of expert attention as the prototype design itself. As in hardware systems, this is a key area for systems engineering oversight.

Software Product Design

In typical hardware system developments, product design consists of the transformation of development prototype hardware components, which might be called "breadboards," into reliable, maintainable, and producible units. In this process, the functional performance is preserved, while the physical embodiments may be changed quite radically. Much of this work is carried out by engineers particularly skilled in the problems of production, environmental packaging, materials, and their fabrication methods, with the objective that the final product can be produced efficiently and reliably.

In the software elements of the system, the product design process is very different. There is no "production" process in software. However, other aspects of a production article are still present. Maintainability continues to be a critical characteristic due to the numerous interfaces inherent within software. Repair by replacement of a failed component—a standby in hardware—does not work in software. An effective user interface is another crucial characteristic of operational software that is often not achieved in the initial version of the system.

Thus, considerable effort is usually required to make a working computer program into a software product usable by others. Fred Brooks has postulated this effort to be three times the effort required to develop a working program. However, there is no professional group in software engineering comparable to the hardware production and packaging engineer. Instead, the "productionization" must be incorporated into the software by the same designers responsible for its basic functionality. Such breadth of

TABLE 11.8. Comparison of Computer Interface Modes

Mode	Description	Advantages	Disadvantages
Menu interaction	Choice from a list of actions	• User preference • Accurate	• Limited choices • Limited speed
Command mode	Abbreviated action commands	• Flexible • Fast	• Long training • Subject to errors
Object manipulation	Click or drag icon	• Intuitive • Accurate	• Moderate flexibility • Moderate speed
Graphical user interface (GUI)	Click graphical buttons	Visual Basic and Java support	• Moderate flexibility • Moderate speed
Touch screen and character recognition	Touch or write on screen	• Simple • Flexible	Easy to make errors

expertise is often not present in the average software designer, with the result that maintainability of software products is frequently less than satisfactory.

Computer User Interfaces. As noted previously, a critical part of engineering operational software systems is the design of the user interface. A computer interface should display information in a form giving the user a clear and well-organized picture of the system status so as to assist the decision process effectively and to provide simple and rapid modes of control. The selection of the appropriate interface mode, display format, interactive logic, and related factors most often requires prototype design and testing with representative users.

The most common control modes offered by computer interfaces are menu interactions, command languages, and object manipulation. A summary of some comparative characteristics of these is given in Table 11.8.

The most rapidly growing computer interface mode is that of object manipulation, the objects being usually referred to as "icons." In addition to the characteristics listed in Table 11.8, graphical presentations of information can often present relationships and can convey meaning better than text. They enable the user to visualize complex information and form inferences that can lead to faster and more error-free decisions than can be achieved by other methods. GUIs are most commonly seen in PC operating systems such as Macintosh OS and Microsoft Windows. The power of the World Wide Web owes a great deal to its GUI formats.

To the systems engineer, GUIs offer both opportunities and challenges. The opportunities are in the virtually infinite possibilities of presenting information to the user in a highly enlightening and intuitive form. The challenges come from the same source, namely, the sheer number of choices that tempt the designer to continue to optimize, unrestrained by an inherent limit. Since GUIs involve a complex software design, there is a risk of cost and schedule impact if the systems engineer is not alert to this hazard.

Advanced Modes. In designing user interfaces for computer-controlled systems, the rapidly advancing technology in this area makes it necessary to consider less conventional modes that offer special advantages. Three examples are briefly described below:

1. *Voice Control.* Spoken commands processed by speech recognition software provide a form of rapid and easy input that leaves the hands free for other actions. Currently, reliable operation is somewhat limited to carefully enunciated words selected from a fixed vocabulary. Capabilities to understand sentences are gradually being evolved.

2. *Visual Interaction.* Computer graphics are being used to aid decision makers by generating displays modeling the results of possible actions, enabling "what-if" simulations in real time. Visual interactive simulation (VIS) is an advanced form of visual interactive modeling (VIM).

3. *Virtual Reality.* A form of 3-D interface in which the user wears stereo goggles and a headset. Head movements generate a simulated motion of the image corresponding to what the eyes would see in the virtual scene. Such displays are used for a growing variety of tasks, such as design of complex structures and pilot training. They are used in battlefield situations and games.

Unit Testing

The engineering design phase of system development begins with the engineering of the individual system components whose functional design has been defined and the technical approach validated in the previous phase. Before the resulting engineered component is ready to be integrated with the other system components, its performance and compatibility must be tested to ensure that they comply with requirements. In software development, this test phase is called "unit testing" and is focused on each individual software component.

Unit tests are generally performed as "white box" tests, namely, those based on the known configuration of the component. Such tests deliberately exercise the critical parts of the design, such as complex control structures, external and internal interfaces, timing or synchronization constraints, and so on.

A compensating characteristic of software for the added testing problems is that the test equipment itself is almost wholly software and can usually be designed and built correspondingly quickly. However, the effort of test design must be as carefully planned and executed as is the system design.

Unit tests for a given component or major module consist usually of a series of test cases, each designed to test a control path, a data structure, a complex algorithm, a timing constraint, a critical interface, or some combination of these. Test cases should be designed to test each function that the unit is required to perform. Since there are typically too many paths to test them all, the selection of test cases requires systems engineering judgment.

Errors uncovered in unit testing should be documented and decisions made as to when and how they should be corrected. Any corrective changes must be carefully considered before deciding which previous test cases should be repeated.

11.6 SOFTWARE INTEGRATION AND TEST

The subject of system integration and evaluation is discussed in detail in Chapter 13, and the general techniques and strategies apply equally to the software components of software-embedded systems and to the software-intensive systems themselves. The discussion makes clear that this aspect of a system's development process is critically important, that it must be carefully planned, expertly executed, and rigorously analyzed, and that the magnitude of the effort required is a large fraction of the entire development effort.

At the system level, the test objectives and strategies of software-dominated systems are similar to those described in Chapter 13. At the software component level, it is necessary to use testing approaches more nearly designed to test software units. The balance of this chapter is devoted to methods of integration and testing complex software programs and software-intensive systems.

The objectives of testing hardware components and subsystems are many—from reducing technical and programmatic risks to verifying specifications. Additional objectives related to politics, marketing, and communications are also part of a system test program. At the lower element level, however, the objectives of testing hardware and software converge.

For software, the objective of testing generally falls into a single category: verification or validation of the software. Moreover, the general method to accomplish this objective is to discover and identify all instances where the program fails to perform its designated function. These range all the way from a case where it fails to meet an essential requirement to where a coding error causes it to crash. Contrary to popular belief, the most valuable test is one that finds a hitherto undiscovered error, rather than one in which the program happens to produce the expected result. Because of the large variety of input scenarios characteristic of the environment of a complex system, the latter result may simply mean that the program happens to handle the particular conditions imposed in that test.

Verification and Validation

Although the terms verification and validation are not for software only, they apply equally to hardware and systems—they are often used more within a software context than any other. *Verification* is simply the process of determining whether the software implements the functionality and features correctly and accurately. These functions and features are usually found in a software specifications description. In other words, verification determines whether we implemented the product right.

Validation, in contrast, is the process of determining whether the software satisfies the users' or customers' needs. In other words, validation determines whether we implemented the right product.

Testing is typically a primary method used to perform verification and validation, though not the only method. However, a robust test program can satisfy a large portion of both evaluation types.

Differences in Testing Software

While the general objectives of testing software may be the same as testing hardware system elements, the basic differences between hardware and software described at the beginning of this chapter make software testing techniques and strategies considerably different.

Test Paths. The unconstrained use of control structures (branches, loops, and switches) may create a multitude of possible logical paths through even a relatively small program. This makes it impractical to test all possible paths and forces the choice of a finite number of cases.

Interfaces. The typically large number of interfaces between software modules, and their depth and limited visibility, makes it difficult to locate strategic test points and to identify the exact sources of discrepancies encountered during testing.

Abstraction. The design descriptions of software are more abstract and are less intuitively understandable than hardware design documentation. This complicates test planning.

Changes. The apparent ease of making changes in software requires correspondingly more frequent retesting. Local changes often require repetition of system-level tests.

Failure Modes. The catastrophic nature of many software errors has two critical consequences. One is the severity of the impact on system operation. The other is that prompt diagnosing of the source of the failure is often frustrated by the inoperability of the system.

Integration Testing

Integration testing is performed on a partially assembled system as system components are progressively linked together. The integration of a complex system is described in Chapter 13 to be a process that must be carefully planned and systematically executed. This is no less true with software systems. The principles and general methods discussed in that chapter apply equally.

Regression Testing

In an integration test sequence, the addition of each component creates new interactions among previously integrated components, which may change their behavior and invalidate the results of earlier successful tests. Regression testing is the process of repeating a selected fraction of such tests to ensure the discovery of newly created discrepancies. The more numerous, complex, and less visible interactions typical of software make it necessary to resort to regression testing more often than for primarily hardware systems.

A problem with regression testing is that unless it is used judiciously, the number of tests can grow beyond practical bounds. For this reason, the test strategy should include careful selectivity of the test cases to be repeated. A balance must be struck between insufficient and excessive rigor to achieve a usable yet affordable product; a systems engineering approach to planning and carrying out integration testing is required.

Validation Testing

Validation testing is intended to determine whether or not a system or a major subsystem performs the functions required to satisfy the operational objectives of the system. Validation testing consists of a series of test scenarios, which collectively exercise the critical system capabilities.

The planning of validation testing and design of test cases also demands a systems engineering approach. The same is true of the analysis of test results, which requires a thorough knowledge of system requirements and of the impact of any significant deviations from nominally required performance. At this stage of system development, decisions on how to handle test discrepancies are critically important. The choice between embarking on a corrective change or seeking a deviation requires an intimate knowledge of the impact of the decision on program cost, schedule, and system performance. Often the best course of action is to investigate the operation of the test equipment, which is itself occasionally at fault, and to repeat the test under more controlled conditions.

Black Box Testing. The section on unit testing described white box testing as addressing the known design features of the component. Validation and other system-level tests consider the system under test as an input-to-output transfer function, without any assumption of its internal workings. As such, black box testing is complementary to white box testing and is likely to uncover interface errors, incorrect functions, initialization errors, as well as critical performance errors.

Alpha and Beta Testing. For software products built for many users, as in the case of much commercial software, most producers have a number of potential customers operate the software before releasing the product for distribution. Alpha testing is typically conducted in a controlled environment at the developer's site, often by employees of a customer. The developer records errors and other problems. Beta testing is conducted at a customer's site without the developer's presence. The customer records the perceived errors and operating problems and reports these to the developer. In both cases, the advantage to the customer is the opportunity to become acquainted with an advanced new product. The developer gains by avoiding the risk of fielding a product containing user deficiencies that would significantly curtail the product's marketability.

11.7 SOFTWARE ENGINEERING MANAGEMENT

The basic elements of managing the development of complex systems were discussed in Chapter 5, and specific aspects in Chapters 6–10. This section deals with some aspects of the management of software-dominated systems that are particularly influenced by the distinguishing character of software, of which systems engineers should be cognizant.

Computer Tools for Software Engineering

Software support tools are software systems that assist the development and maintenance of software programs. In any major software development effort, the availability and quality of the support tools may spell the difference between success and failure. Support tools are used in all aspects of the product life cycle and are becoming more widely available in the commercial marketplace. For these reasons, and the fact that tools for a major software development project require very significant investment, the subject is a proper concern of systems engineers and project managers.

The more specific subject of programming support tools was described briefly in Section 11.5. The paragraphs below discuss the subject of integrated computer-aided software engineering (CASE) tools and some of their typical applications.

CASE. CASE is a collection of tools that are designed to standardize as much of the software development process as possible. Modern CASE tools revolve around graphics-oriented diagramming tools that let the designer define the structure, program and data flow, modules or units, and other aspects of an intended software application. By the use of well-defined symbology, these tools provide the basis for the requirements analysis and design phases of the development cycle.

Requirements Management Tools. The derivation, analysis, quantification, revision, tracing, verification, validation, and documentation of operational, functional, performance, and compatibility system requirements have been seen to extend throughout the system life cycle. For a complex system development, it is a critical and exacting task that involves operational, contractual, as well as technical issues. Several computer-based tools are commercially available that assist in creating an organized database and provide automatic consistency checks, traceability, report preparation, and other valuable services.

Software Metrics Tools. Several commercial tools and tool sets are available to produce automatically measures of various technical characteristics of computer programs, relating to their semantic structure and complexity. (See later section on metrics.)

Integrated Development Support Tools. Several tools have become available that provide a set of compatible integrated support functions, and, in some cases, the capability of importing and exporting data from and to complementary tools from other manufacturers. For example, some tools integrate project management, UML diagramming, requirements analysis, and metrics acquisition capabilities. Such tools simplify the problem of maintaining information consistency among the related domains of software development.

Software Configuration Management (CM). CM in system development was discussed at some length in Chapter 10. Its importance increases with system complexity and criticality. In software systems, strict CM is the most critical activity during and after the engineering development stage. Some of the reasons for this may be inferred from the section on the differences between hardware and software:

1. Software's abstractness and lack of well-defined components makes it difficult to understand.
2. Software has more interfaces; their penetration is deeper and hence is difficult to trace.
3. Any change may propagate deep into the system.
4. Any change may require retesting of the total system.
5. When a software system fails, it often breaks down abruptly.
6. The flexibility of software renders making a software change deceptively easy.

Capability Maturity Model Integration (CMMI)

The abstract nature of software, and its lack of inherent limits on functionality, complexity, or size, makes software development projects considerably more difficult to manage than hardware projects of comparable scope.

Organizations whose business is to produce software-intensive systems or components and to meet firm schedules and costs have often failed to meet their goals because their management practices were not suited to the special needs of software. To help such organizations produce successful products, the Carnegie Mellon University Software Engineering Institute (SEI), operating under government sponsorship, devised a model representing the capabilities that an organization should have to reach a given level of "maturity." This is called a capability maturity model (CMM). A maturity model defines a set of maturity levels and prescribes a set of key process areas that characterize each level. This model provides a means for assessing a given organization's capability maturity level through a defined set of measurements. CMM has been accepted as a standard of industry. It is related to but not equivalent to the International Standard ISO 9000 for software.

Software and systems engineering had separate maturity models until the SEI published the first integrated CMM, combining several previous models into a single, integrated model known as CMMI. Today, CCMI addresses three specific areas of interest: (1) product and service development; (2) service establishment, management, and delivery; and (3) product and service acquisition. As of this writing, CMMI, Version 1.2 is the latest version of the model.

At its core, CMMI is a process improvement methodology. Understanding the current maturity of an organization's processes and identifying the objective maturity level for the future are keys concepts behind the model. Therefore, one aspect of CMMI is the formal definition of maturity levels. These apply to organizations, not projects, although as projects grow in size and complexity, the lines of demarcation between an organization and a project can become blurred.

Capability Maturity Levels. The CMM defines six capability and five maturity levels as summarized in Tables 11.9 and 11.10. The CMMI process is fully institutionalized. Key performance areas (KPAs) are defined for each level and are used in determining an organization's maturity level. Each KPA is further defined by a set of goals

TABLE 11.9. Capability Levels

Capability level 0: incomplete
An "incomplete process" is a process that either is not performed or partially performed. One or more of the specific goals of the process area are not satisfied, and no generic goals exist for this level since there is no reason to institutionalize a partially performed process.

Capability level 1: performed
A performed process is a process that satisfies the specific goals of the process area. It supports and enables the work needed to produce work products.

Capability level 2: managed
A managed process is a performed (capability level 1) process that has the basic infrastructure in place to support the process. It is planned and executed in accordance with police; employs skilled people who have adequate resources to produce controlled outputs; involves relevant stake holders; is monitored, controlled, and reviewed; and is evaluated for adherence to its process description.

Capability level 3: defined
A defined process is a managed (capability level 2) process that is tailored from the organization's set of standard processes according to the organization's tailoring guidelines and contributes work products, measures, and other process improvement information to the organizational process assets.

Capability level 4: quantitatively managed
A quantitatively managed process is a defined (capability level) process that is controlled using statistical and other quantitative techniques. Quantitative objectives for quality and process performance are established and used as criteria in managing the process. Quality and process performance is understood in statistical terms and is managed throughout the life of the process.

Capability level 5: optimizing
An optimizing process is a quantitatively managed (capability level 4) process that is improved based on an understanding of the common causes of variation inherent in the process. The focus of an optimizing process is on continually improving the range of process performance through both incremental and innovative improvements.

TABLE 11.10. Maturity Levels

Maturity level 1: initial
Processes are usually ad hoc and chaotic.

Maturity level 2: managed
The projects of the organization have ensured that processes are planned and executed in
  accordance with policy; the projects employ skilled people who have adequate resources to
  produce controlled outputs; involve relevant stakeholders; are monitored, controlled, and
  reviewed; and are evaluated for adherence to their process descriptions.

Maturity level 3: defined
Processes are well characterized and understood, and are described in standards, procedures,
  tools, and methods. The organization's set of standard processes, is established and improved
  over time. These standard processes are used to establish consistency across the organization.
  Projects establish their defined processes by tailoring the organization's set of standard
  processes according to tailoring guidelines.

Maturity level 4: quantitatively managed
The organization and projects establish quantitative objectives for quality and process
  performance and use them as criteria in managing processes. Quantitative objectives are
  based on the needs of the customer, end uses, organization, and process implementers.
  Quality and process performance is understood in statistical terms and is managed through
  cut the life of the processes.

Maturity level 5: optimizing
An organization continually improves its processes based on a quantitative understanding of
  the common causes of variation inherent in processes.

---

and key practices that address these goals. SEI also defines key indicators that are
designed to determine whether or not the KPA goals have been achieved. These are
used in CMM assessments of an organization's capability maturity level.

CMMI is widely used by industry, especially by large system and software development organizations. The U.S. DoD prescribes a demonstration of CMMI Level 3
capability for major system acquisitions. However, the investment necessary to achieve
CMMI certification is considerable, and it is generally estimated that going from level
1 to level 2 or from level 2 to level 3 requires from 1 to 2 years.

Systems Engineering Implications. Examination of the KPAs reveals that
they address a combination of project management, systems engineering, and process
improvement issues. At level 2, the KPAs addressing requirements management
and CM are clearly systems engineering responsibilities, while project planning,
project tracking and oversight, and subcontract management are mainly project management functions. At level 3, software product engineering, intergroup coordination,
and peer reviews are of direct concern to systems engineers. At higher levels, the focus
is largely on process improvement based on quantitative measurements of process
results.

Software Metrics

Metrics are quantitative measures used to assess progress, uncover problems, and provide a basis for improving a process or product. Software metrics can be classified as project metrics, process metrics, or technical metrics.

Project Metrics. Software project metrics are concerned with measures of the success of project management—stability of requirements, quality of project planning, adherence to project schedules, extent of task descriptions, quality of project reviews, and so on. These are basically the same as would be used on any comparable project to track management practices. A reason for greater attention to project metrics on a software development is the traditionally more difficult task of reliable planning and estimating new software tasks. Project metrics should be tailored to the formality, size, and other special characteristics of the project.

Process Metrics. Software process metrics are fundamental to the practice of establishing process standards as described in the previous section on software capability maturity assessment. Such standards identify a set of process areas that need to be addressed. They do not generally prescribe how they should be handled but require that appropriate practices be defined, documented, and tracked.

Technical Metrics. Technical software metrics are focused largely on assessing the quality of the software product rather than on management or process. In that sense, they are an aid to design by identifying sections of software that are exceptionally convoluted, insufficiently modularized, difficult to test, inadequately commented, or otherwise less than of high quality. Such measures are useful for directly improving the product, and for refining design and programming practices that contributed to the deficiencies. There are numerous commercial tools that are designed to track technical software metrics.

Management of Metrics. Software metrics can be useful in developing good practices and in improving productivity and software quality. However, they can also be misused with negative results for the projects and the software staff. It is important to observe a number of principles in the management of metrics:

1. The purpose of each metric must be clearly understood by all concerned to be beneficial and worth the effort to collect and analyze.
2. The metrics collected on a given project should be appropriate to its character and criticality.
3. The results of metrics collection should be used primarily by the project to increase its probability of success.
4. The results should never be used to threaten or appraise individuals or teams.
5. There should be a transition period for the introduction of new metrics before the data collected are used.

Future Outlook

The continuing growth of information systems is exerting severe pressure to improve software technology in order to keep pace with rising demands and to minimize risks of major software project failures, which have been all too frequent in recent years. Furthermore, the unreliability of much commercial software has frustrated many computer users. Below are some trends that have the potential to meet some of the above needs.

Process Improvement. The establishment and widespread adoption of software process standards, such as CMMI, have significantly strengthened the discipline used in software design. They have introduced engineering practices and management oversight into a culture derived from science and art. For large, well-defined projects, these approaches, which have been found to reduce failure rates, vary significantly. For smaller projects having loosely defined requirements, agile methods have attracted many adherents.

Programming Environment. Computer-aided programming environments, such as that for Visual Basic, are likely to continue to improve, providing better automatic error checking, program visualization, database support, and other features designed to make programming faster and less prone to error. Integration of syntax checking, debugging, and other programming support functions into the environment, along with more powerful user interfaces, is likely to continue to improve productivity and accuracy.

Integrated CASE Tools. Requirements and CM tools are being integrated with modeling and other functions to facilitate the development, upgrading, and maintenance of large software programs. The integration of these tools enables the traceability of program modules to requirements and the management of the massive number of data elements present in complex systems capabilities. While the development of such tools is expensive, their growth and consequent increases in productivity are likely to continue, especially if more emphasis is placed on reducing the time and cost of becoming proficient in their use.

Software Components. Reuse of software components has long been a major goal, but its effective realization has been the exception rather than the rule. One such exception has been the availability of commercial GUI components, supporting features such as windowing and pull-down menus. With the proliferation of automated transactional systems (financial, travel, inventory, etc.), it is likely that numerous other standard components will be identified and made commercially available. The gains in development cost and reliability in automated transactional systems are potentially very large.

Design Patterns. A different approach to reusable components has been the development of design patterns. A seminal work on this subject by Gamma et al. defines

23 basic patterns of OO functions and describes an example of each. The patterns are subdivided into three classes: creational patterns that build various types of objects, structural patterns that operate on objects, and behavioral patterns that perform specified functions. While this approach appears to hold great promise of creating versatile software building blocks, it has thus far not been adopted by a significant fraction of developers.

Software Systems Engineering. Perhaps the most significant advance in the development of software-dominated systems would come from the effective application of systems engineering principles and methods to software system design and engineering. Despite the many differences between the nature of software and hardware technologies, some avenues to narrowing this gap are being actively explored. The development of the CMMI by SEI, which addresses both systems engineering and software engineering in a common framework, may contribute to a more common outlook. However, real progress in this direction must involve education and extensions of current software methodologies to facilitate modular partitioning, clean interfaces, architectural visibility, and other basic features of well-designed systems. The continuing demand for complex software-dominated systems may accelerate efforts to introduce systems engineering methods into software development.

11.8 SUMMARY

The terms software engineering and software systems engineering are not synonymous, however. The former refers to the development and delivery of software products, stand-alone or embedded. The latter refers to the application of principles to the software engineering discipline. We define software as having three major components: (1) instructions, also referred to as code; (2) data structures; and (3) documentation.

Coping with Complexity and Abstraction

The role of software has changed over the past 20 years—most modern systems are dominated by software. Therefore, software engineering has become a full part of system development.

Nature of Software Development

Software can be categorized as either

    (a) system software, providing services to other software;

    (b) embedded software, providing functions, services, or features within a larger system; or

    (c) application software, providing services as a stand-alone system.

Systems that utilize software can be categorized in one of three ways:

1. *Software-Embedded Systems* are a hybrid combination of hardware and software. Although predominantly hardware, these systems use software to control the action of hardware components. Examples are most vehicles, spacecraft, robotics, and military systems.
2. *Software-Intensive Systems* consist of computers and networks, controlled by software. These systems use software to perform virtually all of the systems' functionality, including all automated complex information functionality. Examples are financial management, airline reservations, and inventory control.
3. *Data-Intensive Computing Systems* are large-scale computing resources dedicated to executing complex computational tasks. Examples are weather analysis and prediction centers, nuclear effects prediction systems, advanced information decryption systems, and other computationally intensive operations.

Software has intrinsic differences from hardware, including

- near-infinite variability of software structural units
- few commonly occurring software components;
- software is assigned most critical functions;
- interfaces are more numerous, deeper, and less visible; software functionality and size have almost no inherent limits; software is easily changeable;
- simple software changes may require extensive testing; software often fails abruptly, without warning signs; and
- software is abstract and difficult to visualize.

Software Development Life Cycle Models

The life cycles of software-dominated systems are generally similar to the systems engineering life cycle described in Chapter 4. While there are a plethora of life cycle models, we can define four basic types:

1. *Linear*—a sequence of steps, typically with feedback;
2. *Incremental*—a repetition of a sequence of steps to generate incremental capabilities and functionality until the final increment, which incorporates full capabilities;
3. *Evolutionary*—similar to incremental, except early increments are intended to provide functionality for experimentation, analysis, familiarization, and demonstration. Later increments are influenced heavily from experience with early increments.
4. *Agile*—the typical steps for software development are combined in various forms to enable rapid yet robust development.

Software Concept Development: Analysis and Design

Performance requirements for software-embedded systems are developed at the system level and should be verified by software developers.

Performance requirements for software-intensive systems should be established with close interaction with customers/users and may need to be verified by rapid prototyping. They should not unreasonably stress software extensibility.

Software requirements are typically developed using four steps: elicitation from users, customers and stakeholders, analysis and negotiation with customers, documentation, and validation.

Two prevailing methodologies for designing software systems are structured analysis and design and OOAD. Structured analysis focuses on functional architecture, using functional decomposition, and defines program modules as the primary structural units. This methodology proceeds with top-down functional allocation. In contrast, OOAD focuses on "classes" of objects as program units and encapsulates data variables with operations. This methodology uses an iterative rather than a top-down development.

Other methodologies include robustness analysis, which focuses on initial OO architectural design, FCD, and combined structured and OO approaches.

UML supports all phases of OO development. UML provides 13 types of diagrams, presenting different views of the system, and is widely used. UML has been adopted as an industry standard.

Software Engineering Development: Coding and Unit Test

The engineering design phase of software development implements software architectural design and the computer instructions to execute the prescribed functionality. The phase produces computer programs written in a high-level language (source code) and subjects each program unit to a "unit test" before acceptance.

The programming language must be suited to the type of software and compiler availability. It must conform with the design methodology and requires that staff experienced with the language be available.

Prototyping an iterative development comes in two forms: (1) purely exploratory and is to be discarded once its purpose is fulfilled, and (2) evolutionary, and is to be built upon. In the latter case, high quality must be built in from the beginning.

Human–computer interfaces are critical elements in all software-intensive systems. These types of interfaces usually use interactive graphics formats and may include voice activation and other advanced techniques.

Software Integration and Test

Testing software systems involves many more test paths and interfaces than hardware and requires special test points for diagnosing failures and their sources. Testing often requires end-to-end system-level retesting after eliminating a failure.

Alpha and beta testing subject the new system to tests by the customer and expose user problems before wide product distribution.

Software Engineering Management

CM for software-dominated systems is critical in that software is inherently complex and has numerous and deep interfaces. Since software is responsible for controlling some of the most critical system functions, software tends to be subject to frequent changes.

The CMMI establishes six levels of capability and five levels of maturity for an organization. CMMI establishes KPAs for each level and provides a basis for assessing an organization's overall systems and software engineering capability.

PROBLEMS

11.1 With reference to Figure 11.1, list two specific examples of each of the blocks shown in the diagrams. For one case of each block, describe the kind of data that flows along the paths shown by the lines between the blocks.

11.2 Look up (if necessary) the principal *subcomponents* of the data processor (CPU) of a personal computer. Draw a block diagram of the subcomponents and their interconnections. Describe in your own words the functions of each subcomponent.

11.3 Extend the examples of the three types of software-dominated systems shown in Table 11.1 by listing two more examples of each type. Briefly indicate why you placed each example into the selected category.

11.4 Using the example of an automated supermarket grocery inventory and management system, draw the system context diagram. Assume that the master-pricing database comes from a central office. Neglect special discounts for store card carriers.

11.5 For the same example, define the functions performed by the automated grocery system in processing each individual grocery item. Differentiate between those carrying bar codes and those sold by weight.

11.6 Draw a functional flow diagram for the processing of a grocery item showing the two alternate branches mentioned in Problem 11.5.

11.7 Identify the objects involved in the above automated grocery system and their attributes. Draw an activity diagram corresponding to the processes described in Problem 11.6.

For Problems 11.8–11.12, suppose you have been asked to develop the software for an elevator system for a multistory building. The system will contain three elevators and will have five floors and a basement-level parking garage.

11.8 Develop 20–25 functional and performance requirements for this software system. Please perform analysis on your list to ensure your final list is robust, consistent, succinct, nonredundant, and precise.

11.9 (a) Identify 8–12 top-level functions for this software system.

(b) Draw an FFBD for this system using the functions in (a).

11.10 (a) Identify 8–12 classes for this software system. Each class should have a title, attributes, and operations.

(b) Draw a class diagram showing the associations between the classes in (a).

11.11 (a) Identify the 8–12 top-level hardware components of the elevator system.

(b) Identify the interfaces between the software and hardware components of this system in (a). Please construct a table with three columns. In the first column, labeled "hardware component," identify the component in which the software will need to interface. In the second column, labeled "input/output," identify whether the interface is an input, an output, or both. In the third column, labeled "what is passed," identify what is passed between the software and hardware.

11.12 Develop an operational test plan for this software system. The test plan should include a purpose, a description of no more than five tests, and a linkage between each test and the requirement(s) that are being tested.

FURTHER READING

G. Booch, J. Rumbaugh, and J. Jacobson. *The Unified Modeling Language User Guide.* Addison-Wesley, 1999.

F. P. Brooks, Jr. *The Mythical Man Month—Essays on Software Engineering.* Addison-Wesley, 1995, Chapter 8.

B. Bruegge and A. H. Dutoit. *Object-Oriented Software Engineering.* Prentice Hall, 2000, Chapters 1–7.

P. DeGrace and L. H. Stahl. *Wicked Problems, Righteous Solutions.* Yourdon Press, Prentice Hall, 1990, Chapter 3.

A. Denis, B. H. Wixom, and R. M. Roth. *Systems Analysis Design,* Third Edition. John Wiley & Sons, Inc., 2006, Chapters 4, 6, and 8–10.

G. Eisner. *Computer-Aided Systems Engineering.* Prentice Hall, 1988, Chapters 8 and 14.

H. Eisner. *Essentials of Project and Systems Engineering Management.* John Wiley & Sons, Inc., 1997, Chapters 10 and 12.

E. Gamma, R. Helm, R. Johnson, and J. Dlissides. *Design Patterns.* Addison-Wesley, 1995.

K. E. Kendall and J. E. Kendall. *Systems Analysis and Design,* Sixth Edition. Prentice Hall, 2005, Chapters 6, 7, 14, and 18.

M. Maier and E. Rechtin. *The Art of Systems Architecting.* CRC Press, 2009, Chapter 6.

R. S. Pressman. *Software Engineering: A Practitioner's Approach,* Sixth Edition. McGraw-Hill, 2005, Chapters 20–24.

E. Rechtin. *Systems Architecting: Creating and Building Complex Systems.* Prentice Hall, 1991, Chapter 5.

N. B. Reilly. *Successful Systems for Engineers and Managers.* Van Nostrand Reinhold, 1993, Chapters 13 and 14.

D. Rosenberg. *Use Case Driven Object Modeling with UML*. Addison-Wesley, 1999, Chapters 1–4.

J. Rumbaugh, M. Blaha, W. Premerlani, F. Eddy, and W. Lorenson. *Object-Oriented Modeling and Design*. Prentice Hall, 1991, Chapters 1–3.

Sommerville. *Software Engineering*, Eighth Edition. Addison-Wesley, 2007, Chapters 2, 4, 6, 7, and 11.

12

ENGINEERING DESIGN

12.1 IMPLEMENTING THE SYSTEM BUILDING BLOCKS

The engineering design phase is that part of the development of a new system that is concerned with designing all the component parts so that they will fit together as an operating whole that meets the system operational requirements. It is an intensive and highly organized effort, focused on designing components that are reliable, maintainable, and safe under all conditions to which the system is likely to be subjected, and that are producible within established cost and schedule goals. While the general design approach required to meet the above objectives presumably has been established in previous phases, the engineering design phase is where detailed internal and external interfaces are established and the design is first fully implemented in hardware and software.

It was noted in Chapter 10 that during the advanced development phase, any previously unproven components should be further developed to the point where all significant issues regarding their functional and physical performance have been resolved. However, experience in developing complex new systems has shown that some

Systems Engineering Principles and Practice, Second Edition. Alexander Kossiakoff, William N. Sweet, Samuel J. Seymour, and Steven M. Biemer
© 2011 by John Wiley & Sons, Inc. Published 2011 by John Wiley & Sons, Inc.

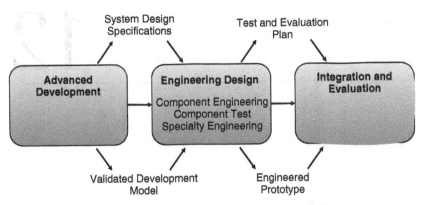

Figure 12.1. Engineering design phase in a system life cycle.

"unknown unknowns" (unk-unks) almost always escape detection until later, revealing themselves during component design and integration. Such eventualities should therefore be anticipated in contingency planning for the engineering design phase.

Place of the Engineering Design Phase in the System Life Cycle

As shown in Figure 12.1, the place of the engineering design phase in the systems engineering life cycle follows the advanced development phase and precedes the integration and evaluation phases. Its inputs from the advanced development phase are seen to be system design specifications and a validated development model of the system. Other inputs, not shown, include applicable commercial components and parts, and the design tools and test facilities that will be employed during this phase. Its outputs to the integration and evaluation phase are detailed test and evaluation plans and a complete set of fully engineered and tested components. Program management planning documents, such as the work breakdown structure (WBS) and the systems engineering management plan (SEMP), as well as the test and evaluation master plan (TEMP), or their equivalents, are utilized and updated in this process. Figure 12.2 shows that the integration and evaluation phase usually begins well before the end of engineering design to accommodate test planning, test equipment design, and related activities.

Design Materialization Status

The change in system materialization status during the engineering design phase is schematically shown in Table 12.1. It is seen that the actions "visualize," "define," and "validate" in previous phases are replaced by the more decisive terms "design," "make," and "test," representing implementation decisions rather than tentative proposals. This is characteristic of the fact that in this phase, the conceptual and developmental results of the previous phases finally come together in a unified and detailed system design.

At the beginning of the engineering design phase, the design maturity of different components is likely to vary significantly; and these variations will be reflected in dif-

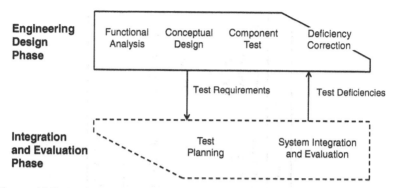

Figure 12.2. Engineering design phase in relation to integration and evaluation.

ferences in component materialization status. For example, some components that were derived from a predecessor system may have been fully engineered and tested in substantially the same configuration as that selected for the new system, while others that utilize new technology or innovative functionality may have been brought only to the stage of experimental prototypes. However, by the end of the engineering design phase, such initial variations in component engineering status must be eliminated and all components fully "materialized" in terms of detailed hardware and software design and construction.

A primary effort in this phase is the definition of the interfaces and interactions among internal components and with external entities. Experience has shown that aggressive technical leadership by systems engineering is essential for the expeditious resolution of any interface incompatibilities that are brought to light during engineering design.

Systems Engineering Method in Engineering Design

The principal activities in each of the four steps in the systems engineering method (see Chapter 4) during engineering design are briefly stated below and are illustrated in Figure 12.3. Steps 3 and 4 will constitute the bulk of the effort in this phase.

1. *Requirements Analysis.* Typical activities include
   - analyzing system design requirements for consistency and completeness and
   - identifying requirements for all external and internal interactions and interfaces.
2. *Functional Analysis and Design (Functional Definition).* Typical activities include
   - analyzing component interactions and interfaces and identifying design, integration, and test issues;
   - analyzing detailed user interaction modes; and
   - designing and prototyping user interfaces.

TABLE 12.1. Status of System Materialization at the Engineering Design Phase

Phase	Concept development				Engineering development	
Level	Needs analysis	Concept exploration	Concept definition	Advanced development	Engineering design	Integration and evaluation
System	Define system capabilities and effectiveness	Identify, explore, and synthesize concepts	Define selected concept with specifications	Validate concept		Test and evaluate
Subsystem		Define requirements and ensure feasibility	Define functional and physical architecture / Allocate functions to components	Validate subsystems		Integrate and test
Component				Define specifications / Allocate functions to subcomponents	Design and test	Integrate and test
Subcomponent		Visualize			Design	
Part					Make or buy	

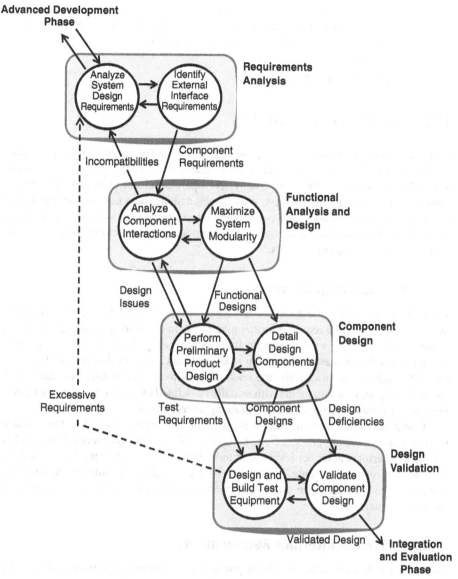

Figure 12.3. Engineering design phase flow diagram.

3. *Component Design (Physical Definition).* Typical activities include
   - laying out preliminary designs of all hardware and software components and interfaces,
   - implementing detailed hardware designs and software code after review, and
   - building prototype versions of engineered components.

4. *Design Validation.* Typical activities include
   - conducting test and evaluation of engineered components with respect to function, interfaces, reliability, and producibility;
   - correcting deficiencies; and
   - documenting product design.

12.2   REQUIREMENTS ANALYSIS

In the advanced development phase, the system functional specifications were translated into a set of system design specifications that defined the design approach selected and validated as fully addressing the system operational objectives. As in previous phases of the development process, these specifications must now be analyzed again for relevance, completeness, and consistency to constitute a sound basis for full-scale engineering. In particular, the analysis must consider any changes occurring due to the passage of time or external events.

System Design Requirements

It will be recalled that the focus of the advanced development phase was on those system components that required further maturation in terms of analysis, design, development, and/or testing to demonstrate fully their validity. These are the components that represent the greatest development risks, and hence their design approach must be carefully analyzed to ensure that the residual risks have been reduced to manageable levels. For example, components with initially ill-defined external interface descriptions must be reexamined to resolve any remaining uncertainties.

Components that were identified as involving some risk, but not to the extent of requiring special development effort, and previously proven components that will be required to perform at higher levels or in more stressful environments must be particularly scrutinized at this stage. The results of these analyses should be inputs to the planning of risk management during engineering design. (See section on risk management in Chapter 5.)

External System Interface Requirements

Since the whole system has not been physically assembled in previous phases, it is likely that the design of its interfaces with the environment has been considered less than rigorously. Hence, a comprehensive analysis of system-level environmental interfaces must be carried out prior to the initiation of engineering design.

User Interfaces. As noted previously, the functional interactions and physical interfaces of the system with the user(s) are not only often critical but also difficult to define adequately. This situation is aggravated by the fact that potential users of a new system do not really know how they can best operate it before they first physically interact with it. Thus, except for very simple human–machine interfaces, a prototype model of the user consoles, displays, and controls should be constructed at the earliest practicable time to enable the user(s) to examine various responses to system inputs and to experiment with alternative interface designs. If this has not been done adequately in the advanced development phase, it must be done early in engineering design.

The user interfaces related to system maintenance involve fault isolation, component replacement, logistics, and a host of related issues. Interface design is often given only cursory attention prior to the engineering design phase, an omission that is likely to lead to the need for a significant redesign of previously defined component interfaces.

Environmental Interfaces. In defining external interfaces subject to shock, vibration, extreme temperatures, and other potentially damaging environments, it is essential to again consider all stages of a system's life, including production, shipment, storage, installation, operation, and maintenance, and to anticipate all of the interactions with the environment during each step. Interface elements such as seals, joints, radiation shields, insulators, shock mounts, and so on, should be reviewed and redefined if necessary to ensure their adequacy in the final design. Some of the above subjects are treated more fully in a later section of this chapter, which discusses interface design.

Assembly and Installation Requirements

In addition to the usual design requirements, the system design must also take into consideration all special requirements for system assembly and installation at the operational site. This is especially important for large systems that must be shipped in sections. An example is a shipboard system, subsystems of which are to be installed below decks in an existing ship. In this case, the size of hatches and passageways will dictate the largest object that can pass through. System installation aboard aircraft is another example. Even buildings have load and size limits on freight elevators. In any case, when on-site assembly is required, the system design must consider where the system will be "cut" and how it will be reassembled. If physical mating is implemented by bolts, for example, then the location and size of these fasteners must take into account the size and position of the wrenches needed for assembly. Many developers have been embarrassed when they realize there is not enough elbowroom to perform a prescribed assembly procedure.

Another on-site problem can occur when the assembly process is found to be difficult and slow to perform. A classic example concerns a suspended walkway that was installed in a large Midwestern hotel lobby. During a large evening dance party, a number of people were dancing on the elevated walkway, causing it to collapse with attendant loss of life. The investigation of this accident revealed that a design change had been made at the assembly site because the originally specified long, threaded supporting rods were difficult to install. A so-called trivial design change was made to permit easier assembly, but it increased the load on the rod structure by a factor of two. The fault was attributed to those involved in the design change. However, it can be argued that if the original designers had given more attention to the difficult assembly process, this problem and the resulting accident might not have occurred.

Risk Mitigation

As in the previous chapters, a necessary step in the planning of the development and engineering process is the consideration of program risk. In the advanced development phase, risk assessment was used to refer to the process of identifying components that required further maturation to eliminate or greatly reduce the potential engineering problems inherent in the application of new technology or complex functionality. By the beginning of the engineering design phase, those risks should have been resolved through further development. This in turn should have reduced the remaining program risks to a level that could be tolerated through the application of risk management, a process that identifies and seeks to mitigate (abate or minimize) the likelihood and impact of residual risks. Methods for mitigating risks are discussed briefly in the section on component design (Section 12.4) and in greater detail in Chapter 5.

Critical Design Requirements

To the extent that previous analysis has shown that a particular requirement places undue stress on the engineering design, this is the last opportunity seriously to explore the possibility of its relaxation and thus to reduce the risk of an unsuccessful design.

12.3 FUNCTIONAL ANALYSIS AND DESIGN

The principal focus of the engineering design phase is on the design of the system components. Insofar as the functional definition of the components is concerned, it may be assumed that the primary allocation of functions has been accomplished in previous phases, but that the definition of their mutual interactions has not been finalized. A primary objective of the functional analysis and design step is to definitize the interactions of components with one another and with the system environment in such a way as to maximize their mutual independence, and thus to facilitate their acquisition, integration, and maintenance and the ease of future system upgrading.

This section stresses three important areas of functional analysis and design:

1. *Modular Configuration:* simplifying interactions among system components and with the environment
2. *Software Design:* defining a modular software architecture
3. *User Interfaces:* defining and demonstrating effective human–machine interfaces.

Modular Configuration

The single most important objective of the functional analysis and design step in the engineering design phase is to define the boundaries between the components and subsystems so as to minimize their interactions (i.e., their dependence on one another). This is essential to ensure that

1. each component can be specified, developed, designed, manufactured, and tested as a self-contained unit;
2. when assembled with the other components, a component will perform its functions properly and without further adjustment;
3. a faulty component can be replaced directly by an equivalent interchangeable component; and
4. a component can be upgraded internally without affecting the design of other components.

A system design with the above characteristics is referred to as "modular" or "sectionalized." These characteristics apply to both hardware and software components. They depend on physical as well as functional interactions, but the latter are fundamental and must be defined before the physical interfaces can be established.

Functional Elements. The system functional elements defined in Chapter 3 are examples of highly modular system building blocks. These building blocks were selected using three criteria:

1. *Significance:* each functional element performs a distinct and significant function, typically involving several elementary functions.
2. *Singularity:* each functional element falls largely within the technical scope of a single engineering discipline.
3. *Commonality:* the function performed by each element is found in a wide variety of system types.

Each of the functional elements was seen to be the functional embodiment of a type of component element (see Table 3.3), which is a commonly occurring building block of modern systems. Their characteristic of "commonality" results from the fact that each is highly modular in function and construction. It follows that the functional elements of a new system should use standard building blocks whenever practicable.

Software Design

As noted previously, the development and engineering of software components are sufficiently different from that of hardware components that a separate chapter is devoted to the special systems engineering problems and solutions of software (Chapter 11). The paragraphs below contain a few selected subjects relevant to this chapter.

Prototype Software. The previous chapters noted that the extensive use of software throughout most modern complex systems usually makes it necessary to design and test many software components in prototype form during the advanced development phase. Common instances of this are found in embedded real-time programs and user interfaces. The existence of such prototype software at the beginning of the engineering design phase presents the problem of whether or not to reuse it in the engineered system and, if so, just how it should be adapted for this purpose.

Redoing the prototype software from scratch can be extremely costly. However, its reuse requires careful assessment and revision, where necessary. The following conditions are necessary for successful reuse:

1. The prototype software must be of high quality, that is, designed and built to the same standards that are established for the engineered version (except perhaps for the degree of formal reviews and documentation).
2. Changes in requirements must be limited.
3. The software should be either functionally complete or compatible with directly related software.

Given the above conditions, modern computer-aided software engineering (CASE) tools are available to facilitate the necessary analysis, modification, and documentation to integrate the prototype software into the engineered system.

Software Methodologies. Chapter 11 identifies many of the key aspects of software engineering that are of direct interest to systems engineers. Two principal methodologies are used in software analysis and design: *structured analysis* and *design and object-oriented analyses and design*. The former and more mature methodology is organized around functional units generally called procedures or functions and is assembled in modules or packages. In good structured design, data values are passed between procedures by means of calling parameters, with a minimum of externally addressed (global) data. Object-oriented analysis (OOA) and object-oriented design (OOD) are more recent methods of software system development and are widely believed to be inherently superior in managing complexity, which is a critical problem in all large, information-rich systems. Using object-oriented methods in developing hardware and combined hardware/software systems has become more commonplace and is usually referred to as object-oriented systems engineering (OOSE). The particulars of this method will be described below in a separate section. Accordingly, today's systems engineers need to know the basic elements and capabilities of these methodologies in order to evaluate their appropriate place in system development.

User Interface Design

Among the most critical elements in complex systems are those concerned with the control of the system by the user—analogous to the steering wheel, accelerator, shift lever, and brakes in an automobile. In system terminology, those elements are collectively referred to as the "user interface." Their criticality is due to the essential role they play in the effective operation of most systems, and to the inherent problem of matching a specific system design to the widely variable characteristics of the many different human operators who will use the system during its lifetime. If several individuals operate different parts of the system simultaneously, their mutual interactions present additional design issues.

The principal elements involved in user control include

1. *Displays:* presentations provided to the user containing information on system status to indicate need for possible user action. They may be dials, words, numbers, or graphics appearing on a display screen, or a printout, sound, or other signals.

2. *User Reaction:* user's interpretation of the display based on knowledge about system operation and control, and consequent decisions on the action to be taken.

3. *User Command:* user's action to cause the system to change its state or behavior to that desired. It may be movement of a control lever, selection of an item from a displayed menu, a typed command, or another form of signal to which the system is designed to respond.

4. *Command Actuator:* device designed to translate the user's action into a system response. This may be a direct mechanical or electrical link or, in automated systems, a computer that interprets the user command and activates the appropriate response devices.

The design of a user–system interface is truly a multidisciplinary problem, as the above list implies, and hence is the domain of systems engineering. Even human factors engineering, considered to be a discipline in itself, is actually fragmented into specialties in terms of its sensory and cognitive aspects. While much research has been carried out, quantitative data on which to base engineering design are sparse. Thus, each new system presents problems peculiar to itself and often requires experimentation to define its interface requirements.

The increasing use of computer automation in modern systems has brought with it the computer-driven display and controls as the preferred user interface medium. A computer interface has the facility to display information in a form processed to give the user a clearer and better organized picture of the system status, to simplify the decision process, and to offer more simple and easier modes of control.

Chapter 11 contains a brief description of computer control modes, graphical user interfaces (GUIs), and advanced modes of user–computer interactions, such as voice control and visual reality.

Functional System Design Diagrams. As the components of a complex system are integrated, it becomes increasingly important to establish system-wide representations of the functional system architecture to ensure its understanding by all those concerned with designing the interacting system elements. Functional diagrams are discussed in more detail in Chapter 8.

12.4 COMPONENT DESIGN

The object of the component design step of the engineering design phase is to implement the functional designs of system elements as engineered hardware and software components with compatible and testable interfaces. During this phase, the system components that do not already exist as engineered items are designed, built, and tested as units, to be integrated into subsystems and then assembled into an engineering prototype in the integration and evaluation phase. The associated engineering effort during this phase is more intense than at any other time during the system life cycle. During the design of any complex new system, unexpected problems inevitably occur; their timely resolution depends on quick and decisive action. This high level of activity, and the potential impact of any unforeseen problems on the successful conduct of the program, tends to place a severe stress on systems engineering during this period.

In the development of major defense and space systems, the engineering design effort is performed in two steps: designated preliminary design and detailed design, respectively. Although preliminary design is typically started under systems architecting, many official programs continue to establish a subphase where the initial architecture is translated into a preliminary design. Each step is followed by a formal design review by the customer before the succeeding step is authorized. The purpose of this highly controlled process is to ensure very thorough preparation prior to commitment to the costly full-scale implementation of the design into hardware and software. This general methodology, without some of its formality, may be applied to any system development.

The level of system subdivision on which the above design process is focused is called a "configuration item" (CI). This level corresponds most closely to that referred to here as "component." It should be noted that in common engineering parlance, the term component is used much more loosely than in this book and sometimes is applied to lower-level system elements, which are identified here as subcomponents. CIs and "configuration baselines" are discussed further in the section on configuration management (CM) (Section 9.6).

Preliminary Design

The objective of preliminary design is to demonstrate that the chosen system design conforms to system performance and design specifications and can be produced by existing methods within established constraints of cost and schedule. It thereafter provides a framework for the next step, detailed design. The bulk of the functional design effort, as described in the previous section, is properly a part of preliminary design.

Typical products of preliminary design include

- design and interface specifications (B specs);
- supporting design and effectiveness trade studies;
- mock-ups, models, and breadboards;
- interface design;
- software top-level design;
- development, integration, and verification test plans; and
- engineering specialty studies (RMA, producibility, logistic support, etc.).

Major systems engineering input and review is essential for all of the above items. Of particular importance is the manner in which the functional modules defined in the functional design process are implemented in hardware and software. Often this requires detailed adjustments in the boundaries between components to ensure that physical interfaces, as well as functional interactions, are as simple as practicable. To the extent that the advanced development phase has not resolved all significant risks, further analyses, simulations, and experiments may have to be conducted to support the preliminary design process.

Preliminary Design Review (PDR). In government programs, the PDR is normally conducted by the acquisition agency to certify the completion of the preliminary design. For major commercial programs, company management acts in the role of the customer. The process is frequently led or supported by a commercial or nonprofit systems engineering organization. The review may last for a few or many days and may require several follow-on sessions if additional engineering is found to be required.

The issues on which PDR is usually centered include major (e.g., subsystem and external) interfaces, risk areas, long-lead items, and system-level trade-off studies. Design requirements and specifications, test plans, and logistics support plans are reviewed. Systems engineering is central to the PDR process and must be prepared to deal with any questions that may arise in the above areas.

Prior to the formal PDR, the development team should arrange for an internal review to ensure that the material to be presented is suitable and adequate. The preparation, organization, and qualification of the review process is critical. This is no less important for commercial systems, even though the review process may be less formal, because success of the development is critically dependent on the quality of design at this stage.

The completion of preliminary design corresponds to the establishment of the allocated baseline system configuration (see Section 12.6).

Detailed Design

The objective of detailed design is to produce a complete description of the end items constituting the total system. For large systems, a massive engineering effort is required to produce all the necessary plans, specifications, drawings, and other documentation necessary to justify the decision to begin fabrication. The amount of effort to produce a detailed design of a particular component depends on its "maturity," that is, its degree of previously proven design. For newly developed components, it is usually necessary to build prototypes and to test them under simulated operating conditions to demonstrate that their engineering design is valid.

Typical products of detailed design include

- draft C, D, and E specs (production specifications);
- subsystem detailed engineering drawings;
- prototype hardware;

- interface control drawings;
- configuration control plan;
- detailed test plans and procedures;
- quality assurance plan; and
- detailed integrated logistic support plans.

Systems engineering inputs are especially important to the interface designs and test plans. Where necessary, detailed analysis, simulation, component tests, and prototyping must be performed to resolve risk areas.

Critical Design Review (CDR). The general procedures for the CDR of the products of detailed design are similar to those for the PDR. The CDR is usually more extensive and may be conducted separately for hardware and software CIs. The CDR examines drawings, schematics, data flow diagrams, test and logistic supply plans, and so on, to ensure their soundness and adequacy. The issues addressed in the CDR are partly predicated on those identified as critical in the PDR and are therefore scheduled for further review in light of the additional analysis, simulations, breadboard or brassboard, or prototype tests conducted after the PDR.

As in the case of PDR, systems engineering plays a crucial role in this process, especially in the review of interfaces and plans for integration and testing. Similarly, internal reviews are necessary prior to the official CDR to ensure that unresolved issues do not arise in the formal sessions. But if they do, systems engineering is usually assigned the responsibility of resolving the issues as quickly as possible.

The completion of detailed design results in the product baseline (see Section 12.6).

Computer-Aided Design (CAD)

The microelectronic revolution has profoundly changed the process of hardware component design and fabrication. It has enabled the development and production of increasingly complex systems without corresponding increases in cost and degradations in reliability. The introduction of CAD of mechanical components has completely changed how such components are designed and built. Even more dramatic has been the explosive development of electronics in the form of microelectronic chips of enormous capacity and power, and their principal product, digital computing.

Mechanical Components. CAD permits the detailed design of complex mechanical shapes to be performed by an engineer at a computer workstation without making conventional drawings or models. The design takes form in the computer database and can be examined in any position, at any scale of magnification, and in any cross section. The same database can be used for calculating stresses, weights, positions relative to other components, and other relevant information. When the design is completed, the data can be transformed into fabrication instructions and transferred to digitally driven machines for computer-aided manufacture (CAM) of exact replicas of the design. It can also generate production documentation in whatever form may be required.

One of the dramatic impacts this technology has had on the design and manufacturing process is that once a part has been correctly designed and built, all subsequent copies will also be correct within the tolerances of the production machines. An equally major impact is on the ease of integrating mechanical components with one another. Since the physical interfaces of components can be specified precisely in three dimensions, two adjacent components made to a common interface specification will match exactly when brought together. Today, a complex microwave antenna can be designed, fabricated, and assembled into a finely tuned device without the months of cut-and-try testing that used to characterize antenna design. This technique also largely eliminates the need for the elaborate jigs and fixtures previously used to make the parts fit a given pattern, or specially built gages or other inspection devices to check whether or not the parts conform to the established tolerances.

Electronic Components. The design of most electronic components has been revolutionized by modern technology even more than that of mechanical components. Processing is almost entirely digital, using standard memory chips and processors. All parts, such as circuit cards, card cages, connectors, equipment racks, and so on, are purchased to strict standards. All physical interfaces fit because they are made to standards. Further, in digital circuits, voltages are low; there is little heat generation; and electrical interfaces are digital streams. Inputs and outputs can be generated and analyzed using computer-based test equipment.

Most circuits are assembled on standard circuit cards and are interconnected by programmed machines. Instead of being composed of individual resistors, capacitors, transistors, and so on, most circuit functions are frequently incorporated in circuit chips. The design and fabrication of chips represents a still higher level of automation than that of circuit boards. The progressive miniaturization of the basic components (e.g., transistors, diodes, and capacitors) and of their interconnections has resulted in a doubling of component density and operating speed every 18 months (Moore's law) since the early 1980s—a trend that has not yet diminished. However, the cost of creating an assembly line for a complex new chip has progressively mounted into hundreds of millions of dollars, restricting the number of companies capable of competing in the production of large memory and processing chips. On the other hand, making smaller customized chips is not prohibitive in cost and offers the advantages of high reliability at affordable prices.

Components that handle high power and high voltage, such as transmitters and power supplies, generally do not lend themselves to the above technology, and for the most part must still be custom built and designed with great care to avoid reliability problems (see later section).

Systems Engineering Considerations. To the systems engineer, these developments are vital because of their critical impact on component cost, reliability, and often design feasibility. Thus, systems engineers need to have first-hand knowledge of the available automated tools, their capabilities and limitations, and their effect on component performance, quality, and cost. This knowledge is essential in judging whether or not the estimated performance and cost of proposed components are realistic, and whether their design takes adequate advantage of such tools.

It is also important that systems engineers be aware of the rate of improvement of automated tools for design and manufacture, to better estimate their capabilities at the time they will be needed later in the system development cycle. This is also important in anticipating competitive developments, and hence the likely effective life of the system prior to the onset of obsolescence.

Example: The Boeing 777. The development of the Boeing 777 airliner has received a great deal of publicity as a pioneer in large-scale automated design and manufacture. It was claimed by Boeing to be the first major aircraft that was designed and manufactured without one or more stages of prototype ground and flight testing. This achievement was made possible mainly because of four factors: (1) the use of automated design and manufacture for all parts of the aircraft structure, (2) the high level of knowledge of aerodynamics and structures of aircraft obtained through years of development and experimentation, (3) the application of computer-based analysis tools, and (4) highly integrated and committed engineering teams. Thus, aircraft body panels were designed and built directly from computer-based design data and fit together perfectly when assembled. This approach was used for the entire airplane body and associated structures.

It should be noted that the 777 engines, whether built by Pratt and Whitney, General Electric, or Rolls Royce, were thoroughly ground tested before delivery because the degree of knowledge and predictability for jet engines is not at the level of that for airframes. Also, the 777 design did not embody radical departures from previous aircraft experience. Thus, the development cycle of the 777 as a total system did not depart as widely from the traditional sequence as it may have appeared to. However, it was a major milestone and a dramatic illustration of the power of automation in certain modern systems.

Reliability

The reliability of a system is the probability that the system will perform its functions correctly for a specified period of time under specified conditions. Thus, the total reliability (P_R) of a system is the probability that every component on which its function depends functions correctly. Formally, reliability is defined as one minus the failure distribution function of a system or component:

$$R(t) = 1 - F(t) = \int_{t}^{\infty} f(t)dt,$$

where $F(t)$ is the failure distribution function and $f(t)$ is the probability density function of $F(t)$. $f(t)$ can follow any number of known probability distributions. A common representation for a failure function is the exponential distribution

$$\text{pdf: } f_X(x) = \begin{cases} \lambda e^{-\lambda} & \text{if } x \geq 0 \\ 0 & \text{otherwise} \end{cases}; \quad \text{cdf: } F_X(x) = \begin{cases} 1 - e^{-\lambda} & \text{if } x \geq 0 \\ 0 & \text{otherwise} \end{cases}$$

$$\text{Expectation: } E[X] = \frac{1}{\lambda}; \quad \text{variance: } Var[X] = \frac{1}{\lambda^2}.$$

This distribution is used quite extensively for common component reliability approximations, such as those relating to electrical and mechanical devices. An advantage of using the exponential distribution is its various properties relating to reliability:

$f(t) = \dfrac{1}{\theta} e^{-t/\theta}$,   where θ = mean life and t is the time period of interest;

$\lambda - \dfrac{1}{\theta}$,   where λ = failure rate; and

$R = e^{-t/M}$,   where R = reliability of the system

M = MTBF.

MTBF is "mean time between failure" and is explained below. By using the exponential distribution, we can calculate individual reliabilities fairly easily and perform simple mathematics to obtain reliability approximations, described below.

Calculating the probability depends on the configuration of the individual system components. If the components are arranged in a series, each one depending on the operation of the others, the total system probability is equal to the product of the reliabilities of each component (Pr):

$$P_R = Pr_1 \times Pr_2 \ldots Pr_n.$$

For example, if a system consisting of 10 critical components in series is required to have a reliability of 99%, then the average reliability of each component must be at least 99.9%.

If a system contains components that are configured in parallel, representing redundancy in operations, a different equation is used. For example, if two components are operating in parallel, the overall reliability of the system is

$$P_R = Pr_1 + Pr_2 - (Pr_1 \times Pr_2).$$

For pure parallel components, such as the example above, at least one component operating would allow the total system to operate effectively. Redundancy is discussed further below.

In most cases, a system consists of both parallel and series components. Keep in mind that for both examples above, time is considered integral to the definition of probability. Pr_i would be defined and calculated from the failure distribution function, which contains t. For the exponential distribution, Pr_i would be expressed as $1/e^{-t/M}$.

For systems that must operate continuously, it is common to express their reliability in terms of the MTBF. In the 10-component system just mentioned, if the system MTBF must be 1000 hours, the component MTBF must average 10,000 hours. From these considerations, it is evident that the components of a complex system must meet extremely stringent reliability standards.

Since system failures almost always occur at the level of components or below, the main responsibility for a reliable design rests on design specialists who understand the details of how components and their subcomponents and parts work and are manufactured. However, the difficulty of achieving a given level of reliability differs widely among the various components. For example, components composed largely of integrated microcircuits can be expected to be very reliable, whereas power supplies and other high-voltage components are much more highly stressed and therefore require a greater fraction of the overall reliability "budget." Accordingly, it is necessary to allocate the allowable reliability requirements among the various components so as to balance, insofar as practicable, the burden of achieving the necessary reliabilities among the components. This allocation is a particular systems engineering responsibility and must be based on a comprehensive analysis of reliability records of components of similar functionality and construction.

A number of specific reliability issues must not be left entirely to the discretion of the component designers; these issues should not only be examined at formal reviews but should also be subject to oversight throughout the design process. Such issues include

1. *External Interfaces:* Surfaces exposed to the environment must be protected from corrosion, leakage, radiation, structural damage, thermal stress, and other potential hazards.
2. *Component Mounting:* Systems subjected to shock or vibration during operation or transport must have suitable shock mountings for fragile components.
3. *Temperature and Pressure:* Systems subjected to extremes of temperature and pressure must provide protective controls at either the system or component level.
4. *Contamination:* Components susceptible to dust or other contaminants must be assembled under clean room conditions and sealed if necessary.
5. *High-Voltage Components:* Components using high voltage, such as power supplies, require special provisions to avoid short circuits or arcing.
6. *Workmanship:* Parts requiring precise workmanship should be designed for easy inspection to detect defects that could lead to failures in operation.
7. *Potential Hazards:* Components that may present operating hazards if not properly made or used should be designed to have large reliability margins. These include rocket components, pyrotechnics, hazardous chemicals, high-pressure containers, and so on.

Software Reliability. Software does not break, short-circuit, wear out, or otherwise fail from causes similar to those that lead to most hardware failures. Nevertheless, complex systems do fail due to malfunctioning software as often as and sometimes even more often than from hardware faults. Anyone whose computer keyboard has "locked up," or who has tried to buy an airline ticket when the "computer is down" has experienced this phenomenon. With systems increasingly dependent on complex software, its reliability is becoming ever more crucial.

Software operating failures occur due to imperfect code, that is, computer program deficiencies that allow the occurrence of unintended conditions, causing the system to produce erroneous outputs, or in extreme cases to abort ("crash"). Examples of conditions that cause such events are infinite loops (repeated sequences that do not always terminate, thereby causing the system to hang up), overflows of memory space allocated to data arrays (which cause excess data to overwrite instruction space, producing "garbage" instructions), and mishandling of external interrupts (which cause losses or errors of input or output).

As described in Chapter 11, there is no possibility of finding all the deficiencies in complex code by inspection, nor is it practical to devise sufficiently exacting tests to discover all possible faults. The most effective means of producing reliable software is to employ experienced software designers and testers in combination with disciplined software design procedures, such as

1. highly modular program architecture,
2. disciplined programming language with controlled data manipulation,
3. disciplined coding conventions requiring extensive comments,
4. design reviews and code "walk-throughs,"
5. prototyping of all critical interfaces,
6. formal CM,
7. independent verification and validation, and
8. endurance testing to eliminate "infant mortality"

Redundancy. Complex systems that must operate extremely reliably, such as air traffic control systems, telephone networks, power grids, and passenger aircraft, require the use of redundant or backup subsystems or components to achieve the required levels of uninterrupted operation. If a power grid line is struck by lightning, its load is switched to other lines with a minimum disruption of service. If an aircraft landing gear's motors fail, it can be cranked down manually. Air traffic control has several levels of backups to maintain safe (though degraded) operation in case of failure of the primary system.

The equation for calculating the reliability of parallel components was presented above. Another perspective on parallel component reliability is to understand that the failure probability is a product of the failure probabilities of the individual system modes. Since the reliability (P_R) is one minus the failure probability (P_F), the reliability of a system with two redundant (parallel) subsystems is

$$P_R = 1 - (P_{F1} \times P_{F2}).$$

The reader is encouraged to prove to himself that the two reliability equations presented are indeed equivalent. As an example, if a system reliability of 99.9 is required for a system, and a critical subsystem cannot be designed to have a reliability better than

99%, providing a backup subsystem of equal reliability will raise the effective reliability of the parallel subsystem to 99.99% ($1 - 0.01 \times 0.01 = 0.9999$).

Systems that must reconfigure themselves by automatically switching over to a backup component in place of a failed one must also incorporate appropriate failure sensors and switching logic. A common example is the operation of an uninterruptible power supply for a computer, which automatically switches to a battery power supply in the event of an interruption in external power. Telephone networks switch paths automatically not only when a link fails but also when one becomes overloaded. An inherent problem with such automatic switching systems is that the additional sensors and switches add further complexity and are themselves subject to failure. Another is that complex automatic reconfiguration systems may overreact to an unexpected set of conditions by a catastrophic crash of the whole system. Such events have occurred in a number of multistate power grid blackouts and telephone outages. Automatically reconfigurable systems require extremely comprehensive systems engineering analysis, simulation, and testing under all conceivable conditions. When this has been expertly done, as in the manned space program, unprecedented levels of reliability have been achieved.

Techniques to Increase Reliability. Several techniques exist to increase, or even maximize, reliability within a system design. Several have been discussed already:

- *System Modularity.* Increase the modularity of system components to achieve loose coupling among components. This will minimize the number of components that are in series and thus could cause a system failure.
- *Redundancy.* Increase component redundancy either with parallel operating components or through the use of switches that automatically transfer control and operations to backup components.
- *Multiple Functional Paths.* A technique to increase reliability without necessarily adding redundant components involves including functional multiple paths within the system design. This is sometimes known as "channels of operation."
- *Derating Components. Derating* refers to the technique of using a component under stress conditions considerably below the rated performance value to achieve a reliability margin in the design.

Several methods and formal techniques exist to analyze failure modes, effects, and mitigation strategies. Five common techniques (not described here) are failure mode, effects, and criticality analysis (FMECA), fault tree analysis, critical useful life analysis, stress–strength analysis, and reliability growth analysis. The reader is encouraged to explore any or all of the techniques as effective analyses strategies.

Maintainability

The maintainability of a system is a measure of the ease of accomplishing the functions required to maintain the system in a fully operable condition. System maintenance takes two forms: (1) repair if a system fails during operation and (2) scheduled periodic servicing including testing to detect and repair failures that occur during standby. High maintainability requires that the system components and their physical configurations be designed with an explicit and detailed knowledge of how these functions will be carried out.

Since to repair a system failure it is first necessary to identify the location and nature of the fault, that is, to carry out a failure diagnosis, system design should provide for means to make diagnosis easy and quick. In case repair is needed, the design must be dovetailed with logistic support plans to ensure that components or component parts that may fail will be stocked and will be replaceable in minimum time.

Unlike hardware faults, replacement of the failed component is not an option for software because software failures result from faults in the code. Instead, the error in the code must be identified and the code modified. This must be done with great care and the change in configuration documented. To prevent the same fault from causing failures in other units of the system, the correction must be incorporated in their programs. Thus, software maintenance can be a critical function.

A measure of system maintainability during operation is the mean time to repair/restore (MTTR). The "time to repair" is the sum of the time to detect and diagnose the fault, the time to secure any necessary replacement parts, and the time to effect the replacement or repair. The "time to restore" also includes the time required to restore the system to full operation and to confirm its operational readiness.

Built-In Test Equipment (BITE). A direct means for reducing the MTTR of a system is to incorporate auxiliary sensors that detect the occurrence of faults that would render the system inoperable or ineffective when called upon, then to signal an operator that repairs are required, and indicate the location of the fault. Such built-in equipment effectively eliminates the time to detect the fault and focuses the diagnosis on a specific function. Examples of such built-in fault detection and signaling devices are present in most modern automobiles, which sense and signal any faulty indications of air bag or antilock brake status, low oil level, or low battery voltage, and so on. In controlling complex systems, such as in aircraft controls, power plant operations, and hospital intensive care units, such devices are absolutely vital. In automatically reconfigurable systems (see section on redundancy, above) the built-in sensors provide signals to automatic controls rather than to a system operator.

The use of BITE presents two important system-level problems. First, it adds to the total complexity of the system and hence to potential failures and cost. Second, it is itself capable of false indications, which can in turn impact system effectiveness. Only when these problems are examined in detail can a good balance be struck between not enough and too much system self-testing. Systems engineering bears the principal responsibility for achieving such a balance.

Design for Maintainability. The issues that must be addressed to ensure a maintainable system design begin at the system level and range all the way down to component parts. They include

1. *Modular System Architecture:* A high degree of system modularity (self-contained components with simple interfaces) is absolutely vital to all three forms of maintenance (repair of operational failures, periodic maintenance, and system upgrading).
2. *Replaceable Units:* Because it is often impractical to repair a failed part in place, the unit that contains the part must be replaced by an identical spare unit. Such units must be accessible, simply and safely replaceable, and part of the logistic support supply.
3. *Test Points and Functions:* To identify the location of a failure to a specific replaceable unit, there must be a hierarchy of test points and functions that permits a short sequence of tests to converge on the failed unit.

To achieve the above, there must be an emphasis on design for maintainability throughout the system definition, development, and engineering design process. In addition to the design, comprehensive documentation and training are essential.

Availability

An important measure of the operational value of a system that does not operate continuously is referred to as system availability, that is, the probability that it will perform its function correctly when called upon. Availability can be expressed as a simple function of system reliability and maintainability for relatively short repair times and low failure rates:

$$P_A = 1 - \frac{\text{MTTR}}{\text{MTBF}},$$

where

P_A = probability that the system will perform when called upon;
MTBF = mean time between failure; and
MTTR = mean time to restore.

This formula shows that system maintainability is just as critical as reliability and emphasizes the importance of rapid failure detection, diagnosis, and repair or parts replacement. It also points to the importance of logistic support to ensure the immediate availability of necessary replacement parts.

Producibility

For systems that are produced in large quantities, such as commercial aircraft, automotive vehicles, or computer systems, reducing the costs associated with the manufacturing process is a major design objective. The characteristic that denotes relative system production costs is called "producibility." The issue of producibility is almost wholly associated with hardware components since the cost of replicating software is only that of the medium in which it is stored.

Design for producibility is the primary province of the design specialist. However, systems engineers need to be sufficiently knowledgeable about manufacturing processes and other production cost issues to recognize characteristics that may inflate costs and to guide design accordingly. Such understanding is necessary for the systems engineer to achieve an optimum balance between system performance (including reliability), schedule (timeliness), and cost (affordability).

Some of the measures that are used to enhance producibility are

1. maximum use of commercially available parts, subcomponents, and even components (referred to as commercial off-the-shelf ["COTS"] items); this also reduces development cost;
2. setting dimensional tolerances of mechanical parts well within the normal precision of production machinery;
3. design of subassemblies for automatic manufacture and testing;
4. maximum use of stampings, castings, and other forms suitable for high-rate production;
5. use of easily formed or machined materials;
6. maximum standardization of subassemblies, for example, circuit boards, cages, and so on; and
7. maximum use of digital versus analog circuitry.

As noted in previous chapters, the objective of producibility, along with other specialty engineering features, should be introduced into the system design process early in the life cycle. However, the application of producibility to specific design features occurs largely in the engineering design phase as part of the design process. Chapter 14 is devoted to the subject of production and its systems engineering content.

Risk Management

Many of the methods of risk mitigation listed in Chapter 5 are pertinent to the component design step in the engineering design phase. Components containing residual risk factors must be subjected to special technical and management oversight, including analysis and testing to ensure the early discovery and resolution of any design problems. Where the acceptability of a given design requires testing under operational conditions, as in the case of user interfaces, rapid prototyping and user feedback may be in order. In exceptional circumstances, where the risk inherent in the chosen approach remains unacceptably high, it may be necessary to initiate a backup effort to engineer a more conservative replacement in case the problems with the first line design cannot be resolved when the design must be frozen. Alternatively, it may be wise to seek relaxation of stringent requirements that would produce only marginal gains in system effectiveness. All of the above measures require systems engineering leadership.

12.5   DESIGN VALIDATION

Design validation proceeds at various levels throughout the engineering development stage of the system life cycle. This section focuses on the validation of the physical implementation of the component system building blocks.

Test Planning

Planning the testing of components to validate their design and construction is an essential part of the overall test and evaluation plan. It covers two types of tests: development testing during the component design process and unit qualification testing to ensure that the final production design meets specifications.

Component test planning must be done during the early part of the engineering design phase for several reasons. First, the required test equipment is often complex and requires a time to design and build comparable to that required for the system components themselves. Second, the cost of test tools usually represents a very significant fraction of the system development costs and must be provided for in the total cost equation. Third, test planning must involve design engineers, test engineers, and systems engineers in a *team effort*, often across organizational and sometimes across contractual lines. From these detailed plans, test procedures are derived for all phases of the test operations.

As in system-level test planning, systems engineering must play a major role in the development of component test plans, that is, what should be tested, at what stage in the development, to what degree of accuracy, what data should be obtained, and so on. An important systems engineering contribution is to ensure that component features that were identified as potential risks are subjected to tests to confirm their elimination or mitigation.

Component Fabrication

In the previous sections, the design process has been discussed in terms of its objectives and has been related to design decisions defined in terms of drawings, schematics, specifications, and other forms of design representation as expressed on paper and in computer data. To determine the degree to which a design will actually result in the desired component performance, and whether or not the component will properly interface with the others, it is necessary to convert its design to a physical entity and to test it. This requires that hardware elements be fabricated and individual software components be coded. Prior to fabrication, reviews are held between the designers and fabrication personnel to assure that what has been designed is within the capabilities of the facility that has to build it.

The implementation process is seldom unidirectional (i.e., noniterative). Design deficiencies are often discovered and corrected during implementation, even before testing, especially in hardware components. Even though CAD has greatly reduced the probability of dimensional and other incompatibilities, it has to be anticipated that some changes will need to be made in the design to achieve a successful functioning product.

At this stage of component engineering, the tools that are to be used in production (such as computer-driven, metal-forming machines and automatic assembly devices) are seldom available for use, so that initial fabrication must often be carried out using manually operated machines and hand assembly. It is important, however, that a realistic experimental replica of the fabrication process be employed for any component parts that are to be built using unconventional manufacturing processes. This is essential to ensure that the transition to production tooling will not invalidate the results of the prototyping process. Involving the production people during sign-off, prior to the time the article reaches the manufacturing facility, will greatly expedite production.

In the case of complex electronic circuits, significant alterations in the initially fabricated model are to be expected before a completely suitable design is finally achieved. Accordingly, it has been customary to first construct and test these circuits in a more open "breadboard" or "brassboard" form (with rudimentary packaging constraints) so as to facilitate circuit changes before packaging the component in its final form. However, with modern automated tools, it is often more efficient to go directly to a packaged configuration, even though this may dictate the fabrication of several such packages before a suitable design is finally achieved.

Development Testing

The objective of engineering development testing is different from production acceptance testing in that the latter is mainly concerned with whether the component should be accepted or rejected, while the former must not only quantify each discrepancy but must also help diagnose its source. It should be anticipated that design discrepancies will be found and design changes will be needed in order to comply with requirements. Thus, component testing is very much a part of the development process. Changes at this point must be introduced via an "engineering change notice" agreed to by all cognizant parties to avoid chaotic, noncoordinated change.

Development testing is concerned with validating the basic design of the component, focusing on its performance, especially on features that are critical to its operation within the system or that represent characteristics that are highly stressed, newly developed, or are expected to operate at levels beyond those commonly attained in previous devices of this type. These tests also focus on the features of the design that are subject to severe environmental conditions, such as shock, vibration, external radiation, and so on.

For components subject to wear, such as those containing moving parts, development tests can also include endurance testing, usually performed under accelerated conditions to simulate years of wear in a matter of months.

Reliability and Maintainability Data. Whereas during development components may not be built from the identical parts used in the production article, it is good practice to begin collecting reliability statistics as early as possible by recording all failures during operation and test and by identifying their source. This will reduce the likelihood of incipient failures carrying on into the production article. This is particularly important where the number of units to be built is too small to collect adequate statistical samples of production components. Involvement of quality assurance engineers in this process is essential.

Development testing must also examine the adequacy and accessibility of test points for providing failure diagnosis during system maintenance. If maintenance of the system will require disassembling the component and replacing subcomponents such as circuit boards, this feature must also be evaluated.

Test Operations. Component development tests are part of the design process and are usually conducted within the design group by a team headed by the lead design engineer and composed of members of the design team as well as other staff experienced in testing the type of component under development. The team should be intimately familiar with the use of test tools and special test facilities that may be required. The validity and adequacy of the test setup and analysis procedures should be overseen by systems engineering.

An important lesson that systems engineers (and test engineers) must learn is that the apparent failure of a component to meet some test objective may not be due to a defective design but rather due to a deficiency in the test equipment or test procedure. This is especially true when a component is first tested in a newly designed test setup. The need for testing the test equipment occurs all too frequently. This is a direct result of the difficulty of ensuring perfect compatibility between two or more interacting and interfacing components, whether they are system elements or test equipment units (hardware or software). Thus, a period of preliminary testing should be scheduled to properly integrate a new component with its test equipment, and unit testing should not begin until all the test bugs have been eliminated.

Change Control. It will be recalled that after the CDR, the detailed design of a complex system is frozen and placed under formal CM (see Section 9.6). This means that thereafter, any proposed design change requires justification, evaluation, and formal approval, usually from a "configuration control board" or an equivalent. Such approval is usually granted only on the basis of a written engineering change request containing a precise definition of the nature of the deficiency revealed by the test process and a thorough analysis of the impact of the proposed change on system performance, cost, and schedule. The request should also contain trade-offs of alternative remedies, including possible relaxation of requirements, and an in-depth assessment of risks and costs associated with making (and not making) the change. This formal process is not intended to prevent changes but to ensure that they are introduced in an orderly and documented manner.

Qualification Testing

Testing a productionized component ("first-unit" testing) prior to its delivery to the integration facility is very much like the acceptance testing of units off the production line. Qualification tests are usually more limited than development tests, but are frequently more quantitative, being concerned with the exact conformance of the unit to interface tolerances so that it will fit exactly with mating system components.

Accordingly, equipment used for this purpose should be much like production test equipment. Qualification tests are generally more severe than the conditions to which the article is subjected in operational use.

The validation of the design of an individual system component can be rigorously accomplished only by inserting it into an environment identical to that in which it will operate as part of the total system. In the case of complex components, it is seldom practicable to reproduce exactly its environment. Therefore, a test setup that closely approximates this situation has to be used.

The problem is made more difficult by the fact that components are almost always developed and built by different engineering groups, often by independent contractors. In the case of software programs, the designers may be from the same company but generally do not understand each other's designs in detail. The system developer thus has the problem of ensuring that the component designers test their products to the identical standards to be used during system integration. The critical point, of course, is that each component's interfaces must be designed to fit exactly with their connecting components and with the environment.

Tolerances. The specification of component interfaces to ensure fit and interchangeability involves the assignment of tolerances to each dimension or other interface parameters. Tolerances represent the positive and negative deviation from a nominal parameter value to ensure a proper fit. The assignment of tolerances requires striking a balance between ease of manufacturing on one hand and assurance of satisfactory fit and performance on the other. Whenever either producibility or reliability is significantly affected, the systems engineer needs to enter the process of setting the preferred balance.

Computer-Aided Tools. The widespread use of CAD and CAM has greatly simplified the above problems in many types of equipment. With these tools, component specifications can be converted into a digital form and can be directly used in their design. The CAD database can be shared electronically between the system developer and the component designer and producer. The same data can be used to automate test equipment.

In the area of electronic equipment, the widespread use of standard commercial parts, from chips to boards to cabinets to connectors, has made interfacing much easier than with custom-built components. These developments have produced economies in test and integration, as well as in component costs. Miniaturization has resulted in a greater number of functions being performed on a circuit board, or encapsulated in a circuit chip, thereby minimizing interconnections and numbers of boards.

Test Operations. Component qualification tests are performed to ensure that the final production component design meets all of its requirements as part of the overall system. Hence, they are much more formal than development tests and are conducted by the test organization, sometimes with oversight by the system contractor. Design engineering supports the test operations, especially during test equipment checkout and data analysis.

Test Tools

A set of test tools for verifying the performance and compatibility of a system component must be designed to provide an appropriate set of inputs and to compare the
resulting outputs with those prescribed in the specifications. In effect, they constitute
a simulator, which models the physical and functional environments of the component,
both external and internal to the system, and measures all significant interactions and
interfaces. Functionally, such a simulator may be as complex as the component that it
is designed to test, and its development usually requires a comparable level of analysis
and engineering effort. Moreover, the assessment of a component's adherence to specified parameter tolerance values usually requires the test equipment precision to be
several times better than the allowable variations in component parameters. This
requirement sometimes calls for precision greater than that readily available, involving
a special effort to develop the necessary capability.

Development test tools often may be available or may be adaptable from other
programs. In addition, standard measuring instruments, such as signal generators, spectrum analyzers, displays, and so on, are readily available in a form that can be incorporated as part of a computer-driven test setup. On the other hand, highly specialized
and complex components, such as a jet engine, may require the provision of dedicated
and extensively instrumented test facilities to be used to support testing during component development and sometimes also during production.

In any event, such special tools as are required to support design and testing during
component development must be designed and built early in the engineering design
phase. Moreover, since similar tools will also be needed to test these same components
during production, efforts should be made to assure that the design and construction of
engineering and production test equipment are closely coordinated and mutually supporting. To keep the cost of such test tools within acceptable bounds, significant systems
engineering effort is usually needed to support the planning and definition of their
design and performance requirements.

Role of Systems Engineering

From the above discussion, it should be evident that systems engineering plays an
essential part in the component validation process. Systems engineers should define the
overall test plan, specify what parameters should be tested and to what accuracy, how
to diagnose discrepancies, and how the test results should be analyzed. Systems engineering must also lead the change initiation and control process. The proper balance
between "undertesting" and "overtesting" requires knowledge of the system impact of
each test, including overall cost. This, in turn, depends on a first-hand knowledge of the
interactions of the component with other parts of the system and with its environment.

12.6   CM

The development of a complex new system has been seen to be resolvable into a series
of steps or phases in which each of the characteristics of the system is defined in terms

of successively more specific system requirements and specifications. The systems engineering process that maintains the continuity and integrity of the system design throughout these phases of system development is called "CM."

The CM process generally begins incrementally during the concept exploration phase, which first defines the selected top-level system configuration in terms of functional requirements after a process of trade-offs among alternative system concepts. It then progresses throughout the phases of the engineering development stage, culminating in system production specifications. The CM process is described more fully in this chapter because the intensity and importance of CM is greatest during the engineering design phase. The terminology of formal CM includes two basic elements, CIs and configuration baselines. Each of these is briefly described below.

CIs

A CI is a system element that is the basis of describing and formally controlling the design of a system. In early phases of system definition, it may be at the level of a subsystem. In later phases, it usually corresponds to that of a component in the hierarchy defined in this book (see Chapter 3). Like the component, the CI is considered as a basic building block of the system, designed and built by a single organization, whose characteristics and interfaces to other building blocks must be defined and controlled to ensure its proper operation within the system as a whole. It is customary to distinguish between hardware configuration items (HWCIs) and computer software configuration items (CSCIs) because of the basically different processes used in defining and controlling their designs.

Configuration Baselines

An important concept in the management of the evolving system design during the system life cycle is that of configuration baselines. The most widely used forms are called functional, allocated, and product baselines. Table 12.2 shows the phase in which each is usually defined, the type of specification that describes it, and the primary characteristics that are specified.

TABLE 12.2. Configuration Baselines

Baseline	Phase defined	Type of specification	Characteristics	Element specified
Functional	Concept definition	A	Functional specifications	System
Allocated	Engineering design	B	Development specifications	Configuration item
Product	Engineering design	C, D, E	Product, process specifications	Configuration item

The functional baseline describes the system functional specifications as they are derived from system performance requirements during the concept definition phase and serves as an input to the advanced development phase.

The allocated baseline is defined during the engineering design phase as the allocations of functions to system components (CIs) are validated by analyses and tests. The resulting development specification defines the performance specifications for each CI, as well as the technical approaches developed to satisfy the specified objective.

The product baseline is established during the engineering design phase in terms of detailed design specifications. It consists of product, process, and material specifications and engineering drawings.

Interface Management

It has been stressed throughout this book that the definition and management of the interfaces and interactions of the system's building blocks with one another and with the system environment is a vital systems engineering function. This function is embodied in the concept of CM, irrespective of whether or not it is formally defined in terms of CIs and baselines as described above. It is therefore incumbent on project management with the aid of systems engineering to organize the necessary people and procedures to carry out this function.

A primary condition for the effective definition and management of a given interface is to ensure the involvement of all key persons and organizations responsible for the designs of the CIs. This is generally accomplished by means of interface configuration working groups (ICWGs), or their equivalents, whose members have the technical knowledge and authority to represent their organizations in negotiating a complete, compatible, and readily achievable definition of the respective interfaces. In large systems, formal sign-off procedures have been found to be necessary to ensure commitment of all parties to the agreed-upon interface coordination documents (ICDs). The form of these documents is a function of the type of interface being documented, but during the engineering design phase, it must be sufficiently specific in terms of data and drawings to specify completely the interface conditions, so that the individual component developers may design and test their products independently.

Change Control

Change is vital to the development of a new and advanced system, especially to take advantage of evolving technology to achieve a sufficient advance in system capability to provide a long useful life. Thus, during the formative stages of system development, it is desirable to maintain sufficient design flexibility to accommodate relevant technological opportunities. The price of such flexibility is that each change inevitably affects related system elements and often requires a series of adaptations extending far beyond the initial area of interest. Thus, a great deal of systems engineering analysis, test, and evaluation is required to manage the system evolution process.

The effort and cost associated with accommodating changes increases rapidly as the design matures. By the time the system design is formulated in detail during the engineering design phase, the search for opportunities for further enhancement can no longer be sustained. Accordingly, the system design is frozen, and formal change control procedures are imposed to deal with necessary modifications, such as those required by incompatibilities, external changes, or unexpected design deficiencies. This usually happens after successful completion of the CDR or its equivalent.

It is customary to categorize proposed changes as class I, or class changes have system- or program-level impact, such as cost, schedule, major interfaces, safety, performance, reliability, and so on. Formal change control of system-level changes is usually exercised by a designated group composed of senior engineers with recognized technical and management expertise capable of making judgments among performance, cost, and schedule. For large programs, this group is called a change control board. It is of necessity led by systems engineering but usually reports at the topmost program level.

12.7 SUMMARY

Implementing the System Building Blocks

The objectives of the engineering design phase are to design system components to performance, cost, and schedule requirements. This phase also establishes consistent internal and external interfaces.

Engineering design culminates in materialization of components of a new system focused on the final design of the system building blocks. Activities constituting engineering design are

- *Requirements Analysis:* identifying all interfaces and interactions,
- *Functional Analysis and Design:* focusing on modular configuration,
- *Component Design:* designing and prototyping all components, and
- *Design Validation:* testing and evaluating system components.

Requirements Analysis

External system interface requirements are especially important at this point in development. User interfaces and environmental interactions require particular attention.

Functional Analysis and Design

Functional design stresses three areas:

- *Modular Configuration:* simplified interactions
- *Software Design:* modular architecture
- *User Interfaces:* effective human interaction.

Modular partitioning groups "tightly bound" functions together into "loosely bound" modules.

Component Design

Major defense and space systems engineering is performed in two steps: preliminary design followed by a PDR detailed design followed by a CDR.

The engineering design process is focused on CIs. These are substantially equivalent to components as defined in this book.

A preliminary design has the objective to demonstrate that chosen designs conform to system performance and design requirements that can be produced within cost and schedule goals. The PDR centers on major interfaces, risk areas, long-lead items, and system-level trade studies.

A detailed design has the objective to produce a complete description of the end items (CIs) constituting the total system. The CDR examines drawings, plans, and so on, for soundness and adequacy. Within the detailed design, CAD has revolutionized hardware implementation—mechanical component design can now be analyzed and designed in software. Digital electronics is miniaturized, standardized, and does not need breadboarding. The Boeing 777 development illustrates the power of automated engineering.

Reliability must be designed at the component level where interfaces, environment, and workmanship are vulnerable areas. Additionally, software must be built to exacting standards and prototyped. Where extreme reliability is required, it is typically achieved by redundancy. Measuring reliability usually includes the MTBF.

Maintainability requires rapid fault detection diagnosis and repair. MTTR is used as a typical measure of maintainability. BITE is used to detect and diagnose faults.

Availability measures the probability of the system being ready when called in: availability increases with MTBF and decreases with MTTR.

Producibility measures the ease of production of system components and benefits from use of commercial components, digital circuitry, and broad tolerances.

Design Validation

Test planning must be done early since test equipment requires extensive time to design and build. Additionally, test costs must be allocated early to ensure sufficient resources. Finally, test planning is a team effort.

Development testing is part of the design process and should start accumulating reliability statistics on failures. These test failures are often due to test equipment or procedures and should be planned for since changes after CDR are subject to formal CM.

Qualification testing validates component release to integration and focuses on component interfaces. Regardless of the testing phase, test tools must be consistent with the system integration process.

CM

CM is a systems engineering process that maintains the continuity and integrity of system design. Configuration baselines defined in major system developments include

- *Functional Baseline:* system functional specifications,
- *Allocation Baseline:* system development specifications, and
- *Product Baseline:* product, process, and material specifications.

The CI is a system element used to describe and formally control system design.

PROBLEMS

12.1 In spite of the effort devoted to develop critical system components during advanced development, unknown unknowns can be expected to appear during engineering design. Discuss what contingency actions a systems engineer should take in anticipation of these "unk-unks." Your answer should include the consideration of the potential impact on cost, schedule, personnel assignments, and test procedures. If you have knowledge of a real-life example from your work, you may use that as the basis for your discussion.

12.2 External system interfaces are especially important during engineering design. Using the design of a new subway system as an example, list six types of external interfaces that will require critical attention. Explain your answer.

12.3 Modular or sectionalized system design is a fundamental characteristic of good system design practice. Using a passenger automobile as an example, discuss its main subsystems from the standpoint of modularity. Describe those that are modular and those that are not. For the latter, state how and why you think they depart from modular design.

12.4 A PDR is an important event during engineering design and the systems engineer has a key role during this review. Assume you (the systems engineer) have been given the assignment to be the principal presenter for an important PDR. Discuss what specific actions you would take to prepare for this meeting. How would you prepare for items that could be considered controversial?

12.5 The personal laptop computer is a product that has proven to be very reliable in spite of the fact that it has many interfaces, is operated by a variety of people, operates nearly continuously, and includes a number of internal moving parts (e.g., floppy disk drive, hard drive, and CD-ROM drive). It is a portable device that operates in a wide range of environments (temperature, shock, vibration, etc.). List six design features or characteristics that contribute to the laptop reliability. For each item in your list, estimate the contributions this item has on the overall computer cost. A ranking of high, medium, and low is sufficient.

12.6 There are six methods of dealing with program risks listed in the section labeled "Risk Management Methods." For four of these six methods, give two examples of situations where that method could be used for risk reduction and explain how.

12.7 Design changes are vital to the development of new and advanced systems, especially to take advantage of evolving technology. Thus, during system development, some degree of design flexibility must be maintained. However, design changes come with a price that increases as the design matures. Assuming you are the systems engineer for the development of a new commercial jet aircraft, give two types of design changes you would support in each of the early part, middle part, and late part of the engineering design phase.

FURTHER READING

C. Alexander. *The Timeless Way of Building*. Oxford University Press, 1979.

A. B. Badiru. *Handbook of Industrial and Systems Engineering*. CRC Press, 2006, Chapters 8, 9, 11, and 15.

B. Blanchard and W. Fabrycky. *System Engineering and Analysis*, Fourth Edition. Prentice Hall, 2006, Chapters 4 and 5.

F. P. Brooks, Jr. *The Mythical Man Month—Essays on Software Engineering*. Addison-Wesley, 1995.

G. E. Dieter and L. C. Schmidt. *Engineering Design*, Fourth Edition. McGraw-Hill, 2009, Chapters 1, 2, 9, 13, and 14.

C. E. Ebeling. *An Introduction to Reliability and Maintainability Engineering*. Waveland Press, Inc., 2005, Chapters 1, 2, 5, 8, 9, and 11.

H. Eisner. *Computer-Aided Systems Engineering*. Prentice Hall, 1988, Chapters 14 and 15.

B. Hyman. *Fundamentals of Engineering Design*, Second Edition. Prentice Hall, 2003, Chapters 1, 5, 6, and 10.

J. A. Lacy. *Systems Engineering Management: Achieving Total Quality, Part II*. McGraw Hill, 1992.

P. D. T. O'Connor. *Practical Reliability Engineering*, Fourth Edition. John Wiley & Sons, Inc., 2008, Chapters 1, 2, and 7.

R. S. Pressman. *Software Engineering: A Practitioner's Approach*, McGraw Hill, 1982.

E. Rechtin. *Systems Architecting: Creating and Building Complex Systems*, Prentice Hall, 1991, Chapter 6.

N. B. Reilly. *Successful Systems for Engineers and Managers*, Van Nostrand Reinhold, 1993, Chapters 8–10.

A. P. Sage. *Systems Engineering*, McGraw Hill, 1992, Chapter 6.

R. M. Shinners. *A Guide for Systems Engineering and Management*, Lexington Books, 1989, Chapter 3.

Systems Engineering Fundamentals. Defense Acquisition University (DAU Press), 2001, Chapters 6 and 10.

Systems Engineering Handbook: A Guide for System Life Cycle Processes and Activities, INCOSE-TP-2003-002-03.2, Section 4. International Council on Systems Engineering, 2010.

13

INTEGRATION AND EVALUATION

13.1 INTEGRATING, TESTING, AND EVALUATING THE TOTAL SYSTEM

As its name implies, the integration and evaluation phase has the objectives of assembling and integrating the engineered components of the new system into an effectively operating whole, and demonstrating that the system meets all of its operational requirements. The goal is to qualify the system's engineering design for release to production and subsequent operational use.

As previously noted, the systems engineering life cycle model defines integration and evaluation as a separate phase of system development because its objectives and activities differ sharply from those of the preceding portion called the engineering design phase. These differences are also reflected in changes in the primary participants engaged in carrying out the technical effort.

If all of the building blocks of a new system were correctly engineered, and if their design was accurately implemented, their integration and subsequent evaluation would be relatively straightforward. In reality, when a team of contractors develops a complex

Systems Engineering Principles and Practice, Second Edition. Alexander Kossiakoff, William N. Sweet, Samuel J. Seymour, and Steven M. Biemer
© 2011 by John Wiley & Sons, Inc. Published 2011 by John Wiley & Sons, Inc.

system during a period of rapidly evolving technology, the above conditions are never fully realized. Hence, the task of system integration and evaluation is always complex and difficult and requires the best efforts of expert technical teams operating under systems engineering leadership.

The success of the integration and evaluation effort is also highly dependent on the advance planning and preparation for this effort that was accomplished during the previous phases. A detailed test and evaluation master plan (TEMP) is required to be formulated by the end of concept exploration and elaborated at each step thereafter (see Chapter 10). In practice, such planning usually remains quite general until well into the engineering design phase for several reasons:

1. The specific test approach is dependent on just how the various system elements are physically implemented.
2. Test planning is seldom allocated adequate priority in either staffing or funding in the early phases of system development.
3. Simulating the system operational environment is almost always complicated and costly.

Hence, the integration and evaluation phase may begin with very considerable preparation remaining and may therefore proceed considerably slower than originally planned. The purpose of this chapter is to describe the essential activities that are typically required in this phase, a number of the problems that are commonly encountered, and some of the approaches to helping overcome the resulting obstacles.

Place of the Integration and Evaluation Phase in the System Life Cycle

It was seen in previous chapters that the general process of test and evaluation is an essential part of every phase of system development, serving as the validation step of the systems engineering method. It can be generally defined as embodying those activities necessary to reveal the critical attributes of a product (in this case a system element, such as a subsystem or component) and to compare them to expectations in order to deduce the product's readiness for succeeding activities or processes. In the integration and evaluation phase, the process of test and evaluation becomes the central activity, terminating with the evaluation of the total system in a realistic replica of its intended operational environment.

Figures 13.1 and 13.2 show two different aspects of the relation between the integration and evaluation phase and its immediately adjacent phases in the system life cycle. Figure 13.1 is a functional flow view, which shows the integration and evaluation phase to be the transition from engineering design to production and operation. Its inputs from the engineering design phase are an engineered prototype, including components, and a test and evaluation plan, with test requirements. The outputs of the integration and evaluation phase are system production specifications and a validated production system design. Figure 13.2 is a schedule and level-of-effort view, which

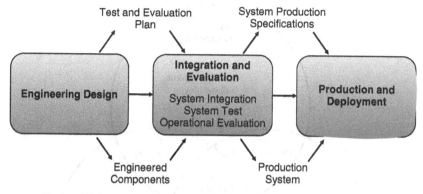

Figure 13.1. Integration and evaluation phase in a system life cycle.

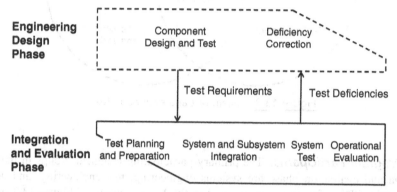

Figure 13.2. Integration and evaluation phase in relation to engineering design.

shows the overlap of the integration and evaluation phase with the engineering design phase.

The differences in the primary objectives, activities, and technical participants of the integration and evaluation phase from those of the engineering design phase are summarized in the following paragraphs.

Program Focus. The engineering design phase is focused on the design and testing of the individual system components and is typically carried out by a number of different engineering organizations, with systems engineering and program management oversight being exercised by the system developer. On the other hand, the integration and evaluation phase is concerned with assembling and integrating these engineered components into a complete working system, creating a comprehensive system test environment and evaluating the system as a whole. Thus, while these activities overlap in time, their objectives are quite different.

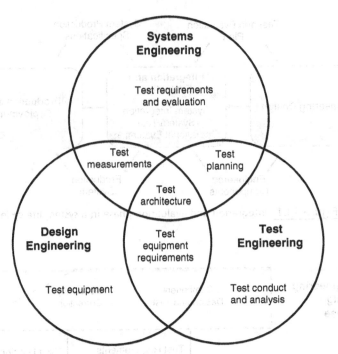

Figure 13.3. System test and evaluation team.

Program Participants. The primary participating technical groups in the integration and evaluation phase are systems engineering, test engineering, and design engineering. Their functions are pictured in the Venn diagram of Figure 13.3, which shows the activities that are primary ones for each technical group and those that are shared. Systems engineering is shown as having the prime responsibility for defining the test requirements and evaluation criteria. It shares the responsibility for test planning with test engineering and the definition of test methodology and data to be collected with design engineering. Test engineering has responsibility for test conduct and data analysis; it usually provides a majority of the technical effort during this period. In many programs, design engineering has the prime responsibility for test equipment design. It is also responsible for component design changes to eliminate deficiencies uncovered in the test and evaluation process.

Critical Problems. The system integration process represents the first time that fully engineered components and subsystems are linked to one another and are made to perform as a unified functional entity. Despite the best plans and efforts, the integration of a system containing newly developed elements is almost certain to reveal unexpected incompatibilities. At this late stage in the development, such incompatibilities must be resolved in a matter of days rather than weeks or months. The same is true when deficiencies are discovered in system evaluation tests. Any crash program to resolve such critical problems should be led by systems engineers working closely with the project manager.

Management Scrutiny. A large-scale system development program represents a major commitment of government and/or industrial funds and resources. When the development reaches the stage of system integration and testing, management scrutiny becomes intense. Any real or apparent failures are viewed with alarm, and temptations to intervene become strong. It is especially important that the program management and systems engineering leadership have the full confidence of top management, and the authority to act, at this time.

Design Materialization Status

The status of system materialization in the integration and evaluation phase is shown in Table 13.1. The table entries identifying the principal activities in this phase are seen to be in the upper right-hand corner, departing sharply from the downward progression of activities in the previous phases. This corresponds to the fact that in the other phases, the activities referred to the stepwise materialization of the individual component building blocks, progressing through the states of visualization, functional definition, and physical definition to detailed design, fabrication, and testing. In contrast, the activities in the integration and evaluation phase refer to the stepwise materialization of the entire system as an operational entity, proceeding through the integration and test of physically complete components into subsystems, and these into the total system.

A very important feature of the materialization status, which is not explicitly shown in Table 13.1, is the characterization of interactions and interfaces. This process should have been completed in the previous phase but cannot be fully validated until the whole system is assembled. The inevitable revelation of some incompatibilities must therefore be anticipated as the new system is integrated. Their prompt identification and resolution is a top priority of systems engineering. Accomplishing the integration of interfaces and interactions may not appear to be a major increase in the materialization of a system, but in reality, it is a necessary (and sometimes difficult) step in achieving a specified capability.

This view of the activities and objectives of the integration and evaluation phase can be further amplified by expanding the activities pictured in the last column of Table 13.1. This is demonstrated in Table 13.2, in which the first column lists the system aggregation corresponding to the integration level as in Table 13.1; the second column indicates the nature of the environment in which the corresponding system element is evaluated; the third column lists the desired objective of the activity; and the fourth defines the nature of the activity, expanding the corresponding entries in Table 13.1. The sequence of activities, which proceeds upward in the above table, starts with tested components, integrates these into subsystems, and then into the total system. The process then evaluates the system, first in a simulated operational environment and finally in a realistic version of the environment in which the system is intended to operate. Thus, as noted earlier, in the integration and evaluation phase, the process of

TABLE 13.1. Status of System Materialization at the Integration and Evaluation Phase

Phase	Concept development			Engineering development		
Level	Needs analysis	Concept exploration	Concept definition	Advanced development	Engineering design	Integration and evaluation
System	Define system capabilities and effectiveness	Identify, explore, and synthesize concepts	Define selected concept with specifications	Validate concept		Test and evaluate
Subsystem		Define requirements and ensure feasibility	Define functional and physical architecture	Validate subsystems		Integrate and test
Component			Allocate functions to components	Define specifications	Design and test	Integrate and test
Subcomponent		Visualize		Allocate functions to subcomponents	Design	
Part					Make or buy	

448

TABLE 13.2. System Integration and Evaluation Process

Integration level	Environment	Objective	Process
System	Real operational environment	Demonstrated operational performance	Operational test and evaluation
System	Simulated operational environment	Demonstrated compliance with all requirements	Developmental test and evaluation
System	Integration facility	Fully integrated system	System integration and test
Subsystem	Integration facility	Fully integrated subsystems	Subsystem integration and test
Component	Component test equipment	Verified component performance	Component test

materialization refers to the system as a whole and represents the synthesis of the total operational system from the previously physically materialized components.

Systems Engineering Method in Integration and Evaluation

Since the structure of the integration and evaluation phase does not conform to the characteristics of the preceding phases, the application of the systems engineering method is correspondingly different. In this phase, the requirements analysis or problem definition step corresponds to test planning—the preparation of a comprehensive plan of how the integration and evaluation tests are to be carried out. Since the functional design of the system and its components has been completed in previous phases, the functional definition step in this phase relates to the test equipment and facilities, which should be defined as a part of test preparation. The physical definition or synthesis step corresponds to subsystem and system integration, the components having been implemented in previous phases. The design validation step corresponds to system test and evaluation.

The organization of the principal sections in this chapter will follow the order of the above sequence. However, it is convenient to combine test planning and test equipment definition into a single section on test planning and preparation and to divide system test and evaluation into two sections: developmental system testing, and operational test and evaluation. These sections will be seen to correspond to the processes listed in the right-hand column of Table 13.2, reading upward from the fourth row.

Test Planning and Preparation. Typical activities include

- reviewing system requirements and defining detailed plans for integration and system testing, and
- defining the test requirements and functional architecture.

System Integration. Typical activities include

- integrating the tested components into subsystems and the subsystems into a total operational system by the sequential aggregation and testing of the constituent elements, and
- designing and building integration test equipment and facilities needed to support the system integration process and demonstrating end-to-end operation.

Developmental System Testing. Typical activities include

- performing system-level tests over the entire operating regime and comparing system performance with expectations,
- developing test scenarios exercising all system operating modes, and
- eliminating all performance deficiencies.

Operational Test and Evaluation. Typical activities include

- performing tests of system performance in a fully realistic operational environment under the cognizance of an independent test agent and
- measuring degree of compliance with all operational requirements and evaluating the readiness of the system for full production and operational deployment.

13.2 TEST PLANNING AND PREPARATION

As described earlier, planning for test and evaluation throughout the system development process begins in its early phases and is continually extended and refined. As the system design matures, the test and evaluation process becomes more exacting and critical. By the time the development nears the end of the engineering design phase, the planning and preparation for the integration and evaluation of the total system represents a major activity in its own right.

TEMP

It was noted in Chapter 10 that acquisition programs often require the preparation of a formal TEMP. Many of the principal subjects covered in the TEMP are applicable to the development of commercial systems as well. For reference purposes, the main elements of the TEMP format, described more fully in Chapter 10, are listed below:

1. *System Introduction:* describes the system and its mission and operational environment and lists measures of effectiveness;
2. *Integrated Test Program Summary:* lists the test program schedule and participating organizations;

3. *Developmental Test and Evaluation:* describes objectives, method of approach, and principal events;

4. *Operational Test and Evaluation:* describes objectives, test configuration, events, and scenarios; and

5. *Test and Evaluation Resource Summary:* lists test articles, sites, instrumentation, and support operations.

Elements 3 and 4 will be referred to in somewhat greater detail in the final sections of this chapter.

Analogy of Test and Evaluation Planning to System Development

The importance of the test and evaluation planning process is illustrated in Table 13.3, which shows the parallels between this process and system development as a whole. The left half of the table lists the principal activities involved in each of four major steps in the system development process. The entries in the right half of the table list the corresponding activities in developing the test and evaluation plan. The table shows that the tasks comprising the test and evaluation planning process require major decisions regarding the degree of realism, trade-offs among test approaches, definition of objectives, and resources for each test event, as well as development of detailed procedures and test equipment. In emphasizing the correspondence between these activities, the table also brings out the magnitude of the test and evaluation effort and its criticality to successful system development.

As may be inferred from Table 13.3, specific plans for the integration and evaluation phase must be developed before or concurrently with the engineering design

TABLE 13.3. Parallels between System Development and Test and Evaluation (T&E) Planning

System development	T&E planning
Need: Define the capability to be fielded.	Objective: Determine the degree of sophistication required of the test program.
System concept: Analyze trade-offs between performance, schedule, and cost to develop a system concept.	Test concept: Evaluate trade-offs between test approaches, schedule, and cost to develop a test concept.
Functional design: Translate functional requirements into two level specification for the (sub)system(s).	Test plan: Translate test requirements into a description of each test event and the resources required.
Detailed design: Design the various components that comprise the system.	Test procedures: Develop detailed test procedures and test tools for each event.

process. This is necessary in order to provide the time required for designing and building special test equipment and facilities that will be needed during integration and system testing. Costing and scheduling of the test program is an essential part of the plan since the costs and duration for system testing are very often underestimated, seriously impacting the overall program.

Review of System Requirements

Prior to the preparation of detailed test plans, it is necessary to conduct a final review of the system-level operational and functional requirements to ensure that no changes have occurred during the engineering design phase that may impact the system test and evaluation process. Three potential sources for such changes are described below:

1. *Changes in Customer Requirements.* Customer needs and requirements seldom remain unchanged during the years that it takes to develop a complex new system. Proposed changes to software requirements seem deceptively easy to incorporate but frequently prove disproportionately costly and time-consuming.

2. *Changes in Technology.* The rapid advances in key technologies, especially in solid-state electronics, accumulated over the system development time, offer the temptation to take advantage of new devices or techniques to gain significant performance or cost savings. The compulsion to do so is heightened by increases in the performance of competitive products that utilize such advances. Such changes, however, usually involve significant risks, especially if made late in the engineering design phase.

3. *Changes in Program Plans.* Changes that impact system requirements and are unavoidable may come from programmatic causes. The most common is funding instability growing out of the universal competition for resources. Lack of adequate funds to support the production phase may lead to a slip in the development schedule. Such events are often beyond the control of program management and have to be accommodated by changes in schedules and fund allocations.

Key Issues

There are several circumstances that require special attention during test planning and preparation for system integration and evaluation. These include the following:

1. *Oversight.* Management oversight is especially intense during the final stages of a major development. System tests, especially field tests, are regarded as indicators of program success. Test failures receive wide attention and invite critical investigation. Test plans must provide for acquisition of data that are necessary to be able to explain promptly and fully any mishaps and remedial measures to program management, the customer, and other concerned authorities.

2. *Resource Planning.* Test operations, especially in the late stages of the program, are costly in manpower and funds. Too frequently, overruns and slippages in the development phases cut into test schedules and budgets. Serious problems of this type can be avoided only through careful planning to assure that the necessary resources are made available when required.

3. *Test Equipment and Facilities.* Facilities for supporting test operations must be designed and built concurrently with system development to be ready when needed. Advance planning for such facilities is essential. Also, the sharing of facilities between developmental and operational testing, wherever practicable, is important in order to stay within program funding limits.

Test Equipment Design

As noted in Chapter 11, the testing of system elements, as well as the system as a whole, requires test equipment and facilities that can stimulate the element under test with external inputs and can measure the system responses. This equipment must meet exacting standards:

1. *Accuracy.* The inputs and measurements should be several times more precise than the tolerances on the system element inputs and responses. There must be calibration standards available for ensuring that the test equipment is in proper adjustment.

2. *Reliability.* The test equipment must be highly reliable to minimize test discrepancies due to test equipment errors. It should be either equipped with self-test monitors or subjected to frequent checks.

3. *Flexibility.* To minimize costs where possible, test equipment should be designed to serve several purposes, although not at the expense of accuracy or reliability. It is frequently possible to use some of the equipment designed for component tests also for.

Before designing the test equipment, it is important to define fully the test procedures so as to avoid later redesign to achieve compatibility between test equipment and the component or subsystem under test. This again emphasizes the importance of early and comprehensive test planning.

The paragraphs below discuss some of the aspects of test preparation peculiar to the integration, system test, and operational evaluation parts of the test and evaluation process.

Integration Test Planning

Preparing for the system integration process is dependent on the manner in which the system components and subsystems are developed. Where one or more components of a subsystem involve new technical approaches, the entire subsystem is often developed by the same organization and integrated prior to delivery to the system contractor. For example, aircraft engines are usually developed and integrated as units before delivery to the airplane developer. In contrast, components using mature technologies are often acquired to a specification and delivered as individual building blocks. The integration process at the system contractor's facility must deal with whatever assortment of components, subsystems, or intermediate assemblies is delivered from the respective contractors.

As stated previously, it is important to support the integration process at both the subsystem and system levels by capable integration facilities. These must provide the necessary test inputs, environmental constraints, power and other services, output measurement sensors, as well as test recording and control stations. Many of these must be custom designed for each specific use. The facilities must be designed, built, and calibrated before integration is to begin. A typical physical test configuration for is described in Section 13.3, System Integration.

Developmental System Test Planning

Preparing for system-level tests to determine that the system performance requirements are met and that the system is ready for operational evaluation is more than a normal extension of the integration test process. Integration testing is necessarily focused on ensuring that the system's components and subsystems fit together in form and function. System performance tests go well beyond this goal and measure how the system as a whole responds to its specified inputs and whether its performance meets the requirements established at the outset of its development.

The success or failure of a test program is critically dependent on the extent to which the total effort is thoughtfully planned and precisely detailed, the test equipment is well engineered and tested, and the task is thoroughly understood by the test and data analysis teams. Problems in system testing are at least as likely to be caused by faults in the test equipment, poorly defined procedures, or human error as by improper system operation. Thus, it is necessary that the test facilities be engineered and tested under the same rigorous discipline as that used in system development. Many programs suffer from insufficient time and effort being assigned to the testing process, and pay for such false economy by delays and excessive costs during system testing. To minimize the likelihood of such consequences, the test program must be planned early and in sufficient detail to identify and estimate the cost of the required facilities, equipment, and manpower.

Operational Evaluation Planning

Because operational evaluation is usually conducted by the customer or a test agent, its planning is necessarily done separately from that for integration and development testing. However, in many large-scale system developments, the costs of system-level testing compel the common use of as much development test equipment and facilities as may be practicable.

In some cases, a joint developer–customer test and evaluation program is carried out, in which the early phases are directed by the developer and the later phases by the customer or the customer's agent. Such collaborative programs have the advantage of providing a maximum exchange of information between the developer and customer, which is to their mutual benefit. This also helps to avoid misunderstandings, as well as to quickly resolve unanticipated problems encountered during the process.

At the other extreme are operational test and evaluation programs that are carried out in a very formal manner by a special system evaluation agent and with maximum independence from the developer. However, even in such cases, it is important for both the developer and the system evaluation agent to establish channels of communication to minimize misinformation and unnecessary delays.

13.3 SYSTEM INTEGRATION

In the engineering of new complex systems with many interacting components, testing at the system level cannot begin until the system has been fully assembled and demonstrated to operate as a unified whole. The likelihood that some of the interfaces among the elements may not fit or function properly, or that one or more interactions among them may fall outside prescribed tolerances, is usually high. It is only the very simplest systems that are assembled without testing at several intermediate levels of aggregation. Thus, experience has shown that no matter how thoroughly the individual components have been tested, there almost always remain unforeseen incompatibilities that do not reveal themselves until the system elements are brought together. Such discrepancies usually require changes in some components before the integrated system works properly. These changes, in turn, frequently require corresponding alterations in test equipment or procedures and must be reflected in all relevant documentation. This section describes the general process and problems involved in integrating a typical complex system.

The successful and expeditious integration of a complex system depends on how well it has been partitioned into subsystems that have simple interactions with one another and are themselves subdivided into well-defined components. The integration process can be thought of as the reverse of partitioning. It is normally accomplished in two stages: (1) the individual subsystems are integrated from their components, and (2) the subsystems are assembled and integrated into the total system. At intervals during both stages, the assembled elements are tested to determine whether or not they fit and interact together in accordance with expectations. In the event that they do not, special test procedures are instituted to reveal the particular design features that need to be corrected. Throughout the entire process, system integration proceeds in an orderly, stepwise manner with system elements added one or two at a time and then tested to demonstrate proper operation before proceeding to the next step. This procedure maintains control of the process and simplifies diagnosis of discrepancies. The price for this stepwise integration of the system is that at every step, the test equipment must simulate the relevant functions of the missing parts of the system. Nevertheless, experience in the development of large systems has repeatedly demonstrated that the provision of this capability is, in the long run, quite cost-effective. In the integration of large software programs, this is frequently done by connecting the "program executive" to "stubbed-off" or nonfunctioning modules, which are successively replaced one at a time by functioning modules.

Determining the most effective order of assembly and selecting the optimum test intervals are critical to minimizing the effort and time needed to accomplish the integration process. Since both system-level knowledge and test expertise are essential to the definition of this process, the task is normally assigned to a special task team composed of systems engineers and test specialists.

Physical Test Configuration

Integration testing requires versatile and readily reconfigurable integration facilities. To understand their operation, it is useful to start with a generic model of a system element test configuration. Such a model is illustrated in Figure 13.4 and is described below.

The *system element* (component or subsystem) under test is represented by the block at the top center of the figure. The *input generator* converts test commands into exact replicas, functionally and physically, of the inputs that the system element is expected to receive. These may be a sequence of typical inputs covering the range expected under operational conditions. The input signals in the same or simulated form are also fed to the element model. The *output analyzer* converts any outputs that are not already in terms of quantitative physical measures into such form. Whether or not the data obtained in the tests are compared in real time with predicted responses from the element model, they should also be recorded, along with the test inputs and other conditions, for subsequent analysis. In the event of discrepancies, this permits a more detailed diagnosis of the source of the problem and a subsequent comparison with results of

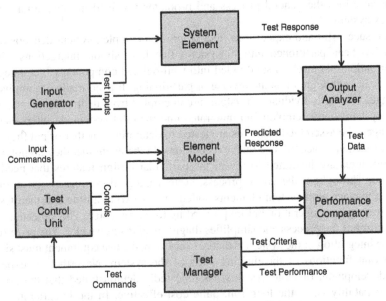

Figure 13.4. System element test configuration.

suitably modified elements. The physical building blocks in the top row of Figure 13.4 may be seen to implement the corresponding functional elements of Figure 13.3.

The *element model*, pictured in the center of the Figure 13.4, has the function of reproducing very precisely the response that the component or subsystem under test is expected to produce to each input, according to its performance specifications. The element model may take several forms. At one extreme, it may be a specially constructed and validated replica of the system element itself. At the other, it may be a mathematical model of the element, perhaps as simple as a table lookup if the predicted performance is an explicit function of the input. How it is configured determines the form of input required to drive it.

The *test manager* introduces a function not represented in the basic test architecture of Figure 13.3. Because the testing of most elements of complex systems is a complicated process, it requires active supervision by a test engineer, usually supported by a control console. This allows critical test results to be interpreted in real time in terms of required performance so that the course of testing can be altered if significant deviations are observed.

The *performance comparator* matches the measured system element outputs with the expected outputs from the element model in accordance with test criteria provided by the test manager. The comparison and assessment is performed in real time whenever practicable to enable a rapid diagnosis of the source of deviations from expected results, as noted previously. The evaluation criteria are designed to reflect the dependence of the operational performance on individual performance parameters.

Most actual test configurations are considerably more complex than the simplified example in Figure 13.4. For example, tests may involve simultaneous inputs from several sources involving various types of system elements (e.g., signal, material, and mechanical), each requiring a different type of signal generator. Similarly, there are usually several outputs, necessitating different measuring devices to convert them into forms that can be compared with predicted outputs. The tests may also involve a series of programmed inputs representing typical operating sequences, all of which must be correctly processed.

It is clear from the above discussion that the functionality embodied in the test configuration of a system element is necessarily comparable to that of the element itself. Hence, designing the test equipment is itself a task of comparable difficulty to that of developing the system element. One factor that makes the task somewhat simpler is that the environment in which the test equipment operates is usually benign, whereas the system operating environment is often severe. On the other hand, the precision of the test equipment must be greater than that of the system element to ensure that it does not contribute significantly to measured deviations from the specified element performance.

Subsystem Integration

As noted previously, the integration of a subsystem (or system) from its component parts is normally a stepwise assembly and test process in which parts are systematically aggregated, and the assembly is periodically tested to reveal and correct any faulty interfaces or component functions as early in the process as practicable. The time and effort required to conduct this process is critically dependent on the skillful organization of the test events and the efficient use of facilities. Some of the most important considerations are discussed below.

The order in which system components are integrated should be selected to avoid the need to construct special input generators for simulating components within the subsystem, that is, other than those simulating inputs from sources external to the subsystem being integrated. Thus, at any point in the assembly, the component that is to be added should have inputs that are derivable from either generators of external inputs or the outputs of components previously assembled.

The above approach means that subsystem integration should begin with components that have only external inputs, either from the system environment or from other subsystems. Examples of such components include

1. subsystem support structures,
2. signal or data input components (e.g., external control transducers), and
3. subsystem power supplies.

The application of the above approach to the integration of a simple subsystem is illustrated in Figure 13.5. The figure is an extension of Figure 13.4, in which the subsystem under test is composed of three components. The configuration of components in the figure is purposely chosen so that each component has a different combination of inputs and outputs. Thus, component A has a single input from an external subsystem and two outputs—one an internal output to B and the other to another subsystem. Component B has no external interfaces—getting its input from A and producing an output to C. Component C has two inputs—one external and the other internal, and a single output to another subsystem.

The special features of the test configuration are seen to be

1. a compound input generator to provide the two external inputs to the subsystem —one to A and the other to C;
2. internal test outputs from the interfaces between A and B, and between B and C; these are needed to identify the source of any observed deviation in the overall performance and are in addition to the external subsystem outputs from A and C; and
3. a compound element model containing the functions performed by the constituent components and providing the predicted outputs of the test interfaces.

Following the integration sequence approach described above, the configuration in Figure 13.5 would be assembled as follows:

1. Start with A, which has no internal inputs. Test A's outputs.
2. Add B and test its output. If faulty, check if input from A is correct.
3. Add C and test its output.

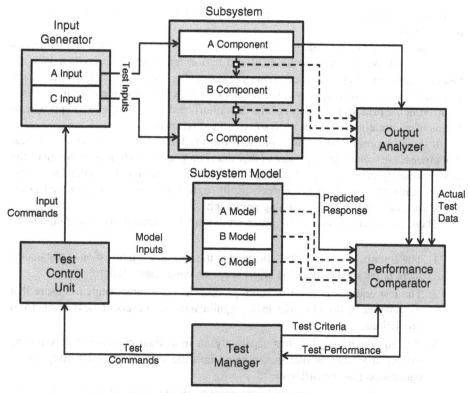

Figure 13.5. Subsystem test configuration.

The above integration sequence does not require the construction of input genera-tors to provide internal functions and should rapidly converge on the source of a faulty component or interface.

The approach described above works in the great majority of cases but must, of course, be carefully reviewed in the light of any special circumstances. For example, there may be safety issues that make it necessary to leave out or add steps to circumvent unsafe testing conditions. The temporary unavailability of key components may require a substitution or simulation of elements. Particularly critical elements may have to be tested earlier than in the ideal sequence. Systems engineering judgment must be applied in examining such issues before defining the integration sequence.

Test Conduct and Analysis. The determination of whether or not a given step in the integration process is successful requires matching the outputs of the partially assembled components against their expected values as predicted by the model. The effort required to make this comparison depends on the degree of automation of the test configuration and of the analytical tools embodied in the performance comparator block in Figure 13.5. The trade-off between the sophistication of the test and analysis tools and the analysis effort itself is one of the critical decisions to be made in planning the integration process.

In scheduling and costing the integration effort, it must be expected that numerous deviations will be observed in the measured performance from that predicted by the model, despite the fact that all components presumably have previously passed qualification tests. Each discrepancy must be dealt with by first documenting it in detail, identifying the principal source(s) of the deviations, and devising the most appropriate means of eliminating or otherwise resolving the discrepancy.

It should be emphasized that in practice, most failures observed during the integration process are usually due to causes other than component malfunctions. Some of the most frequently occurring problem areas are faulty test equipment or procedures, misinterpretation of specifications, unrealistically tight tolerances, and personnel error. These are discussed in the succeeding paragraphs.

There are several reasons why faults are frequently found in the test equipment:

1. The amount of design effort allocated to the design and fabrication of test equipment is far smaller than the effort spent on component design.
2. The test equipment must be more precise than the components to ensure that its tolerances do not contribute significantly to observed deviations from predictions.
3. The equipment used to test separately an individual component may not be exactly the same as that incorporated into the integration test facility, or its calibration may be different.
4. The predictions of expected performance of the element under test by the element model may be imperfect due to the impossibility of modeling exactly the behavior of the test element.

Not infrequently, the specifications of interfaces and interactions among components permit different interpretations by the designers of interfacing components. This can result in significant mismatches when the components are assembled. There is no practical and foolproof method of entirely eliminating this source of potential problems. Their number can, however, be minimized through critical attention to and review of each interface specification prior to its release for design of the associated hardware or software. In most cases, establishing an interface coordination team, including all involved contractors, has proven to be advantageous.

To ensure that interfacing mechanical, electrical, or other elements fit together and interact properly, the specifications for each separate element must include the permitted tolerances (deviations from prescribed values) in the interacting quantities. For example, if the interfacing components are held together by bolts, the location of the holes in each component must be specified within a plus/minus tolerance of their nominal dimensions. These tolerances must allow for the degree of precision of production machinery, as well as normal variations in the size of standard bolts. If the specified tolerances are too tight, there will be excessive rejects in manufacture; if too loose, there will be occasional misalignments, causing fit failures.

Personnel errors are a common source of test failure and one that can never be completely avoided. Such failures may occur because of inadequate training, unclear or insufficiently detailed test procedures, overly complex or demanding test methods, fatigue, or simple carelessness. Errors of this type can occur at any point in the planning, execution, and support of the testing process.

Changes. If the diagnosis of a faulty test traces the problem to a component design feature, it is necessary to undertake a highly expedited effort to determine the most practical and effective means of resolving the problem. At this stage of development, the design should be under strict configuration management. Since any significant change will be costly and potentially disruptive, all means of avoiding or minimizing the change must be explored and several alternatives examined. The final decision will have to be made at the program management level if significant program cost and schedule changes are involved.

If there is no "quick fix" available, consideration may be given to seeking a waiver to deviate from a certain specification for an initial quantity of production units so as to afford adequate time to design and validate the change prior to its release for production. Not infrequently, careful analysis reveals that the effect of the deviation on operational performance is not sufficient to warrant the cost of making the change, and a permanent waiver is granted. Systems engineering analysis is the key to determining the best course of action in such circumstances, and to advocating its approval by management and the customer.

Total System Integration

The integration of the total system from its subsystems is based on the same general principles as those governing the integration of individual subsystems, described in the preceding paragraphs. The main differences are those of relative scale, complexity, and hence criticality. Faults encountered at this stage are more difficult to trace, costly to remedy, and have a greater potential impact on overall program cost and schedule. Hence, a more detailed planning and direction of the test program are in order. Under these conditions, the application of systems engineering oversight and diagnostic expertise are even more essential than in the earlier stages of system development.

System Integration Test Facility. It was noted that specially designed facilities are normally required to support the integration and test of systems and their major subsystems. This is even more true for the assembly and integration of total systems. Often, such a facility is gradually built up during system development to serve as a "test bed" for risk reduction testing and may be assembled in part from subsystem test facilities.

As in the case of subsystem integration test facilities, the system integration facility must provide for extracting data from test points at internal boundaries between subsystems, as well as from the normal system outputs. It should also be designed to be flexible enough to accommodate system updates. Thus, the design of the integration facilities needed to achieve the necessary test conditions, measurements, and data analysis capabilities is itself a major systems engineering task.

13.4 DEVELOPMENTAL SYSTEM TESTING

The system integration process was seen to be focused on ensuring that component and subsystem interfaces and interactions fit together and function as they were designed. Once this is accomplished, the system may, for the first time, be tested as a unified whole to determine whether or not it meets its technical requirements, for example, performance, compatibility, reliability, maintainability, availability (RMA), safety, and so on. The above process is referred to as verification that the system satisfies its specifications. Since the responsibility for demonstrating successful system verification is a necessary part of the development process, it is conducted by the system developer and will be referred to as developmental system testing.

System Testing Objectives

While the primary emphasis of developmental system-level testing is on the satisfaction of system specifications, evidence must also be obtained concerning the system's capability to satisfy the operational needs of the user. If any significant issues exist in this regard, they should be resolved before the system is declared ready for operational evaluation. For this reason, the testing process requires the use of a realistic test environment, extensive and accurate instrumentation, and a detailed analysis process that compares the test outputs with predicted values and identifies the nature and source of any discrepancies to aid in their prompt resolution. In a real sense, the tests should include a "rehearsal" for operational evaluation.

In the case of complex systems, there are frequently several governing entities in the acquisition and validation process that must be satisfied that the system is ready for full-scale production and operational use. These typically include the acquisition or distribution agency (customer), which has contracted for the development and production of the system, and in the case of products to be used by the public, one or more regulatory agencies (certifiers) concerned with conformance with safety or environmental regulations. In addition, the customer may have an independent testing agent who must pass favorably on the system's operational worth. In the case of a commercial airliner, the customer is an airline company and the certification agencies are the Federal Aviation Administration (FAA) and the Civil Aeronautics Board (CAB).

An essential precondition to system-level testing is that component and has been successfully completed and documented. When system test failures occur in components or subsystems because of insufficient testing at lower levels, the system evaluation program risks serious delays. A required "stand-down" at this point in the program is time-consuming, expensive, and may subject the program to a critical management review. It is axiomatic, therefore, that the system test program should not be started unless the developer and customer have high confidence in the overall system design and in the quality of the test equipment and test plans.

Despite careful preparation, the test process should be conducted with the expectation that something may go wrong. Consequently, means must be provided to quickly identify the source of such unexpected problems and to determine what, within the bounds of acceptable costs in money and time, can be done to correct them. Systems engineering knowledge, judgment, and experience are crucial factors in the handling of such "late-stage" problems.

Developmental Test Planning

The provisions of the defense TEMP regarding developmental test and evaluation state that, in part, plans should

- define the specific technical parameters to be measured;
- summarize test events, test scenarios, and the test design concept;
- list all models and simulations to be used; and
- describe how the system environment will be represented.

System Test Configuration

System testing requires that the test configuration be designed to subject the system under test to all of the operational inputs and environmental conditions that it is practical to reproduce or simulate, and to measure all of the significant responses and operating functions that the system is required to perform. The sources for determining which measurements are significant should be found largely in system-level requirements and specifications. The principal elements that must be present in a system test configuration are summarized below and are discussed in the subsequent paragraphs of this section.

- *System Inputs and Environment*
  1. The test configuration must represent all conditions that affect the system's operation, including not only the primary system inputs but also the interactions of the system with its environment.
  2. As many of the above conditions as practicable should be exact replicas of those that the system will encounter in its intended use. The others should be simulated to realistically represent their functional interactions with the system.
  3. Where the real operational inputs cannot be reproduced or simulated as part of the total test configuration (e.g., the impact of rain on an aircraft flying at supersonic speed), special tests should be carried out in which these functions can be reproduced and their interaction with the system measured.
- *System Outputs and Test Points*
  1. All system outputs required for assessing performance should be converted into measurable quantities and recorded during the test period.

2. Measurements and recordings should also be made of the test inputs and environmental conditions to enable correlation of the variations in inputs with changes in outputs.

3. A sufficient number of internal test points should be monitored to enable tracing the cause of any deviations from expected test results to their source in a specific subsystem or component.

- *Test Conditions*

  1. To help ensure that contractor system testing leads to successful operational evaluation by the customer, it is important to visualize and duplicate, insofar as possible, the conditions to which the system is most likely to be subjected during operational evaluation.

  2. Some system tests may intentionally overstress selected parts of the system to ensure system robustness under extreme conditions. For example, it is common to specify that a system degrade "gracefully" when overstressed rather than suddenly crash. This type of test also includes validating the procedures that enable the system to recover to full capability.

  3. Wherever practicable, customer operating and evaluation agent personnel should be involved in contractor system testing. This provides an important mutual exchange of system and operational knowledge that can result in better planned and more realistic system tests and more informed test analyses.

Development of Test Scenarios

In order to evaluate a system over the range of conditions that it is expected to encounter in practice, as defined in top-level system requirements, a structured series of tests must be planned to explore adequately all relevant cases. The tests should seek to combine a number of related objectives in each test event so that the total test series is not excessively prolonged and costly. Further, the order in which tests are conducted should be planned so as to build upon the results of preceding tests, as well as to require the least amount of retesting in the event of an unexpected result.

Composite system tests of the type described above are referred to as test events conducted in accordance with test scenarios, which define a series of successive test conditions to be imposed on the system. The overall test objectives are allocated among a set of such scenarios, and these are arranged in a test event sequence. The planning of test scenarios is a task for systems engineers with the support of test engineers because it requires a deep understanding of the system functions and internal as well as external interactions.

The combination of several specific test objectives within a given scenario usually requires that the operational or environmental inputs to the system must be varied to exercise different system modes or stress system functions. Such variations must be properly sequenced to produce maximum useful data. Decisions have to be made as to whether or not the activation of a given test event will depend on a successful result of the preceding test. Similarly, the scenario test plan must consider what test results outside expected limits would be cause for interrupting the test sequence, and if so, when the sequence would be resumed.

System Performance Model

In describing the testing and integration of system components, a necessary element was stated to be a model of the component that predicted how it is expected to respond to a given set of input conditions. The model is usually either a combination of physical, mathematical and hybrid elements, or wholly a computer simulation.

In predicting the expected behavior of a complex system in its totality, it is usually impractical to construct a performance model capable of reproducing in detail the behavior of the whole system. Thus, in system-level tests, the observed system performance is usually analyzed at two levels. The first is in terms of the end-to-end performance characteristics that are set forth in the system requirements documents. The second is at the subsystem or component level where certain critical behavior is called for. The latter is especially important when an end-to-end test does not yield the expected result and it is required to locate the source of the discrepancy.

Decisions as to the degree of modeling that is appropriate at the system test level are very much a systems engineering function, where the risks of not modeling certain features have to be weighed against the effort required. Since it is impractical to test everything, the prioritization of test features, and hence of model predictions, must be based on a system-level analysis of the relative risks of omitting particular characteristics.

The design, engineering, and validation of system performance models is itself a complex task and must be carried out by the application of the same systems engineering methods used in the engineering of the system itself. At the same time, pains must be taken to limit the cost of the modeling and simulation effort to an affordable fraction of the overall system development. The balance between realism and cost of modeling is one of the more difficult tasks of systems engineering.

Engineering Development Model (EDM)

As mentioned earlier, the system test process often requires that essentially all of the system be subjected to testing before the final system has been produced. For this reason, it is sometimes necessary to construct a prototype, referred to as an "EDM," for test purposes, especially in the case of very large complex systems. An EDM must be as close as possible to the final product in form, fit, and function. For this reason, EDMs can be expensive to produce and maintain, and must be justified on the basis of their overall benefit to the development program.

System Test Conduct

The conduct of contractor system tests is usually led by the test organization, which is also involved in the integration-testing phase, and is intimately familiar with system design and operation. There are, however, numerous other important participants.

Test Participants. As shown in Figure 13.3, systems engineers should have been active in the planning of the test program from its inception and should have approved the overall test plans and test configurations. An equally critical systems engineering function is that of resolving discrepancies between actual and predicted test results. As mentioned previously, those may arise from a variety of sources and must be quickly traced to the specific system or test element responsible; a system-level approach must be taken to devise the most effective and least disruptive remedy.

Design engineers are also key participants, especially in the engineering of test equipment and analysis of any design problems encountered during testing. In the latter instance, they are essential to effect quickly and expertly such design changes as may be required to remedy the deficiency.

Engineering specialists, such as reliability, maintainability, and safety engineers, are essential participants in their respective areas. Of particular importance is the participation of specialists in the testing of human–machine interfaces, which are likely to be of critical concern in the operational evaluation phase. Data analysts must participate in test planning to ensure that appropriate data are acquired to support performance and fault diagnostic analysis.

As noted earlier, while system testing is under the direction of the developer, the customer and/or the customer's evaluation agent will often participate as observers of the process and will use this opportunity to prepare for the coming operational evaluation tests. It is always advantageous for customer test personnel to receive some operation training during this period.

Safety. Whenever system testing occurs, there must be a section of the test plan that specifically addresses safety provisions. This is best handled by assigning one or more safety engineers to the test team, making them responsible for all aspects of this subject. Many large systems have hazardous exposed moving parts, pyrotechnic and/or explosive devices, high voltages, dangerous radiation, toxic materials, or other characteristics that require safeguards during testing. This is particularly true of military systems.

In addition to the system itself, the external test environment may also pose safety problems. The safety engineers must brief all participating test personnel on the potential dangers that may be present, provide special training, and supply any necessary safety equipment. Systems engineers must be fully informed on all safety issues and must be prepared to assist the safety engineers as required.

Test Analysis and Evaluation

Test analysis begins with a detailed comparison of system performance, as a function of test stimuli and environments, with that predicted by the system performance model. Any deviations must trigger a sequence of actions designed to resolve the discrepancies.

Diagnosing the Sources of Discrepancies. In all discrepancies in which the cause is not obvious, systems engineering judgment is required to determine the most promising course of action for identifying the cause. Time is always of the essence, but never as much so as in the middle of system-level evaluation. The cause of a test discrepancy can be due to a fault in (1) test equipment, (2) test procedures, (3) test execution, (4) test analysis, (5) the system under test, or (6) occasionally, to an excessively stringent performance requirement. As noted previously, faults are frequently traceable to one of the first four causes, so that these should be eliminated before contemplating emergency system fixes. However, since there is seldom time to investigate possible causes one at a time, it is usually prudent to pursue several of them in parallel. It is here that the acquisition of data at many test points within the system may be essential to rapidly narrow the search and to indicate an effective priority of investigative efforts. This is also a reason why test procedures must be thoroughly understood and rehearsed well in advance of actual testing.

Dealing with System Performance Discrepancies

If a problem is traced to the system under test, then it becomes a matter of deciding if it is minor and easily corrected, or serious, and/or not understood, in which case delays may be required, or not serious and agreeable to the contractor and customer that corrective action may be postponed.

The above decisions involve one of the most critical activities of systems engineers. They require a comprehensive knowledge of system design, performance requirements, and operational needs, and of the "art of the possible." Few major discrepancies at this stage of the program can be quickly corrected; any design change initiates a cascade of changes in design documentation, test procedures, interface specifications, production adjustments, and so on. In many instances, there may be alternative means of eliminating the discrepancy, such as by software rather than hardware changes. Many changes propagate well beyond their primary location. Dealing with such situations usually requires the mobilization of a "tiger team" charged with quickly reaching an acceptable resolution of the problem.

Any change made to the system raises the question whether or not the change requires the repetition of tests previously passed—another systems engineering issue with a serious impact on program schedule and cost.

In cases where the system performance discrepancy is not capable of being eliminated in time to meet established production goals, the customer has the option of choosing to accept release of the system design for limited production, assuming that it is otherwise operationally suitable. Such a decision is taken only after exhaustive analysis has been made of all viable alternatives and usually provides for later backfitting of the initial production systems to the fully compliant design.

13.5   OPERATIONAL TEST AND EVALUATION

In previous periods of subsystem and system testing, the basis of comparison was a model that predicted the performance expected from an ideal implementation of the functional design. In system operational evaluation, the test results are compared to the operational requirements themselves rather than to their translation into performance requirements. Thus, the process is focused on *validation* of the system design in terms of its operational requirements rather than on *verification* that it performs according to specifications.

The operational evaluation of a new system is conducted by the customer or by an independent test agent acting on the customer's behalf. It consists of a series of tests in which the system is caused to perform its intended functions in an environment identical or closely similar to that in which it will operate in its intended use. The satisfactory performance of the system in meeting its operational requirements is a necessary prerequisite to initiation of production and deployment. In the case of systems built for public use, such as commercial aircraft, there will also be special tests or inspections by government agents responsible for certifying the product's safety, environmental suitability, and other characteristics subject to government regulation.

Operational Test Objectives

Operational test and evaluation is focused on operational requirements, mission effectiveness, and user suitability. The subject of operational evaluation is usually a preproduction prototype of the system. The expectation is that all obvious faults will have been eliminated during development testing, and that any further significant faults may cause suspension of evaluation tests, pending their elimination by the developer. The limitations of time and resources normally available for operational evaluation require careful prioritization of test objectives. A generally applicable list of high-priority areas for testing includes the following:

1. *New Features.* Features designed to eliminate deficiencies in a predecessor system are likely to be the areas of greatest change and hence greatest uncertainty. Testing their performance should be a top priority.

2. *Environmental Susceptibility.* Susceptibilities to severe operational environments are areas least likely to have been fully tested. Operational evaluation is sometimes the first opportunity to subject the system to conditions closely resembling those that it is designed to encounter.

3. *Interoperability.* Compatibility with external equipment, subject to nonstandard communication protocols and other data link characteristics, makes it essential to test the system when it is connected to the same or functionally identical external elements as it will be connected to in its operational condition.

4. *User Interfaces.* How well the system users/operators are able to control its operations, that is, the effectiveness of the system human–machine interfaces, must be determined. This includes assessing the amount and type of training that will be required, the adequacy of training aids, the clarity of displays, and the effectiveness of decision support aids.

Example: Operational Evaluation of an Airliner. The function of a commercial airliner is to transport a number of passengers and their luggage from a given location to remote destinations, rapidly, comfortably, and safely. Its operational con-

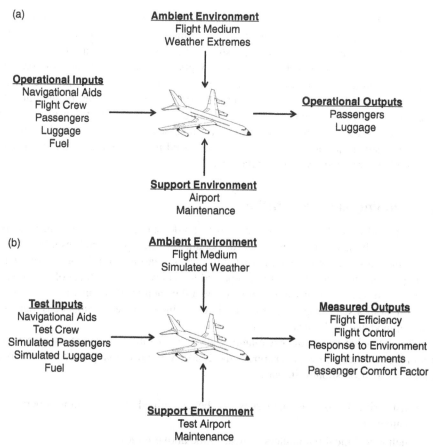

Figure 13.6. (a) Operation of a passenger airliner. (b) Operational testing of an airliner.

figuration is illustrated by a so-called context diagram in Figure 13.6a. The diagram lists the principal operational inputs and outputs, together with the ambient and support environments, that contribute to and affect the operation of the system. The principal inputs besides passengers and luggage are fuel, flight crew, and navigation aids. Numerous secondary but important functions, such as those relating to the comfort of the passengers (food, entertainment, etc.) that must also be considered are omitted from the figure for the sake of clarity. The operational flight environment includes the flight medium, with its variation in pressure, temperature, wind velocity, and weather extremes, which the system must be designed to withstand with minimum effect on its primary functions.

Figure 13.6b is the corresponding diagram of the airliner in its operational test mode. A comparison with Figure 13.6a shows that the test inputs duplicate the operational inputs, except that most of the passengers and luggage are simulated. The measured outputs include data from the plane's instruments and special test sensors to enable the evaluation of performance factors relating to efficiency, passengers comfort, and safety, as well as to permit the reconstruction of the causes of any in-flight abnormalities. The operational test environment duplicates the operational environment, except for conditions of adverse weather, such as wind shear. To compensate for the difficulty of reproducing adverse weather, an airplane under test may be intentionally subjected to stresses beyond its normal operating conditions so as to ensure that sufficient safety margin has been built in to withstand severe environments. In addition, controllable severe flight conditions can be produced in wind tunnel tests, in specially equipped hangars, or in system simulations.

Test Planning and Preparation

Test plans and procedures, which are used to guide operational evaluation, must not only provide the necessary directions for conducting the operational tests but should also specify any follow-up actions that, for various reasons, could not be completed during previous testing, or need to be repeated to achieve a higher level of confidence. It should also be noted that while there are general principles that apply to most system test configurations, each specific system is likely to have special testing needs that must be accommodated in the test planning.

The extensive scope of test planning for the operational evaluation of a major system is illustrated by the provisions of the TEMP. It requires that plans for operational test and evaluation should, in part,

- list critical operational issues to be examined to determine operational suitability,
- define technical parameters critical to the above issues,
- define operational scenarios and test events,
- define the operational environment to be used and the impact of test limitations on conclusions regarding operational effectiveness,
- identify test articles and necessary logistic support, and
- state test personnel training requirements.

Test and Evaluation Scope. Evaluation planning must include a definition of the appropriate scope of the effort, how realistic the test conditions must be, how many system characteristics must be tested, what parameters must be measured to evaluate system performance, and how accurately. Each of these definitions involves trade-offs between the degree of confidence in the validity of the result, and the cost of the test and evaluation effort. Confidence in the results, in turn, depends on the realism with which the test conditions represent the expected operational environment. The general relationship between test and evaluation realism and evaluation program cost is pictured in Figure 13.7. It obeys the classic law of diminishing returns, in which cost escalates as the test sophistication approaches full environmental reality and complete parameter testing.

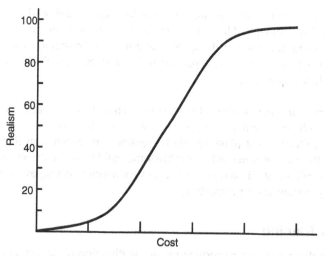

<u>Figure 13.7.</u> Test realism versus cost.

The decision of "how much testing is enough" is inherently a systems engineering issue. It requires a basic knowledge of the operational objectives, how these relate to system performance, what system characteristics are most critical and least well proven, how difficult it would be to measure critical performance factors, and other equally vital elements of the trade-offs that must be made. It also requires the inputs of test engineers, design engineers, engineering specialists, and experts in the operational use of the system.

Test Scenarios. System operational evaluation should proceed in accordance with a set of carefully planned test scenarios, each of which consists of a series of events or specific test conditions. The objective is to validate all of the system requirements in the most efficient manner, that is, involving the least expenditure of time and resources.

The planning of the test events and their sequencing must not only make the most effective use of test facilities and personnel but also must be ordered so that each test builds on the preceding ones. The proper functioning of the links between the system and external systems, such as communications, logistics, and other support functions, is essential for the successful testing of the system itself and must, therefore, be among the first to be tested. At the same time, all test equipment, including data acquisition, should be recalibrated and recertified.

Test Procedures. The preparation of clear and specific test procedures for each test event is particularly important in operational testing because the results are critical for program success. Also, the user test personnel are generally less familiar with the detailed operation of the system under test than development test personnel. The test procedures should be formally documented and thoroughly reviewed for completeness and accuracy. They should address the preparation of the test site, the configuration of the test equipment, the setup of the system, and the step-by-step conduct of each test. The required actions of each test participant should be described, including those involved in data acquisition.

Analysis Plan. An analysis plan must be prepared for each test event specifying how the data obtained will be processed to evaluate the proper performance of the system. The collective test plans should be reviewed to ensure that they combine to obtain all of the measures needed to establish the validity of the system in meeting its operational requirements. This review requires systems engineering oversight to provide the necessary system-level perspective.

Personnel Training

The fact that these tests are performed under the direction of personnel who have not been part of the system development team makes the evaluation task especially challenging. An essential part of the preparation for operational evaluation is, therefore, the transfer of technical system knowledge from the development organization and the acquisition agency to those responsible for planning and executing the evaluation process. This must be started at least during the developmental system test period, preferably by securing the active participation of the evaluation agent's test planning and analysis personnel. The developer's systems engineering staff should be prepared to take the lead in effecting the necessary transfer of this knowledge.

While it is to everyone's benefit to effect the above knowledge transfer, the process is too often inadequate. Significant funding is seldom earmarked for this purpose, and the appropriate personnel are often occupied with other priority tasks. Another common obstacle is an excessive spirit of independence that motivates some evaluation agents to avoid becoming involved in the preevaluation testing phase. Therefore, it usually remains for an experienced program manager or chief systems engineer in either organization to take the initiative to make it happen.

Test Equipment and Facilities

Since the focus of operational evaluation is on end-to-end system performance, only limited data are strictly required regarding the operation of individual subsystems. On the other hand, it is essential that any system performance discrepancy be quickly identified and resolved. To this end, the system developer is often permitted to make auxiliary measurements of the performance of selected subsystems or components. The same equipment as was employed in developmental testing is usually suitable for this purpose. It is to the advantage of both the evaluation agent and the developer to monitor and record the outputs from a sufficient number of system test points to support a detailed posttest diagnosis of system performance when required.

As stated previously, the conditions to which each system is subjected must be representative of its intended operational environment. In the above example of a commercial airliner, the operational environment happens to differ from readily reproducible flying conditions only in the availability of adverse weather conditions that the airliner must be able to handle safely. This fortunate circumstance is not typical of the evaluation of most complex systems. Operational testing of ground transport vehicles requires a specially selected terrain that stresses their performance capabilities over a broad range of conditions. Systems depending on external communications require special auxiliary test instrumentation to provide such inputs and to receive any corresponding output.

Test Conduct

If system developer personnel participate, they do so either as observers, or more commonly, in a support capacity. In the latter role, they assist in troubleshooting, logistic support, and provision of special test equipment. In no case are they allowed to influence the conduct of the tests or their interpretation. Nonetheless, they often can play a key role in helping quickly to resolve unexpected difficulties or misunderstandings of some feature of the system operation.

As a preliminary to conducting each test, the operational personnel should be thoroughly briefed on the test objectives, the operations to be performed, and their individual responsibilities. As noted previously, personnel and test equipment errors are often the most prevalent causes of test failures.

Test Support. Operational and logistic support of evaluation tests is critical to their success and timely execution. Since these tests are in series with key program decisions, such as authorization of full-scale production or operational deployment, they are closely watched by both developer and customer management. Thus, adequate supplies of consumables and spare parts, transportation and handling equipment, and technical data and manuals must be provided, together with associated personnel. Test equipment must be calibrated and fully manned. As noted earlier, support should be obtained from the system developer to provide engineering and technical personnel capable of quickly resolving any minor system discrepancies that may invalidate or delay testing.

Data Acquisition. It was noted in the previous paragraphs that data acquired during operational evaluation are usually much more limited than that which was collected during developer system tests. Nevertheless, it is essential that the end-to-end system performance be measured thoroughly and accurately. This means that the "ground truth" must be carefully monitored by instrumenting all external conditions to which the system is subjected and the measurements recorded for posttest analysis. The external conditions include all functional system inputs as well as significant environmental conditions, especially those that may interfere with or otherwise affect system operation.

Human–Machine Interfaces. In most complex systems, there are human–machine interfaces that permit an operator to observe information and to interact with the system, serving as a critical element in achieving overall system performance. A classic example is an air traffic controller. While data input from various sensors is automatic, the controller must make life-and-death decisions and take action based on information displayed on a control console and received from reporting pilots. A similar operator function is part of many types of military combat systems.

In such operator interactions, system performance will depend on two interrelated factors: (1) effectiveness of operator training and (2) how well the human interface units have been designed. During operational testing, this aspect of system performance will be an important part of the overall evaluation because improper operator action often results in test failures. When such errors do occur, they are often difficult to track down. They can result from slow reaction time of the operator (e.g., fatigue after many hours on station), awkward placement of operator controls and/or display symbology, or many other related causes.

Safety. As in the case of development system tests, special efforts must be exerted to ensure the safety of both test personnel and inhabitants neighboring the test area. In the case of military missile test ranges, instrumentation is provided to detect any indication of loss of control, in which case a command is sent to the missile by the range safety officer to actuate a self-destruct system to terminate the flight.

Test Analysis and Evaluation

The objectives of operational evaluation have been seen to determine whether or not the system as developed meets the needs of the customer, that is, to validate that its performance meets the operational requirements. The depth of evaluation data analysis varies from "go no-go" conclusions to a detailed analysis of the system and all major subsystems.

Under some circumstances, an independent evaluation agent may judge that a new system is deficient in meeting the user's operational goals to a degree not resolvable by a minor system design or procedural change. Such a situation may arise because of changes in operational needs during the development process, changes in operational doctrine, or just differences of opinion between the evaluator and the acquisition agent. Such cases are usually resolved by a compromise, in which a design change is negotiated with the developer through a contract amendment, or a temporary waiver is agreed upon for a limited number of production units.

Test Reports

Because of the attention focused on the results of the operational evaluation tests, it is essential to provide timely reports of all significant events. It is customary to issue several different types of reports during the evaluation process.

Quick-Look Reports. These provide preliminary test results immediately following a significant test event. An important purpose of such reports is to prevent misinterpretation of a notable or unexpected test result by presenting all the pertinent facts and by placing them in their proper perspective.

Status Reports. These are periodic reports (e.g., monthly) of specific significant test events. They are designed to keep the interested parties generally aware of the progress of the test program. There may be an interim report of the cumulative test findings at the conclusion of the test program while the data analysis and final report are being completed.

Final Evaluation Report. The final report contains the detailed test findings, their evaluation relative to the system's intended functions, and recommendations relative to its operational suitability. It may also include recommendations for changes to eliminate any deficiencies identified in the test program.

13.6 SUMMARY

Integrating, Testing, and Evaluating the Total System

The objectives of the integration and evaluation phase are to integrate the engineered components of a new system into an operating whole and to demonstrate that the system meets all its operational requirements. The outputs of the integration and evaluation phase are

- validated production designs and specifications, and
- qualification for production and subsequent operational use.

The activities constituting integration and evaluation are

- *Test Planning:* defining test issues, test scenarios, and test equipment;
- *System Integration:* integrating components into subsystems and the total system;
- *Developmental System Testing:* verifying that the system meets specifications; and
- *Operational Test and Evaluation:* validating that the system meets operational requirements.

Test Planning and Preparation

Integration and evaluation "materializes" the system as a whole and synthesizes a functioning total system from individual components. These activities solve any remaining interface and interaction problems.

Defense systems require a formal TEMP, which covers test and evaluation planning throughout system development.

System requirements should be reviewed prior to preparing test plans to allow for customer requirements changing during system development. Late injection of technology advances always poses risks.

Key issues during system integration and evaluation include

- intense management scrutiny during system testing,
- changes in test schedules and funding due to development overruns, and
- readiness of test equipment and facilities.

System test equipment design must meet exacting standards and accuracy must be much more precise than component tolerances. Reliability must be high to avoid aborted tests. Finally, the design must accommodate multiple use and failure diagnosis.

System Integration

A typical test configuration consists of

- the system element (component or subsystem) under test,
- a physical or computer model of the component or subsystem,
- an input generator that provides test stimuli,
- an output analyzer that measures element test responses, and
- control and performance analysis units.

Subsystem integration should be organized to minimize special component test generators, to build on results of prior tests, and to monitor internal test points for fault diagnosis.

Test failures are often not due to component deficiencies, but test equipment may be inadequate. Additionally, interface specifications may be misinterpreted or interface tolerances may be mismatched. And finally, inadequate test plans, training, or procedures may lead to personnel errors.

Integration test facilities are essential to the engineering of complex systems and represent a significant investment. However, they may be useful throughout the life of the system.

Developmental System Testing

Developmental system testing has the objectives of verifying that the system satisfies all its specifications and of obtaining evidence concerning its capability to meet operational requirements.

The system test environment should be as realistic as practicable—all external inputs should be real or simulated. Conditions expected in operational evaluation should be anticipated. Moreover, effects impractical to reproduce should be exercised by special tests. However, the entire system life cycle should be considered.

Test events must be carefully planned—related test objectives should be combined to save time and resources. Detailed test scenarios need to be prepared with sufficient flexibility to react to unexpected test results.

A predictive system performance model must be developed. This is a major task requiring systems engineering leadership and effort; however, an EDM is excellent for this purpose.

Developmental tests are carried out by a coordinated team consisting of

- systems engineers, who define test requirements and evaluation criteria;
- test engineers, who conduct test and data analysis; and
- design engineers, who design test equipment and correct design discrepancies.

System performance discrepancies during developmental testing must be accounted for in test scheduling, quickly responded to by a remedial plan of action.

Operational Test and Evaluation

System operational test and evaluation has the objectives of validating that the system design satisfies its operational requirement and of qualifying the system for production and subsequent operational use.

Typical high-priority operational test issues are

- new features designed to eliminate deficiencies in a predecessor system,
- susceptibilities to severe operational environments,
- interoperability with interacting external equipment, and
- user system control interfaces.

The essential features of an effective operational evaluation include

- familiarity of the customer's or the customer agent's test personnel with the system;
- extensive preparation and observation of developmental testing;
- test scenarios making effective use of facilities and test results;
- clear and specific test procedures and detailed analysis plans;
- thorough training of test operation and analysis personnel;
- fully instrumented test facilities replicating the operational environment;
- complete support of test consumables, spare parts, manuals, and so on;
- accurate data acquisition for diagnostic purposes;
- special attention to human–machine interfaces;
- complete provisions for the safety of test personnel and neighboring inhabitants;
- technical support by system development staff; and
- timely and accurate test reports.

PROBLEMS

13.1 Figure 13.3 pictures the individual and common responsibilities of design engineers, test engineers, and systems engineers. In addition to differences in their responsibilities, these classes of individuals typically approach their tasks with significantly different points of view and objectives. Discuss these differences, and emphasize the essential role that systems engineers play in coordinating the total effort.

13.2 Figure 13.4 diagrams the test configuration for a component or a subsystem in which it is subjected to controlled inputs and its response is compared in real time with that of a computer model of the element under test. When a real-time simulation of the element is not available, the test configuration records the test response to be analyzed at a later time. Draw a diagram similar to Figure 13.4 representing the latter test configuration, as well as that of the subsequent test analysis operation. Describe the functioning of each unit in these configurations.

13.3 Test failures are not always due to component deficiencies; sometimes, they result from an improper functioning of the test equipment. Describe what steps you would take before, during, and after a test to enable a quick diagnosis in the event of a test failure.

13.4 The systems engineering method in the integration and evaluation phase is outlined in the introduction to this chapter. Construct a functional flow diagram for the four steps in this process.

13.5 In designing system tests, probes are placed at selected internal test points, as well as at system outputs, to enable a rapid and accurate diagnosis of the cause of any discrepancy. List the considerations that must be applied to the selection of the appropriate test points (e.g., what characteristics should be examined). Illustrate these considerations using the example of testing the antilock brake system of an automobile.

13.6 Describe the differences in objectives and operations between developmental test and evaluation and operational test and evaluation. Illustrate your points with an example of a lawn tractor.

13.7 Define the terms "verification" and "validation." Describe the types of tests that are directed at each, and explain how they meet the definitions of these terms.

FURTHER READING

B. Blanchard and W. Fabrycky. *System Engineering and Analysis*, Fourth Edition. Prentice Hall, 2006, Chapters 6, 12, and 13.

W. P. Chase. *Management of Systems Engineering*. John Wiley & Sons, Inc., 1974, Chapter 6.

D. K. Hitchins. *Systems Engineering: A 21st Century Systems Methodology*. John Wiley & Sons, Inc., 2007, Chapters 8, 11, and 12.

International Council on Systems Engineering. *Systems Engineering Handbook: A Guide for System Life Cycle Processes and Activities*, INCOSE-TP-2003-002-03.2, Section 4. July, 2010.

D. C. Montgomery. *Design and Analysis of Experiments*, Sixth Edition. John Wiley & Sons, Inc., 2005, Chapters 1 and 2.

P. D. T. O'Connor. *Test Engineering: A Concise Guide to Cost-effective Design, Development and Manufacture*. John Wiley & Sons, Inc., 2005, Chapters 6–8 and 10.

H. Petroski. *Success through Failure: The Paradox of Design*. Princeton University, 2006.

E. Rechtin. *Systems Architecting: Creating and Building Complex Systems*. Prentice Hall, 1991, Chapter 7.

M. T. Reynolds. *Test and Evaluation of Complex Systems*. John Wiley & Sons, Inc., 1996.

S. M. Shinners. *A Guide for Systems Engineering and Management*. Lexington Books, 1989, Chapter 7.

R. Stevens, P. Brook, K. Jackson, and S. Arnold. *Systems Engineering, Coping with Complexity*. Prentice Hall, 1998, Chapter 5.

Systems Engineering Fundamentals. Defense Acquisition University (DAU Press), 2001, Chapter 7.

PART IV

POSTDEVELOPMENT STAGE

Part IV goes beyond most systems engineering books in examining the role that systems engineering must play in the production, installation, operations, and support of complex systems. It also identifies the knowledge of these phases that systems engineers should acquire to ensure that the system will be affordable and fully effective in its intended operational environment.

The transition of a system from development to production is often a source of serious difficulties and program delays. If the properties of reliability, producibility, and maintainability have not been fully integrated into the system design, the transition is likely to be slow and costly. Chapter 14, Production, discusses these problems and describes the production facilities and operations as a system in its own right. It also discusses what a systems engineer needs to learn about production processes and problems associated with the types of systems he or she is concerned with, to guide effectively the development and engineering of such systems.

As in the case of production, the operations and support of complex systems also requires the participation of systems engineering. Unanticipated problems are the rule rather than the exception in the operation of new complex systems, and they require urgent resolution by system-oriented personnel. Chapter 15 discusses such problems as well as the systems engineering participation in the process of system upgrading and modernization.

Systems Engineering Principles and Practice, Second Edition. Alexander Kossiakoff, William N. Sweet, Samuel J. Seymour, and Steven M. Biemer
© 2011 by John Wiley & Sons, Inc. Published 2011 by John Wiley & Sons, Inc.

PART IV

POSTDEVELOPMENT STAGE

14

PRODUCTION

14.1 SYSTEMS ENGINEERING IN THE FACTORY

The production phase of the system life cycle represents the culmination of the system development process, leading to the manufacture and distribution of multiple units of the engineered and tested system. The objective of this phase is to embody the engineering designs and specifications created during the engineering development stage into identical sets of hardware and software components, and to assemble each set into a system suitable for delivery to the users. Essential requirements are that the produced system performs as required, is affordable, and functions reliably and safely as long as required. To fulfill these requirements, systems engineering principles must be applied to the design of the factory and its operations.

Most of the discussion in this chapter is concerned with the production of hardware system elements. On the other hand, as noted in Chapter 11, almost all modern products are controlled by embedded microprocessors. Thus, production tests necessarily include testing the associated software.

This chapter is organized in four main sections. It begins with Engineering for Production, which describes where production considerations must be applied during

Systems Engineering Principles and Practice, Second Edition. Alexander Kossiakoff, William N. Sweet, Samuel J. Seymour, and Steven M. Biemer
© 2011 by John Wiley & Sons, Inc. Published 2011 by John Wiley & Sons, Inc.

each phase of system development in order to ensure that the end product is both afford-able and satisfies performance and reliability goals. The section Transition from Development to Production describes the problems typically encountered in the transfer of responsibility from the engineering to the manufacturing organizations and the role of systems engineering in their resolution. Production Operations describes the organi-zation of the overall system manufacturing program as a complex system in its own right, especially as it is typically distributed among a team of contractors. The final section, Acquiring a Production Knowledge Base, describes the scope of knowledge that development systems engineers need to acquire in order to lead properly a system development effort, together with some of the means by which it may be best obtained.

Place of the Production Phase in the System Life Cycle

The production phase is the first part of the postdevelopment stage of the system life cycle. This relation is shown in Figure 14.1, which is a functional flow diagram relating the production phase to the preceding integration and evaluation phase and to the suc-ceeding phase, operation, and support. The inputs from integration and evaluation are specifications and the production system design; the outputs are operational documenta-tion and the delivered system.

Figure 14.2 shows the timing of the production phase relative to its preceding and succeeding phases. As in the case of the integration and evaluation phase, there is a considerable overlap between the end of each phase and the beginning of the next. Overlap between the end of integration and evaluation and the beginning of the produc-tion phase is necessary to order long-lead materials, to acquire factory tooling and test equipment, and to prepare production facilities for operations. Similarly, the initial produced systems are expected to be placed in operation while the production of sub-sequent units is continuing.

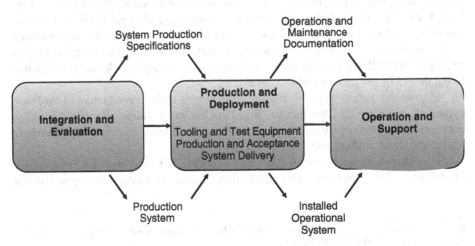

Figure 14.1. Production phase in a system life cycle.

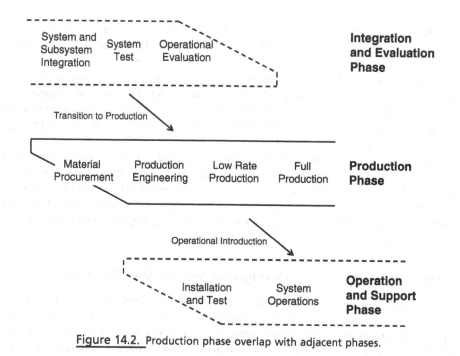

Figure 14.2. Production phase overlap with adjacent phases.

Design Materialization Status

The materialization status of the system would seem to be off the scale of previous diagrams, such as Table 13.1 and its predecessors, because the prior phases of the system development process have essentially fully "materialized" the system components and the system as a whole. However, since the majority of complex systems are made from components produced at a variety of locations, the process of materialization cannot be considered completed until the components are assembled at a central location and are accepted as a total system. This dispersal of manufacturing effort creates stress on vendor coordination, interface control, integration testing, and calibration standards. These will be discussed further in a subsequent section.

14.2 ENGINEERING FOR PRODUCTION

During the development stages of the system life cycle, and especially during concept development, the technical effort is focused primarily on issues related to achieving the performance objectives of the system. However, unless the final product is also affordable and functions reliably, it will not meet its operational need. Since these latter factors are strongly influenced by the choice of system functions and especially by their physical implementation, they must be considered from the beginning and throughout the development process. The process of introducing production considerations during development is generally referred to as "concurrent engineering" or "product development." This section addresses how this process is applied during each phase of system development.

The accepted method of incorporating production considerations into the development process is to include such production specialists and other specialty engineers as members of the system design team. These may include experts in such specialties as reliability, producibility, safety, maintainability, and user interfaces, as well as packaging and shipping.

To make the contributions of these experts effective, it is necessary to bring them into active participation in the system design process. In this connection, it is essential to apply their specialized knowledge to the system requirements, as well as to interpret their specialty languages to other members of the system design team. Without systems engineering leadership, communication skills, and insistence on system balance, the concurrent engineering process is not likely to be effective.

Concurrent Engineering throughout System Development

The following paragraphs describe examples of the application of concurrent engineering in successive phases of system development, as well as the role of systems engineers in making these applications effective. As may be expected, this effort grows in relevancy as the system design progresses; however, it must be initiated at the outset of the program and effectively implemented throughout even the earliest phases.

Needs Analysis. Production and reliability considerations apply in the needs analysis for both needs-driven and technology-driven situations. The decision to begin a new system development must consider its feasibility to be produced as a reliable and affordable entity. Making such a decision relies heavily on systems engineering analyses, together with first-hand knowledge of the postulated development and manufacturing processes.

Concept Exploration. A principal product of the concept exploration phase is a set of system performance requirements that will serve as a basis for selecting the most desirable system concept from among competing candidates. In framing these requirements, a balance must be struck among performance, cost, and schedule goals—a balance requiring a total system perspective in which production processes are essential factors.

As will be discussed in the section on production operations, just as technology is advancing rapidly in solid-state electronics, communications, system automation, materials, propulsion, and many other system component areas, it is similarly revolutionizing production processes. A clear sense of the status and trend of manufacturing technology is a necessary element in the formulation of realistic system requirements. Systems engineering must make informed evaluations of the contributions by production specialists. For example, the selection of materials will be influenced by the difficulty and cost of production processes.

Concept Definition. Perhaps the most critical contribution of systems engineering is in the selection and definition of the preferred system conceptual design. At this point in the development, a clear concept of the implementation of the system in hardware and software is required to develop credible estimates of manufacturing and life cycle costs.

The selection of the proposed system design requires a balance among many factors, and for most of these, the assessment of risk is a central factor. As noted in Chapter 8, taking advantage of advancing technology necessarily involves some degree of risk both in terms of component design and process design. The estimates of risks are influenced by the nature and maturity of the associated manufacturing methods, which must be heavily weighted in trade-off analyses of alternative system configurations. In bringing the experience of production experts to bear on these judgments, systems engineers must serve as informed translators and mediators between them and design engineers and analysts.

Advanced Development. The advanced development phase provides an opportunity to reduce program risks by conducting analyses, simulations, experiments, and demonstrations of critical subsystems or components. Similarly, new production processes and materials must be validated before acceptance. Because of the expense involved in all such activities, especially experiments and demonstrations, the decision as to which ones warrant such validation must be made with full knowledge of the nature and extent of the risks, the magnitude of the gains expected from the use of the proposed processes and materials, and the scope of experimentation necessary to settle the issue. Again, this is a major systems engineering issue requiring expertise in production as well as in system design and performance.

This phase must provide a suitable basis for defining production processes, critical materials, tooling, and so on, through trade studies that consider the risks and costs of alternative approaches. Systems engineering must be intimately involved in the planning and evaluation of such studies to ensure their appropriate integration into the overall plans for the engineering design phase. In this connection, critical attention must be given to the impact of manufacturing methods on the compatibility of component interfaces in order to minimize production, assembly, and testing problems.

Engineering Design. The engineering design phase is where production factors become especially prominent in the detailed design of system components. Component and subcomponent interface tolerance specifications must be compatible with the capabilities of manufacturing processes and allocated costs. The design and construction of factory test equipment must also be accomplished during this phase so as to be ready when production is authorized.

During this phase, design engineers obtain major inputs from specialty engineers applying their experience in the areas of producibility, reliability, maintainability, and safety. In this collective effort, systems engineers serve as mediators, interpreters, analysts, and validators of the final product. To play these roles, the systems engineers must have a sufficient understanding of the intersecting disciplines to effect meaningful communication across technical specialties and to guide the effort toward the best available outcome.

An important part of the engineering design phase is the design and fabrication of production prototypes to demonstrate the performance of the product, as it will be manufactured. The degree to which the prototype fabrication methods are selected to duplicate the actual manufacturing tooling and process control is a matter requiring systems engineering judgment as well as design and manufacturing considerations.

Usually, many components of a complex system are designed and manufactured by subcontractors. The selection of component contractors must involve the evaluation of their manufacturing as well as engineering capabilities, especially when the components involve advanced materials and production techniques. Systems engineers should be able to help judge the proficiency of candidate sources, be key participants in source selection and in setting the requirements for product acceptance, and serve as technical leads in subcontracting.

Such knowledge is also essential for leading the interface definition effort among component suppliers, the specification of interface tolerances, and the definition of component test equipment design and calibration standards for use in both development and production acceptance testing.

The above considerations all affect production cost estimates, which systems engineers must contribute to and evaluate; considerations of uncertainty and risks must also be given due weight in forming the final cost and schedule estimates.

Integration and Evaluation. Unexpected incompatibilities at component interfaces are often first brought to light during the integration of prototype system components and subsequent system testing. These problems are normally corrected through component redesign, refinement of component test equipment, and so on, prior to final release for production. Nevertheless, during the subsequent assembly and test of the production system, design changes introduced to correct these incompatibilities, together with other "minor" changes and adjustments introduced to facilitate production and test activities, may themselves produce new areas of incompatibility. Accordingly, systems engineers should monitor the initial production assembly and test activities so as to alert program management to any problem areas that must be addressed prior to deployment of the product. In order to identify and expeditiously deal with such problems at the earliest possible time, systems engineers must be knowledgeable of both factory production and test acceptance processes. In some cases, acceptance test procedures are written by systems engineers.

Application of Deployment Considerations in System Development

It has been stressed in previous chapters that the system design must consider system behavior throughout the total life cycle. In many systems, the deployment or distribution process subjects the system and its constituent components to substantial environmental stresses during transportation, storage, and installation at the operational site. While these factors are considered during system definition, in many instances, they are not quantitatively characterized until the advanced development phase or sometimes even later. It is therefore mandatory that the deployment of the system be planned in detail as early in the development process as possible. Factors such as the risk of exposure to environments that might affect system performance or reliability must be assessed and reflected either in the system design requirements or in restrictions to be observed during the deployment process. In some cases, protective shipping containers will be required. In those cases where problem areas still exist, provision should be made for their resolution through further analysis and/or experimentation.

In many cases, the predecessor system provides a prime source of information regarding the conditions that a new system may encounter during its transit from producer to user. When the operational site and the system physical configuration are the same or similar to that of the new system, the deployment process can be quantitatively defined.

14.3 TRANSITION FROM DEVELOPMENT TO PRODUCTION

Transition in Management and Participants

As may be inferred from the life cycle model, wholesale changes in program management focus and participants necessarily take place when the production of a new system is initiated. These areas are briefly summarized below.

Management. The management procedures, tools, experience base, and skills needed for successful program direction and control during the production phase differ materially from those needed during system development. Accordingly, the production of a new system is almost always managed by a team different from the one that directed the earlier engineering development, integration, and test efforts. Moreover, the production contract is sometimes completed among several companies, some of which may have been only peripherally involved in the system development. For all these reasons, there is normally little carryover of key personnel from the engineering into the production phase. At best, selected members of the development engineering team may be made available when requested to provide technical assistance to the production organization. Production funding is usually embodied in a contract separate from the one that was in force during engineering development and is administered separately to provide its own audit trail and future costing data.

Program Focus. As noted earlier, the production phase is focused on producing and distributing identical copies of the product design. The stress is on efficiency, economy, and product quality. Automated manufacturing methods are employed where practicable. Configuration control is extremely tight.

Participants. The participants in this phase are very different from those who were involved in the development effort. Specifically, the great majority of participants in this phase are technicians, many of whom are highly skilled as automatic equipment and factory test operators. The engineering participants are chiefly concerned with process design, tool and test equipment design and calibration, quality control, factory supervision, and troubleshooting. Most are specialists in their respective disciplines. However, as stated earlier, to effect a successful transition into production, there must also be an experienced team of systems engineers guiding the process.

Problems in the Transition Process

The transition of a new system from development to production can be a particularly difficult process. Many of the associated problems can be ascribed to the factors that were first cited in Chapter 1 (i.e., advancing technology, intercompany competition, and technical specialization) as dictating the need for a special systems engineering activity. These factors are further discussed below.

Advancing Technology. It was seen that while the incorporation of technological advances in the design of new systems is often necessary to achieve the desired gain in capability and thus preclude premature obsolescence, this also incurs the risk of introducing unanticipated complications in both the development and production processes. Although the development process provides methods for the identification and reduction of performance problems, production-related difficulties are frequently not revealed until production prototypes have been fabricated and tested. By that time, remedial action is likely to cause severe and very expensive delays in production schedules. Systems engineering expertise is crucial, both for anticipating such unintended results insofar as possible and for quickly identifying and resolving those that still do unexpectedly occur.

An example of technological advances that must be considered in the transition process is that of the speed of digital processors, accompanied by reductions in size and cost. The pressure to install the latest products can be irresistible but comes at a price of changes in packaging, testing, and sometimes software revision.

Competition. Competition puts stresses on the transition process from several directions. Competition for funds often results in insufficient effort being budgeted for production preparation; moreover, it almost always eliminates the availability of reserve funding to deal with unexpected problems, which always arise in the development of complex systems. This results in too little testing of production prototypes, or delay of their fabrication until after the time that decisions on tooling, materials, and other production factors have to be made. Despite slippages in production preparation, production schedules are frequently held firm to avoid the external appearance of program problems, which are likely to cause customer concern and possibly even direct intervention. Competition for experienced staff within the organization can also result in reassignment of key engineers to a higher priority activity, even though they may have been counted on for continued commitment to the project. Competition for facilities may delay the availability of the facilities needed for the start of production. These are only examples of the competing forces that must be dealt with in managing the transition process.

Specialization. The transition from development to production also involves transfer of prime technical responsibility for the system from specialists in engineering and development to specialists in manufacturing. Moreover, at this point, the primary location of activity also shifts to the manufacturing facilities and their supporting organizations, which typically are separated physically from and managerially independent of engineering—an arrangement that can and often does severely attenuate essential communications between the engineering and production organizations. Systems engineers with some knowledge of production are frequently the only individuals who can communicate effectively between engineering and production personnel.

The above problems are rendered still more difficult by the usual dispersion of development and production of major components and subsystems among several specialized subcontractors. Coordination during the production phase in such cases is many times more complicated than during development because of the need to closely synchronize the timing and tempo of fabrication and testing with system assembly and delivery schedules. For these reasons, successful prototypes do not necessarily guarantee successful production systems.

Product Preparation

The importance of the above transition process in commercial development and production has led the National Society of Professional Engineers (NSPE) to dedicate a separate phase in their system life cycle to "commercial validation and production preparation." The engineering activities during this phase of development are stated to include the following:

- Complete a preproduction prototype.
- Select manufacturing procedures and equipment.

Demonstrate the effectiveness of

- final product design and performance;
- installation and start-up plans for the manufacturing process, selection of production tools and technology;
- selection of materials, components, and subsystem vendors and logistics; design of a field support system; and
- preparing a comprehensive deployment/distribution plan.

Either as part of the production plan or separately, other associated activities must also be defined or refined at this time. These include

- logistic support plans,
- configuration control plans, and
- document control plans and procedures.

Production Configuration Management

The forces of advancing technology, competition, and specialization all exert pressure to make changes in the engineering design of the system, especially at the component and subcomponent levels. As noted previously, new technology offers opportunities to introduce higher performance or cheaper elements (e.g., new materials, commercial off-the-shelf [COTS]). Moreover, competition presses for less costly designs, and engineers at component producers may petition to adapt designs to fit their particular production tooling. All of these factors tend to produce numerous engineering change proposals (ECPs), each of which must be analyzed and accepted, modified, or rejected. The system contractor's systems engineers play a crucial role in analyzing these proposals, planning and overseeing test efforts where required and recommending the appropriate action on change proposals. The time available for such action is very short and the stakes are very high. Intercontractual pressures often complicate the decision process.

Viewed in this light, the transition from engineering design to production is the most critical period in the configuration management process and calls for effective analytical, engineering, and communication skills on the part of systems engineers and project managers. Above all, documentation must not be allowed to lag significantly behind the change process, and all concerned must be kept in the communication loop. Systems engineering is the keeper of the integrity of the design.

It follows that the configuration management process does not stop when production begins; it continues even more intensively throughout the production process. At the beginning of production, component interface incompatibilities that have not been previously detected and eliminated (or inadvertently have been created in product design) will be revealed and must be dealt with quickly. Each incompatibility requires a decision as to whether it can be remedied in parallel with continuing production or if production should be interrupted, and if so, at what point. Because of their impact on cost and schedule, such decisions are made at management levels, but the most critical inputs are provided by systems engineering. These inputs come from close teamwork between the configuration management team and supporting systems and production engineering staffs. If, as often happens, communication between the production and engineering organization is poor, the above process will be inefficient and costly.

14.4 PRODUCTION OPERATIONS

Planning the development and evaluation of a major new system requires well thought-out and documented plans, such as the systems engineering management plan (SEMP) and the test and evaluation master plan (TEMP), which are promulgated widely to coordinate the efforts of the system development. For the same reasons, the production phase must have a formal system production plan to provide a blueprint of the organization, tasks, and schedules for system production.

Production Planning

The key elements of a production plan include the following subplans and sections:

- responsibility and delivery schedule for each major subassembly (component);
- manufacturing sites and facilities;
- tooling requirements, including special tools;
- factory test equipment;
- engineering releases;
- component fabrication;
- components and parts inspection;
- quality control;
- production monitoring and control assembly;
- acceptance test;
- packaging and shipping;
- discrepancy reports;
- schedule and cost reports; and
- production readiness reviews.

Preparation of the production plan should begin during engineering design and forms the basis for initiating production. It must be a living document and must evolve during the production process. Lessons learned should be documented and passed on to future programs. Systems engineers not only contribute to the plan but, in the process, also benefit by learning about the diverse activities that must be managed during production.

Production Organization as a Complex System

The manufacture of a new complex system typically requires the coordinated efforts of a team of contractors with extensive facilities, equipment, and technical personnel, usually distributed geographically but working to unified specifications and schedules. As in an engineered system, all these subsystems and their elements must work together effectively and efficiently to perform their collective mission—the production of units of a system of value to its users. The planning, design, implementation, and operation of this production system are tasks of comparable complexity to that required to develop the system itself.

Figure 14.3 is a schematic representation of the configuration of the facilities for producing a typical new complex system. The large blocks correspond to the engineering and production facilities of the prime contractor. The blocks on the left represent suppliers of newly developed components, while those at the top represent suppliers of standard components. The suppliers of developed components are shown to have engineering elements that operate under the technical direction of the prime contractor.

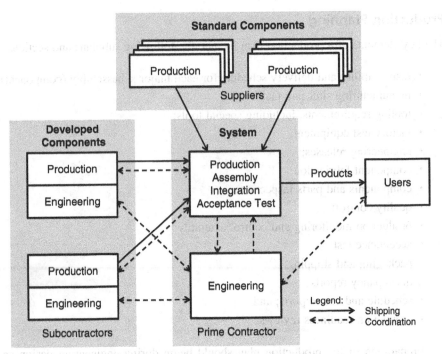

Figure 14.3. Production operation system.

Whether or not the component suppliers are owned by the prime contractor, they are to all intents separate organizations that have to be technically coordinated by the prime contractor's engineering organization. It is seen that this combination of facilities must itself be managed as an integrated system, with strict control of all of the interfaces with respect to product performance, quality, and schedule.

The overall task of bringing this entity into being is usually led by a management team assembled by the prime production contractor. While systems engineers do not lead this effort, they must be important contributors because of their broad knowledge of the system requirements, architecture, risk elements, interfaces, and other key features.

The "architecting" of the production system is complicated by a number of factors, including

1. *Advancing Technology*, especially of automated production machinery, which raises issues as to when to introduce the most recent development and into which processes; similar decisions are required on the extent and timing of introducing advanced materials;

2. *Requirement to Ensure Compatibility of New Processes with Workforce Organization and Training*—in numerous cases, technology-driven changes have resulted in decreases in productivity;

3. *Design of Communications among Distributed Production Facilities*—a balance between lack of information exchange and information overload is crucial;

4. *Factory and Acceptance Test Equipment*—in a distributed system, there must be a highly coordinated set of component test equipment that ensures identical acceptance criteria at component manufacturers and at the integration and assembly facility, as well as conformity of system-level acceptance test equipment with the integrated component tolerance structure;

5. *Manufacturing Information Management*—in any complex system, an enormous amount of data must be collected at all system levels in order to effectively govern and control the manufacturing and assembly process; the database management system required to deal with this information is a large software system in its own right and therefore requires an expert staff for its implementation and operation;

6. *Provisions for Change*—for production operations expected to extend for a period of years, the facilities need to be designed to adapt to variations in rates of production and to the introduction of design changes; many systems are first produced at low rates to validate the process on when production is stretched out for funding reasons; accommodation of the process to these changes while maintaining an efficient operation is an important goal.

All of the above problems require the application of systems engineering principles to obtain effective solutions.

Component Manufacture

We have seen that the building blocks of complex systems are components representing different specialized product lines. These are integrated from subcomponents into complete units, tested, and shipped to a system assembly plant or to spare parts distribution facilities. Thus, the manufacturing process takes place at a number of separate facilities, many of which are usually under different company managements. As noted in the previous section, the management of such a distributed operation poses special problems. An example is the necessity for extremely tight coordination between the component manufacturers and the system producer of production schedules, testing, inspection, and quality control activities. The difficulty of managing a distributed production process for a new and complex system necessitates an integrated team effort in which systems engineers play an essential role in helping to deal rapidly and efficiently with such inadvertent incompatibilities as may be encountered.

Component manufacturing is the place where most special tooling, such as automatic material forming, joining, and handling machinery, is required. The use of automation can substantially reduce the cost of production, but at the same time may involve large development costs and extensive worker training. If newly introduced, it can also cause start-up delays. Thus, the introduction of special tooling for component manufacture must be closely coordinated by the production contractor to minimize scheduling problems.

Production tolerances require special attention because they are directly affected by tooling, as well as by any minor changes that may be made to reduce production costs. Since these may affect both the ability to interface with components made by another contractor and also system performance, systems engineering oversight by the production contractor is necessary.

Usually, the company that produces a given new system component is also the one that developed it. However, the organizational separation of the company's manufacturing from its engineering operation creates the potential for mistakes in the design of production tooling and test equipment resulting from imperfect communications. Incompatibilities inadvertently introduced by design changes made in the interest of cost reduction or other worthy objectives may consequently pass unnoticed until final component testing or even until system assembly. A degree of systems engineering oversight is important, especially to ensure compatibility between the test equipment at component manufacturers and that which will be used at the integration facility for component acceptance. This should also include provisions for and periodic revalidation of common calibration standards.

The establishment of commercial standards at the part and subcomponent levels has greatly simplified many aspects of production and integration of electronic and mechanical components. Economies of scale have reduced costs and have enabled a broad degree of interchangeability, especially in component containers, mounting, and interconnections.

System Acceptance Tests

Before each production system is accepted by the customer for delivery, it must pass a formal systems acceptance test. This is usually an automated end-to-end test with go–no go indications of key system performance.

For a complex system, the design and development of suitable acceptance test procedures and equipment is a major task requiring strong systems engineering leadership. The test must determine that the requirement of ensuring that the product is properly constructed meets the key requirements and is ready for operational use. Its results must be unequivocal, regarding success or failure, and must require minimum interpretation. At the same time, the test must be capable of being performed relatively quickly without adding materially to the total cost of manufacturing. Such a balance requires the application of systems engineering judgment as to what is essential to be tested and what is not.

The system acceptance test is usually witnessed by representative(s) of the customer and is signed off on successful completion.

Manufacturing Technology

The explosive advance of modern technology has had dramatic impacts on products and the process of production. Microelectronic chips, high-speed computing devices, low-cost optics, piezoelectrics, and microelectromechanical devices are but a few of dozens of technological advances that have radically changed the composition of components and the way they are made. Even more changes in manufacturing methods and equipment have been produced by the wholesale replacement of human factory operators by automatic controls and robotics. The new methods have greatly increased speed, precision, and versatility of machining and other processes. Of comparable importance is the reduction of the time to convert a machine from one operation to another from days or weeks to minutes or hours. These changes have resulted in major economies at nearly every aspect of manufacture. They have also made it possible to produce higher-quality and more uniform components.

Prior to the widespread application of computer-aided manufacture (CAM) and component design, control of interfaces had to rely on inspection and testing using a multiplicity of special tools and fixtures. Today's computer-controlled manufacture and assembly, as well as the use of computer-based configuration management tools that can be electronically coordinated among organizations, make the management of interfaces of components built remotely far easier than in the past. However, to effectively implement this degree of automation requires planning, qualified staff, and significant funding. This, in turn, requires systems engineering thinking on the part of those organizing the production system.

14.5   ACQUIRING A PRODUCTION KNOWLEDGE BASE

For inexperienced systems engineers, the acquisition of knowledge regarding the production phase that is both broad enough and sufficiently detailed to influence effectively the development process can appear to be an especially daunting task. However, this task is basically similar to that of broadening the knowledge base in diverse engineering specialties, in the elements of program management, and in the interorganizational communications that every systems engineer must accomplish over time. Some of the most effective means for acquiring this knowledge are summarized below.

Systems Engineering Component Knowledge Base

In order to guide the engineering of a new system, systems engineers must acquire a basic level of knowledge concerning the basic design and production processes of system components. This means that systems engineers must appreciate the impact of production factors on the suitability of particular components to meet the requirements for their use in a specific system application. To make the acquisition of such a knowledge base more achievable, the following considerations may be helpful:

1. Focus on those components that use advanced technology and/or recently developed production processes. This means that attention to mature components and established production processes may be relaxed.
2. Focus on previously identified risk areas as they may affect or be affected by production.

3. For these identified risk areas, identify and establish contact with sources of expert knowledge from key in-house and contractor engineers. This will be invaluable in helping solve problems that may arise later.

The type and extent of the necessary knowledge base will vary with the system and component areas. Some examples are described below.

Electronics Components. Modern electronics is largely driven by semiconductor technology, so familiarity with the nature of circuit chips, circuit boards, solid-state memories, microprocessors, and gate arrays is necessary, though only to the level of understanding what they are, what they do, and how they should, and even more importantly, should not be used. Their development is in turn driven by commercial technology, and in many instances, their capability is multiplying according to Moore's law. It is therefore important to have a feel for the current state of the art (e.g., component densities, processor speeds, chip capabilities) and its rate of change.

Electro-optical Components. In communications and displays, electro-optical components play key roles, thanks to advances in lasers, fiber optics, and solid-state electro-optical elements. Their development is also driven by commercial applications and is advancing rapidly in the above areas.

Electromechanical Components. As their name implies, these components combine the features of electrical and mechanical devices (e.g., antennae, motors). Their characteristics tend to be peculiar to the specific application and can best be learned on a case-by-case basis.

Mechanical Components. Most applications of mechanical components are mature. However, several areas are moving rapidly. These include advanced materials (e.g., composites, plastics), robotics, and micro devices. Their design and production have been revolutionized by computer-aided engineering (CAE) and CAM.

Thermomechanical Components. Most of these components relate to energy sources and thermal controls. For this reason, safety is frequently a key issue in their system applications, as is the related function of control.

Software Components. Software, and embedded firmware derived from it, is rapidly becoming part of virtually every device (e.g., communications, transportation, toys). The process of designing and producing reliable software is also advancing as rapidly. The production aspects of software and firmware are of course very different from hardware. Every systems engineer should understand the general capabilities, including the advantages and limitations, of software quality and software design and implementation, as well as the basic differences between computer-based software and firmware. Software is treated in greater depth in Chapter 11.

Production Processes

Production processes are not the responsibility of systems engineers. Nevertheless, the general nature of these processes and typical problems associated with them must be understood by systems engineers to give them the knowledge to resolve problems that occur in production, especially during start-up.

Observing Production Operations. The factory floor is often the most illuminating source of insight concerning the manufacturing process, especially when observation is supplemented by questioning factory personnel. Opportunities to observe production operations occur naturally during site visits, production planning, and other activities, but these are seldom adequate to provide even a superficial understanding of manufacturing processes. Systems engineers should endeavor to schedule special factory tours to acquire a first-hand feel of how the factory operates. This is especially important because of the rapid advances in manufacturing tools and processes brought about by increasing automation. Because the initial production of new components is likely to run into problems with tools, processes, materials, parts availability, quality control, and so on, it is important to develop a feel for the nature of the associated activities and possible means for problem resolution. Of course, the best opportunity to learn production processes is a short assignment in the manufacturing organization.

Production Organization. It has been previously noted that the organization and management of the production process of a major system is different from the organization and management of the system development process. It is important for systems engineers to be acquainted with the differences, both generically and for the specific system under development. While this is of most immediate concern for program management, it strongly influences how the transition from engineering to production should be planned, including the transfer of design knowledge from design engineers to production engineers. In particular, in many companies, the communications between engineering and production personnel are often formal and largely inadequate. When this is the case, company management should provide special means to establish adequate communication across this critical interface—a function in which systems engineers can play a leadership role. Failure to bridge this potential communication gap properly has been a major contributor to critical delays and near failures in the production of numerous major systems.

Production Standards. Virtually all types of manufacturing are governed by industry or government standards. The U.S. government is replacing most of its own standards by those developed by industry, as well as moving to utilize COTS parts and components insofar as is practicable. These standards are primarily process oriented and define all aspects of production. Systems engineers must be familiar with the standards that are applicable to components and subsystems in their own system domain and with the way in which these standards are applied to the manufacturing process. These standards are often indicative of the quality of the components that are likely to be produced, and hence of the degree to which oversight, special testing, and other management measures will be required. While the decisions regarding such actions are the responsibility of program management, systems engineering judgment is a necessary ingredient.

14.6 SUMMARY

Systems Engineering in the Factory

The objectives of the production phase are to produce sets of identical hardware and software components, to assemble components into systems meeting specifications, and to distribute produced systems to customers.

Essential requirements are that the produced system performs as required, is affordable, and functions reliably and safely as long as required.

Engineering for Production

Concurrent engineering, or product development, has the following features: it is the process of introducing production considerations during development. Production specialists and other specialty engineers are key members of the design team. Therefore, systems engineers must facilitate communications among team members.

The decision to begin new system development must demonstrate its need, technical feasibility, and affordability. The formulation of realistic system requirements must include a clear sense of the status and trend in manufacturing technology. As technology evolves, requirements must also evolve to stay consistent.

Production risks are influenced by the nature and maturity of the associated manufacturing methods and are heavily weighted in trade-off analyses of system alternatives.

Successful production requires that new production processes and materials are validated before acceptance, that component interfaces are compatible with manufacturing processes, and that factory test equipment is validated and ready. The latter is typically demonstrated by production prototypes that have demonstrated product performance.

Unexpected incompatibilities at component interfaces have the following features:

- They are often first discovered during the integration of prototype components.
- Corrections of incompatibilities may themselves produce new areas of incompatibility.

Systems engineers must be knowledgeable of factory production and test acceptance processes. Direction and control of production differs from system development in the following: (1) different tools, experience base, and skills; and (2) a different team of specialists—few key personnel carry over from development.

Production risks are frequently not revealed until production prototypes are fabricated and tested. Remedial action is likely to cause expensive delays; therefore, systems engineering expertise is crucial for resolution.

Transition from Development to Production

Stresses on the transition from development to production result from

- insufficient funding for production preparation,
- little or no reserve funds for unexpected problems,
- too little testing of production prototypes, and
- schedules held firm even though problems exist.

The transition to production is a most critical period for continuity of operations and features must be recognized. The transition transfers responsibility from development to manufacturing specialists. And manufacturing facilities are typically separate and independent of engineering. Therefore, communication is difficult between engineering and production personnel. Consequently, systems engineers are needed for facilitating communications. Finally, a system production plan is required as a blueprint for transition.

The transition to production is critical to the configuration management process because documentation cannot lag behind the change process; systems engineering is the keeper of the integrity of the design.

Production Operations

The planning, design, implementation, and operation of a "production system" is a task of comparable complexity to developing the system itself. Architecting of the production system requires

- acquisition of extensive tooling and test equipment,
- coordination with component manufacturing facilities,
- organization of a tight configuration management capability,
- establishment of an effective information system with enginery organization,
- training the production staff in the use of new tooling,
- accommodation of both low and high production rates, and
- promotion of flexibility to accommodate future product changes.

Specialized components often represent different product lines and pose special problems. Tight coordination is needed between component manufacturing and system producers, in production schedules, testing, inspection, and quality control. Establishment of commercial standards at the part and subcomponent levels leads to greatly simplified production and integration of components.

Computer-controlled manufacturing methods greatly increase speed, precision, and versatility of factory operations. They reduce time to reconfigure machines between operation modes, and produce higher-quality parts and more uniform components. This often leads to major cost savings.

Acquiring a Production Knowledge Base

Systems engineers must acquire a basic knowledge concerning production processes to be capable of guiding the engineering of a new system. They must focus on advanced technology and new production processes, as well as risk areas as they may be affected by production.

PROBLEMS

14.1 Because complex systems contain a large number of subsystems, components, and parts, it is usually necessary to obtain a significant number of them from outside subcontractors and vendors. In many cases, it is possible to make these items either in-house or procure them elsewhere. Both approaches have advantages and disadvantages. Discuss the main criteria that are involved in deciding which approach is best in a given case.

14.2 One of the requirements of a good systems engineer who is engaged in developing systems which have significant components that are manufactured is that he or she be knowledgeable about factory production and acceptance test processes. Give two examples that illustrate the importance of this knowledge in achieving on-time delivery of the final product.

14.3 Configuration management is particularly important during the transition from system development to production. Identify four specific areas where close attention to configuration management is crucial during this phase transition and explain why.

14.4 Discuss how the planning, design, implementation, and operation of a production system is a task of comparable complexity to that required to develop the actual system itself.

14.5 Describe the process referred to as concurrent engineering, its objectives, use of interdisciplinary integrated product teams (IPTs), and its place in the system life cycle. Describe the role of systems engineers on the teams. Describe what problems you would expect to be encountered in assembling an IPT and in making its effort productive and how they might be handled.

14.6 Discuss four typical problems that make the transition from development to production difficult and the approaches to minimizing them.

14.7 Production is typically the responsibility of a division of a company independent of the development organization. It has been stated that the transition to production and the production process itself requires systems engineering expertise in certain critical areas. List some instances where systems engineering expertise in the production organization is required in the production of medical devices (e.g., implantable pacemakers).

14.8 Discuss the principal areas in which CAM has revolutionized the manufacture of automobiles.

FURTHER READING

B. Blanchard and W. Fabrycky. *System Engineering and Analysis*, Fourth Edition. Prentice Hall, 2006, Chapters 16 and 17.

G. E. Dieter and L. C. Schmidt. *Engineering Design*, Fourth Edition. McGraw-Hill, 2009, Chapter 13.

International Council on Systems Engineering. *Systems Engineering Handbook: A Guide for System Life Cycle Processes and Activities*, INCOSE-TP-2003-002-03.2, Sections 4 and 9. July, 2010.

Systems Engineering Fundamentals. Defense Acquisition University (DAU Press), 2001, Chapter 7.

15

OPERATIONS AND SUPPORT

15.1 INSTALLING, MAINTAINING, AND UPGRADING THE SYSTEM

The operations and support phase of the system life cycle is the time during which the products of the system development and production phases perform the operational functions for which they were designed. In theory, the tasks of systems engineering have been completed. In practice, however, the operation of modern complex systems is never without incident. Such systems usually require substantial technical effort in their initial installation and can be expected to undergo significant testing and component replacement during periodic maintenance periods. Occasional operational glitches must also be expected due to operator error, operating stresses, or random equipment failures. In such cases, systems engineering principles must be applied by system operators, maintenance staff, or outside engineering support to identify the cause of the problem and to devise an effective remedy. Further, large complex systems, such as an air traffic control system, are too costly to replace in their entirety and therefore are subject to major upgrades as they age, which introduce new subsystems in place of obsolescent ones. All of these factors are sufficiently significant in the total role of systems engineering in the overall system life cycle to warrant a special place in the study of systems engineering.

Systems Engineering Principles and Practice, Second Edition. Alexander Kossiakoff, William N. Sweet, Samuel J. Seymour, and Steven M. Biemer
© 2011 by John Wiley & Sons, Inc. Published 2011 by John Wiley & Sons, Inc.

The principal sections of this chapter summarize the typical activities that take place in the course of a system's operating life, beginning with the time it is delivered from the production or integration facility to the operational site until it is replaced by a newer system or otherwise rendered obsolete and disposed of. The section on *installation and test* deals with problems associated with integrating the system with its operating site and the successful interconnection of internal and external interfaces. The section covering *in-service support* concerns activities during the normal operations of the system; these include maintenance, field service support, logistics, and dealing with unexpected operational emergencies. The section on *major system upgrades* is concerned with periodic subsystem modifications that may be introduced to maintain system effectiveness in the face of changing user requirements and advancing technology. Such system upgrades require the same type of systems engineering expertise as did the original system development, and may also present new and unique challenges due to added constraints that may be imposed by the process of integrating new and old components. The last section on *operational factors in system development* describes the kinds of information that systems engineers should seek to acquire regarding the operational characteristics of the system being developed, together with the opportunities that they may have for obtaining such knowledge. Such knowledge is just as important to systems engineers who lead the system development as is a firm grounding in factors that affect system production processes and costs.

Place of the Operations and Support Phase in the System Life Cycle

Before discussing the systems engineering activities during the operations and support phase, it should be noted that this phase is the concluding step of the system life cycle. The functional flow diagram of Figure 15.1 shows the inputs from the production phase to be operational documentation and a delivered system, and the outputs to be an obsolete system and a plan for disposing it in an appropriate way.

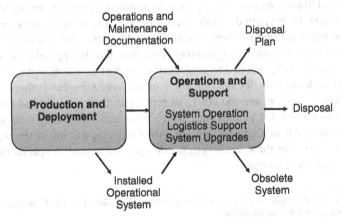

Figure 15.1. Operations and support phase in a system life cycle.

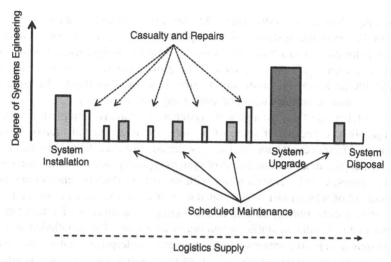

Figure 15.2. System operations history.

Systems Engineering in the Operations and Support Phase

During its operating life, a typical complex system encounters a number of different periods when its operation is interrupted. These incidents are represented in Figure 15.2. The abscissa is time, running from system delivery to its disposal. The ordinate represents the relative level of systems engineering involvement in the various events identified by the captions in the upper part of the figure. At the start, a column is seen symbolizing the installation and test period, which is shown to take substantial time (usually weeks or months) and a relatively large systems engineering effort. The four low, regularly spaced columns represent planned maintenance periods, which may require days of system downtime. The narrow spikes at irregular intervals are meant to correspond to random system breakdowns requiring emergency fault identification and repairs. These are usually fixed quickly but may take considerable systems engineering effort to find a solution that can be effected with minimum downtime. The large column on the right represents a major system upgrade, requiring a relatively long period (many months) and a high level of systems engineering. The latter may rival the effort involved in a new system development and may itself require a multiphase approach.

15.2 INSTALLATION AND TEST

System Installation

The effort required for installing a delivered system at its operating site is strongly dependent on two factors: (1) the degree of physical and functional integration that had been accomplished at the production facility and (2) the number and complexity of the interfaces between the system and the operating site (including other interacting systems). In the case of an aircraft, for example, virtually all significant system elements typically are assembled and integrated at the prime contractor's factory site so that when the aircraft leaves the production facility, it is ready for flight. The same is true for an automobile, a military truck, or almost any kind of vehicle.

The installation of many large-scale systems on a land or ship platform may be a major operation, especially if some of its subsystems are manufactured at separate locations by different contractors and are assembled only after delivery to the operating site. For example, an air terminal control system typically consists of several radars, a computer complex, and a control tower with an array of displays and communication equipment, all of which must be integrated to operate as a system and linked to the en route control system, runway landing control systems, and an array of associated equipment required to handle air traffic in and out of an airport. The installation and test of such a system is in itself a major systems engineering enterprise. Another such example is a ship navigation system, which consists of many subsystems that are manufactured at various sites, frequently by different contractors, and which have complex interfaces with ship elements. After the initial ship systems pass integration tests at a land site, subsequent production subsystems are often assembled and integrated only after they are delivered to the shipyard. The task of interfacing elements of the system with ship structure, power, controls, and communications is usually performed at the shipyard by experts in ship installations.

Internal System Interfaces. As discussed previously, the systems engineer has a responsibility to assure that system integrity is maintained throughout assembly and installation. Installation procedures must be carefully planned and agreed to by all involved organizations. The systems engineer must be a key participant in this planning effort and in seeing it properly carried out. However, regardless of the degree of planning, the proper integration of subsystem interfaces will always be a potential source of trouble and therefore deserves special effort.

As noted previously, interfacing of the subsystems at the operating site is especially complicated when major subsystems are designed and manufactured by different contractors. In shipboard systems, two common examples of such subsystems are propulsion and communications. These subsystems include interfacing elements that employ both low- and high-power digital and analog signals, together with numerous switching and routing processors. Some of these equipments will be new state-of-the-art elements designed specifically for this project, and some will be older, off-the-shelf items.

Under such circumstances, problems during installation and checkout are almost certain to be encountered. Moreover, some problems will be difficult to track down because the necessary resources, such as test specialists and troubleshooting equipment, may not be available at the installation site. In such instances, it is not unusual for the acquisition agency to bring in special "tiger teams" for assistance. If available, the people who developed the original system, particularly the systems engineers who will have the most system-level knowledge and management skills, are best qualified to work on these problems.

System Integration Site. For systems where integration of major subsystems is especially difficult to accomplish at the operational site, it becomes cost-effective to utilize a specially equipped and supported integration site where subsystems and components are assembled and various levels of checkout are performed prior to partial disassembly and shipment to the operational site. This may be the same integration site that is used during development to test and evaluate various elements of prototype equipment, or it may be a separate facility also used for training of operators and maintenance personnel. In either case, such a special site can also be extremely valuable in checking out fixes to problems encountered during initial operations, as well as in supporting the engineering of system upgrades.

External System Interfaces. In addition to numerous internal subsystem interfaces, complex systems have many critical external interfaces. Two examples are prime power, which is usually generated and distributed by an external system, and communication links, which interface through hardwired electronic circuits or by microwave links. Communication links not only have to be electronically compatible but must also have the appropriate set of message protocols, which are usually processed by software.

A further complicating factor is that large systems must often interface with systems that are procured from developers who are not under the control of either the prime system contractor or the system acquisition agent. This means that design changes, quality control, delivery schedules, and so on, can become major coordination issues. This also makes the resolution of problems more troublesome by raising the issue of who may be at fault and should therefore assume responsibility for correcting any resulting problems. Problems of this type emphasize the importance of having a well-planned and executed test program during system assembly and integration.

During system development, special pains must be taken to ensure that the details of external interactions are fully specified early in the design process. In many cases, the documentation of interacting systems is insufficiently detailed and sometimes so far out of date that their interface connections to the newly developed system are no longer valid. Systems engineers who have first-hand experience with system environments can often anticipate many such critical factors relating to external interfaces, thereby ensuring that their characteristics are defined early enough in the development process to avoid problems during system installation.

Communication links other than standard commercial communications are notoriously troublesome. They often employ special connections and message protocols whose detailed specifications are difficult to obtain ahead of time.

The result can lead to surprises during system installation and initial operation, with no clear evidence as to which organization is responsible for the incompatibility and capable of resolving it. In such circumstances, it is usually advisable for the development contractor to take the initiative to at least identify the specific technical problem and to propose the means for a solution. Otherwise, the blame for the lack of system interoperability is commonly placed arbitrarily on the new system and its developer.

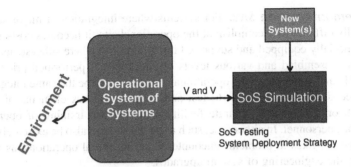

Figure 15.3. Non-disruptive installation via simulation.

Nondisruptive Installation

Some critical systems require continuous operations and cannot be stopped or paused during system installation or upgrades. This tends to be the case when installing a system into a large system of systems. The installation of a new or upgraded system into the system of systems cannot disrupt current operations. Examples include system installation into a city power grid, a complex industrial wide area network, a national communications network, a major defense system of systems, and the national air traffic control system of systems. All of these examples require 24-hour operations without significant disruption.

Installing major systems into a system of systems without disruption requires careful planning and attention to detail. In the recent past, two general approaches have emerged to assist in this area: maintaining a system of systems simulation and maintaining a system of systems test bed. Figure 15.3 depicts the first option.

With this strategy, a system of systems simulation with hardware-in-the-loop is created. This simulation is typically user-in-the-loop as well, as opposed to stand alone. This simulation facility is verified and validated against actual data collected from the operational system of systems, which interacts with the environment. Typically, the simulation would not interact with the environment (although there are exceptions to this).

The system of systems simulation is used as a test bed to determine (1) the impact the new system will have on the system of systems before it is actually installed and (2) an installation strategy that will keep operations at an acceptable level. Once a strategy has been developed and verified using the system of systems simulation facility, knowledge and confidence is gained on how to install the system into the actual system of systems.

The advantage of this nondisruptive installation mode is the cost savings and the ability to model installation procedures and techniques before installing the system into the actual system of systems. The system of systems simulation facility, while expensive and complex, is only a representation of the actual system of systems and can be scoped to desired budget and tolerance levels. Obviously, if the system of systems in

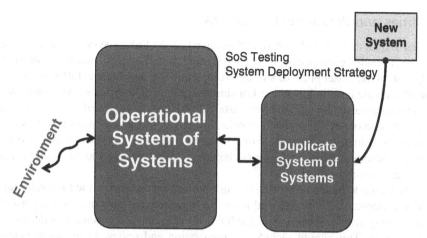

Figure 15.4. Non-disruptive installation via a duplicate system.

question is a defense network responsibility for the survival of a nation, extremely high tolerances would be required. However, if the system of systems is a business information technology (IT) network, tolerance may be relaxed to a comfortable risk level.

The second concept used within nondisruptive installations involves the development of a duplicate system of systems, scaled down from the operational one, and is depicted in Figure 15.4. The concept is similar to the first concept in that the system is installed into the scaled-down version of the system of systems, and testing occurs. During this process, the duplicate system of systems is typically disconnected from the operational system of systems to avoid any interference or disruption. An installation strategy is developed from the experience to apply for installation into the full-scale system of systems.

Once confidence in the risk of disruption is acceptable, the system is installed into the operational system of systems. Many times, the operational system of systems is disconnected from the environment—the duplicate system of systems is used as a surrogate during the installation. This is typically performed during a low demand situation or time frame to ensure the limited capacity of the duplicated system of systems is sufficient.

Although this strategy for nondisruptive installation is expensive (you are basically building a scaled-down version of the operation system of systems), it has two major benefits: (1) the duplicate system of systems is an architecture copy of the operational system of systems and is the closest representation that is possible without duplicating both the architecture and scale of the original; and (2) during peak demand, the duplicate, scaled-down system of systems can be used to augment the operational system of systems. National communications systems use this technique to keep its networks operational continuously, and to allow for unexpected peak demand periods.

Facilities and Personnel Limitations

Neither the facilities nor the personnel assigned to the task of system installation and test are normally equipped to deal with significant difficulties. Funds are inevitably budgeted on the assumption of success. And, while the installation staff may be experienced with the installation and test of similar equipment, they are seldom knowledgeable about the particular system being installed until they have gained experience during the installation of several production units. Moreover, the development contractor staff consists of field test engineers, while systems engineers are seldom assigned until trouble is encountered, and when it is, the time required to select and assign this additional support can be costly.

The lesson to be learned is that the installation and test part of the life cycle should be given adequate priority to avoid major program impact. This means that particular attention to systems engineering leadership in the planning and execution of this process is a necessity. This should include the preparation and review of technical manuals describing procedures to be followed during installation and operation.

Early System Operational Difficulties

Like many newly developed pieces of equipment, new systems are composed of a combination of new and modified components and are therefore subject to an excessive rate of component failure or other operational problems during the initial period of operation, a problem that is sometimes referred to as "infant mortality." This is simply the result of the difficulty of finding all system faults prior to total system operation. Problems of this type are especially common at external system interfaces and in operator control functions that can be fully tested only when the system is completely assembled in an operational setting. During this system shakedown period, it is highly desirable that a special team, led by the user and supported by developer engineers, be assigned to rapidly identify and resolve problems as soon as they appear. Systems engineering leadership is necessary to expedite such efforts, as well as to decide what fixes should be incorporated into the system design and production, when this can best be done, and what to do about other units that may have been already shipped or installed. The need for rapid problem resolution is essential in order to effect necessary changes in time to resolve uncertainties regarding the integrity of the production design. Continuing unresolved problems can lead to stoppages in production and installation, resulting in costly and destructive impact on the program.

15.3 IN-SERVICE SUPPORT

Operational Readiness Testing

Systems that do not operate continuously but that must be ready at all times to perform when called upon are usually subjected to periodic checks during their standby periods to ensure that they will operate at their full capability when required. An aircraft that has been idle for days or weeks is put through a series of test procedures before being released to fly. Most complex systems are subjected to such periodic readiness tests to ensure their availability. Usually, readiness tests are designed to exercise but not to fully stress all functions that are vital to the basic operation of the system or to operational safety.

All systems, sooner or later, will experience unexpected problems during operational use. This can occur when they encounter environmental conditions that were not known or planned for during development. Periodic system tests provide information that helps assess and resolve such problems quickly when they occur.

Periodic operational readiness tests also provide an opportunity to collect data on the history of the system operating status throughout its life. When unexpected problems occur, such data are immediately available for troubleshooting and error correction. System readiness tests have to be designed and instrumented with great skill to serve their purpose effectively and economically—a true systems engineering task.

Readiness tests often must be modified after system installation to conform more fully to the needs and capabilities of system operators and maintenance personnel. Development systems engineers can effectively contribute to such an activity. Location of data collection test points and the characteristics of the data to be collected, for example, data rate, accuracy, recording period, and so on, also represent systems engineering decisions.

Commonly Encountered Operational Problems

Software Faults. Faults in complex software-intensive systems are notoriously difficult to eliminate and tend to persist well past the initial system shakedown period. The difficulties stem from such inherent features as the abstractness and lack of visibility of software functionality, sparseness of documentation, multiplicity of interactions among software modules, obscure naming conventions, changes during fault resolutions, and a host of other factors. This is especially true of embedded real-time software commonly found in dispersed automated systems.

The variety of computer languages and programming methodology further complicates system software support. While most analog circuitry has been replaced by digital circuits in signal processing and many other applications, computer code written in older languages, such as COBOL, FORTRAN, and JOVIAL, is still in widespread use. This "legacy" code, mixed with more recent and modern code (e.g., C++, Java), makes it that much more difficult to maintain and modify operational computer programs.

Remedies for software faults are correspondingly complicated and troublesome. A corrective patch in a particular program module is likely to affect the behavior of several interacting modules. The difficulty of tracing all paths in a program and the mathematical impossibility of testing all possible conditions make it virtually impossible to ensure the validity of changes made to correct faults in operational software.

The relative ease of making software changes often leads to situations where these changes are made too quickly, and without significant analysis and testing. In such cases, documentation of the system changes is likely to be incomplete, causing difficulties in system maintenance.

The only way to prevent serious deterioration of system software quality is to continue to subject all software changes to strict configuration control procedures and formal review and validation as practiced during the engineering design and production phases. As noted elsewhere, proving-in changes at a test facility by experienced software engineers prior to installing these in the operational system is an excellent practice; this procedure will pay for itself by minimizing the inadvertent introduction of additional faults in the course of system repair. Chapter 11 is devoted to a discussion of all of the special aspects of software engineering.

Complex Interfaces. In the section on system installation and test (Section 15.2), it was stated that external system interfaces were always a potential source of problems. During installation, there is always a strong push for accomplishing the process as quickly as possible so that operational schedules are maintained. So, while documented installation procedures are generally followed, insufficient time is often allocated to exercise thoroughly the necessary checkout procedures. As noted earlier, examples of areas where operational problems typically show up in a shipboard system are displays, navigation, and communication subsystems. The control panels for these subsystems are usually distributed among various locations and therefore have a strong functional as well as physical interaction. In such cases, the operational crews should be alerted to the potential problems and should be provided with explicit information on the locations and interfaces of all interacting system elements.

Field Service Support

It is common for deployed complex systems to require field support during the lifetime of the system. In the case of military systems, this is often provided by an engineering support unit within a branch of the service. It is also common for that unit to contract with civilian agencies to provide general engineering support to keep the system operating as intended.

When system operating problems are detected, it is necessary first to determine whether the problem is due to a fault in the operational system or is a result of improper functioning of a built-in fault indicator. For example, the device may be erroneously signaling a failure (false alarm) or may be ascribing it to the wrong function. Therefore, the field engineer who is called upon to troubleshoot a problem should be knowledgeable in system operation, including especially the functioning of built-in test devices.

When any fault is encountered during system operation, the required remedial actions are more difficult to implement than they would have been during development or even during installation and test. This is because (1) user personnel are not technical specialists; (2) special checkout and calibration equipment used during installation will have been removed; (3) most analysis and troubleshooting tools (e.g., simulations) are not available at the operational site; and (4) most knowledgeable people originally assigned to the development project are likely to have been reassigned, to have changed jobs, or to have retired. Because of these factors, for operational fixes to be done reliably, they often have to be developed remotely; that is, data will have to be collected at the operational site and transmitted back to the appropriate development site for analysis; corrective action will have to be formulated; and finally, the required changes will have to be implemented at the operational site by a special engineering team.

As noted previously, facilities at the developer's test site are excellent locations for follow-on system work because of the availability of knowledgeable people, configuration flexibility, extensive data collection and analysis equipment, and the opportunity to carry out disciplined and well-documented tests and analyses.

Scheduled Maintenance and Field Changes

Most complex systems undergo periods of scheduled maintenance, testing, and often revalidation. Nonemergency field changes are best accomplished during such scheduled maintenance periods, where they can be carried out under controlled conditions by expert personnel and can be properly tested and documented. Fortunately, this usually accommodates the majority of significant changes. In most cases, as in that of commercial aircraft, such operations utilize special facilities with a full complement of checkout equipment, have a substantial parts stockpile and an automated inventory system, and are conducted by specially trained personnel.

Any changes, large or small, to an operational system require careful planning. As noted earlier, changes should be made under configuration control and should conform to documentation requirements that specifically state how they will be carried out. All changes should be viewed from a system perspective so that a change in one area does not cause new problems in other areas. Any technical change to an operational system will usually also require changes in hardware–software system documentation, repair manuals, spare parts lists, and operating procedure manuals. In this process, systems engineering is required to see that all issues are properly handled and to communicate these issues to those responsible for the overall operation.

Severe Operational Casualties

The previous paragraphs dealt with operational problems that could be corrected during operations or short periods of scheduled maintenance. It must be assumed that a complex system built to operate for a dozen years or more may accidentally suffer a failure of such magnitude that it is effectively put out of commission until corrected, such as by a fire, a collision, or through other major damage. Such a situation normally calls for the system to be taken out of service for the time necessary to repair and reevaluate it. However, before undertaking the drastic step of an extended interruption of service, a systems engineering team should be assembled to explore all available alternatives and to recommend the most cost-effective course of action for restoring operation. The severe casualty poses a classical system problem where all factors must be carefully weighed and a recovery plan developed that suitably balances operational requirements, cost, and schedule.

Logistics Support

The materials and processes involved in the logistics support of a major operational system constitute a complex system themselves. The logistics for a major fielded system may consist of a chain of stations, extending from the factory to the operational sites, which supplies a flow of spare parts, repair kits, documentation, and, when necessary, expert assistance as required to maintain the operating system in a state of readiness at all times. Technical manuals and training materials should be considered part of system support. The effort of developing, producing, and supporting effective logistics support for a major operating system can represent a substantial fraction of the total system development, production, and operating cost.

A basic problem in logistics support is that it must be planned and implemented on the basis of estimates of which system components (not yet designed) will need the most spare parts, what the optimum replacement levels will be for the different subsystems (not yet completely defined), what means of transportation, and hence time to resupply, will be available in potential (hypothetical) theaters of operation, and many other assumptions. These estimates can benefit from strong systems engineering participation and must be periodically readjusted on the basis of knowledge gained during development and operating experience. This means that logistic plans will need continual review and revision, as will the location and stocking level of depots and transport facilities.

There are also direct connections between the logistics support system and system design and production. The sources of most spare parts are usually the production facilities that manufacture the corresponding components and may include the system production contractor and the producers of system components. Moreover, subcomponents and parts commonly include commercial elements and hence are subject to obsolescence design changes or discontinued availability.

System field changes also directly affect the logistic supply of the affected components and other spare parts. Since the process of reflecting such changes in the logistics inventory cannot be instantaneous, it is essential to expedite it, as well as to maintain complete records of the status of each affected part wherever it is stored.

It can be seen from the above that the quality and timeliness of overall support provided by the logistic system will have direct effects on operability. This is particularly true for systems operating in the field, where the timely delivery of spare parts can be crucial to survival. In the case of commercial airlines, timely delivery of needed parts is also critical to maintaining schedules. Managing such a logistics enterprise is itself an enormous task of vital importance to the successful operational capability of the system.

15.4 MAJOR SYSTEM UPGRADES: MODERNIZATION

In the chapters dealing with the origin of new systems, it was noted that systems are usually developed in response to the forces of advancing technology and competition, which combine to create technical opportunities and generate new needs. Similarly, during the development and operational life of a system, the dynamic influence of these same factors continues, thereby leading to a gradual decrease in the system's effective operational value relative to advances made by its potential competitors or adversaries.

Advances in technology are far from uniform across the many components that constitute a modern complex system. The fastest growth has been in semiconductor technology and electro-optics, with the resultant dramatic impact on computer speed and memory and on sensors. Mechanical technology has also advanced, but mainly in relatively limited areas, such as special materials and computer-aided design and manufacture. For example, in a guided missile system, the guidance components may become outdated, while the missile structure and launcher remain effective.

Thus, obsolescence of a large complex system often tends to be localized to a limited number of components or subsystems rather than affecting the system as a whole. This presents the opportunity of restoring its relative overall effectiveness by replacing a limited number of critical components in a few subsystems at a fraction of the cost of replacing the total system. Such a modification is usually referred to as a system upgrade. Aircraft generally undergo several such upgrades during their operating life, which, among other modifications, incorporate the most advanced computers, sensors, displays, and other devices into their avionics suites. A complication often encountered is discontinued production by manufacturing sources, which requires adjusting system interfaces to fit the replacements.

System Upgrade Life Cycle

The development, production, and installation of a major system upgrade can be considered to have a mini life cycle of its own, with phases that are similar to those of the main life cycle. Active participation by systems engineering is therefore a vital part of any upgrade program.

Conceptual Development Stage. Like the beginning of a new system development, the upgrade life cycle begins with the recognition through a needs analysis process of a need for a major improvement in mission effectiveness because of growth in the mission needs and deficiencies in the current system's response to these needs.

There follows a process of concept exploration, which compares several options of upgrading a portion of the current system with its total replacement by a new and superior system, as well as with options for achieving the objective by different means. If the comparison shows a convincing preference for the strategy of a limited system modification or upgrade, and is feasible both technically and economically, then a decision to inaugurate such a program is appropriate.

The equivalent of the concept definition phase for a system upgrade is similar to that for a new system, except that the scope of system architecture and functional allocation is limited to designated portions of the system and to those components that contain the parts of the system to be replaced. Proportionally greater effort is required to achieve compatibility with the unmodified parts of the system, keeping the original functional and physical architecture unaltered. The above constraints require a high order of systems engineering to accommodate successfully the variety of interfaces and interactions between the retained elements of the system and the new components, and to accomplish this with a minimum of rework while assuming that performance and reliability have not been compromised.

Engineering Development Stage. The advanced development phase of the upgrade program, and most of the engineering design phase, is limited to the new components that are to be introduced. Here again, special effort must be directed toward interfacing the new components with the retained portions of the system.

The integration of the upgraded system faces difficulties well beyond those normally associated with the integration of a new system. This is caused by at least the following two factors.

First, the system being modified will likely have been subjected to numerous repair and maintenance actions over a period of years. During this time, changes may not always have been rigorously controlled and documented, as would have been the case if strict configuration management procedures had then been in force. Accordingly, over time, the deployed systems are likely to become increasingly different from each other. This situation is especially troublesome in the case of software changes, which themselves are often patched to repair coding errors. The above uncertainty in the detailed configuration of each fielded system requires extensive diagnostic testing and adaptation during the integration process.

Second, while vehicles and other portable systems are normally brought to a special integration facility for the installation of the upgrade components, many large land- and ship-based systems must be upgraded at their operating sites, thereby complicating the integration process. The upgrading of the navigation systems on a fleet of cargo vessels with new displays and added automation requires effecting these changes on board ship, using a combination of contractor field engineers and shipyard installation technicians. Installation and integration plans should provide special management oversight, extra support when needed, and generous scheduled time to ensure a successful completion of the task.

System Test and Evaluation. The level and scope of system test and evaluation required after a major system upgrade can range all the way from evaluating only the new capabilities provided by the upgrade to a repeat of the original system evaluation effort. The choice usually rests on the degree to which the modifications affect a distinct and limited part of the system capabilities that can be tested separately. Accordingly, when the upgrade alters the central functions of the system, it is customary to perform a comprehensive reevaluation of the total system.

Operations and Support. Major system upgrades always require correspondingly large changes in the logistics support system, especially in the inventory of spare parts. Operation training, with accompanying manuals and system documentation, must also be provided.

These phases require the same expert systems engineering guidance as did the development of the basic system. While the scope of the effort is less, the criticality of design decisions is no less important.

Software Upgrades

As described in Chapter 11, software is much easier to change than hardware. Such changes usually do not require an extensive system stand-down or special facilities.

With increasing system functionality being controlled by software, the pressure for software upgrades tends to make them considerably more frequent than major hardware upgrades.

However, to ensure that such operations are successful, special systems engineering and project management oversight is required to manage the difficulties inherent in system software changes:

1. It is essential that the proposed changes be thoroughly checked out at the developer's site before being installed into the operational software.
2. The changes must be entered into the configuration management database to document the changed system configuration.
3. An analysis must be performed to determine the degree of regression testing necessary to demonstrate the absence of unintended consequences.
4. Operation and maintenance documentation must be suitably updated.

The above actions are required for any system change but are often neglected for apparently small software changes. It must be remembered that in a complex system, no changes are "small."

Obsolescent legacy programs suffer from two disadvantages. First, the number of software support personnel willing to work on legacy software is diminishing and becoming inadequate. Second, modern high-performance digital processors do not have compilers that handle the legacy languages. On the other hand, the task of rewriting the programs in a modern language is comparable to the task of its original development and is generally prohibitively costly. This presents a difficult system problem for systems in the above position. Some programs have successfully used a software language translation to greatly reduce the cost of converting legacy programs to a modern language.

Preplanned Product Improvement (P³I)

For systems that are likely to require one or more major upgrades, a strategy referred to as P³I is often employed. This strategy calls for the definition during system development of a planned program of future upgrades that will incorporate a specified set of advanced features, thereby increasing system capabilities in particular ways.

The advantage of P³I is that changes are anticipated in advance so that, when needed, the planning is already in place; the design can accommodate the projected changes with minimum reconfiguration; and the upgrade process can proceed smoothly with minimum disruption to system operations. These preplanned changes will vary in magnitude and complexity depending on the need and availability of appropriate technology. Commercial airlines, for example, will often plan for a stretched version of an existing aircraft that will carry more passengers and incorporate larger engines and new control systems. By modifying an existing aircraft instead of developing a new one, the problems of government recertification can often be alleviated. In the military, the planned upgrade process has the advantage of prior mission justification. Since the current system is operational and performing a needed function, the proposed system changes will not affect already approved mission and system objectives.

In the case of future improvements defined during initial system development, the contract for implementing them is usually awarded to the development contractor. This is the most straightforward contractual arrangement for carrying out a major system upgrade. It is also most likely to secure the services of engineers familiar with the current system characteristics to participate in the planning and execution of system changes. While even in this case the original development team may have largely dispersed, that part that remains provides a major advantage by its knowledge of the system. However, as can sometimes occur in government-sponsored programs, the pressure for competition can become especially severe and can even lead to the selection of a different contractor team for the upgrade contract. In such cases, an intensive education program will be required for the new team to learn the finer points of the system environment and detailed operation.

15.5 OPERATIONAL FACTORS IN SYSTEM DEVELOPMENT

In Chapter 14, Production, it was pointed out that systems engineers who guide the development of a new system must have significant first-hand knowledge of relevant production processes, limitations, and typical problems in order to coordinate the introduction of producibility considerations into the system design process. It is likewise important that systems engineers be knowledgeable about the system's operational functions and environment, including its interaction with the user(s), in order to be aware of how the system design can best meet the user's needs and accommodate the full range of conditions under which the system is to be used.

Unfortunately, the kinds of opportunities described in Chapter 14 that exist for systems engineers in a development organization to learn about manufacturing processes frequently are not available for learning about the system's operational environment. The latter is seldom accessible to development contractor personnel, except for those who provide technical support services, and these are more likely to be technicians or equipment specialists rather than systems engineers. Another inhibiting factor is that the operational environment is usually so system specific that acquaintance with the environment of an existing operational system does not necessarily provide insight into the conditions under which the particular system under development will operate.

The type of operational knowledge that systems engineers must acquire can be illustrated by the example of developing a new display for an air traffic control terminal. In this case, it is essential that the systems engineers have an intimate knowledge of how the controllers do their job, such as the data they need, its relative importance in sending messages to aircraft, the expected fluctuations in air traffic, the traffic conditions that are deemed critical, and a host of other data that impact the controller's functions. Engineers developing a control console for a civil air terminal can usually observe the operations at first hand and interview controllers and pilots.

However difficult, it is essential that engineers responsible for system design acquire a solid understanding of the conditions under which the system being developed will operate. Without such knowledge, they cannot interpret the formal requirements that are provided to guide the development since these are seldom complete and fully representative of user needs. As a result, it is possible that deficiencies due to faulty operational interfaces will be discovered only during system operation, when they will be very costly or even impractical to remedy.

The term "operational environment" as used here includes not only the external physical conditions under which a system operates but also other factors such as the characteristics of all systems interfaces, procedures for achieving various levels of system operational readiness, factors affecting human–machine operations, maintenance and logistic issues, and so on. Figure 3.4 illustrates the complex environment in which a passenger airliner routinely operates.

Operational environments can vary radically depending on the type of system under consideration. For example, an information system (e.g., a telephone exchange or airline reservation system) operates in a controlled climate inside a building. In contrast, most military systems (airplanes, tanks, and ships) operate in harsh physical, electronic, and climatic conditions that can severely stress the systems they carry. Systems engineers must understand the key characteristics and effects of these environments, including how they are specified in the system requirements and measured during operations.

Sources of Operational Knowledge

A number of potential sources of operational knowledge may be available in certain situations. These include operational tests of similar systems, integration testing during system installation, system readiness tests, and maintenance operations. These activities all address the problems associated with successfully integrating the system's external interfaces with the site and with associated external systems. These can often expose serious problems that are not adequately revealed by the interface specifications provided to the developer.

To gain the necessary operational background, the systems engineer should endeavor to witness the operation of as many systems of the type under consideration as possible. Serving as an active participant in system test operations, or even by simply acting as an observer, is a good opportunity for learning. When present at such tests, the systems engineer should make the most of the opportunity by asking questions of system operators at appropriate times. Of special importance is information regarding what parts of the system are the sources of most problems and why. Learning about operational human–machine interfaces is particularly valuable because of the difficulty of realistically representing them during development.

System Readiness Tests. A useful source of operational knowledge is observing procedures used to determine the level of system readiness. All complex systems go through some form of checklist or fast test sequence prior to operation, often using automatic test equipment under operator control. A commercial airliner goes through an extensive checklist prior to each takeoff and a much more thorough series of checks prior to and during scheduled maintenance. It is instructive to observe how operators react to fault indications, what remedial action is taken, what level of training these operators have been given, and what type of documentation has been provided.

Operating Modes. Most complex systems include a number of operating modes in order to respond effectively to differences in their environment or operating status. Some systems that must operate in a variety of external conditions, such as a military system, usually have several levels of operational readiness, for example, "threats possible," "threats likely," "full-scale hostilities," as well as periods of scheduled maintenance or standby. There may also be backup modes in case of degraded system operation or power failure. The systems engineer should observe the conditions under which each mode is induced and how the system responds to each mode change.

Assistance from Operational Personnel

In view of the limited opportunities for the developer's systems engineers to acquire an adequate level of operational expertise, it is often advisable to obtain the active participation of experienced operational personnel during system development. A particularly effective arrangement is when the user stations a team designated to be system operators at the development contractor's facility during the period of systems engineering, integration, and test. These individuals bring knowledge of the special circumstances of the system's interaction with the intended operational site, as well as represent the system operator's viewpoint.

Another source of operational expertise comes from system maintenance personnel who are experienced in the problems of servicing similar systems at their operating sites and in their logistics support. Systems engineers can gain considerable knowledge by well-planned interviews with such individuals. As noted earlier, complex systems often have maintenance support facilities that may be excellent sources of operational knowledge.

15.6 SUMMARY

Installing, Maintaining, and Upgrading the System

The application of systems engineering principles and expertise continue to be required throughout the operational life of the system. The operations and support phase includes installation and test, in-service support, and implementation of major system upgrades.

Interface integration and test can be challenging due to a mix of various organizational units, complex external interfaces, and incomplete or poorly defined interfaces.

Installation and Test

Installation and test problems can be difficult to solve because installation staff have a limited system knowledge. Systems engineers are seldom assigned until trouble is encountered. However, periodic operational readiness testing is necessary for systems that do not operate continuously. This can help minimize unexpected system problems.

Where nondisruptive installation is required, care to plan the installation procedures, via a hybrid simulation or a duplicate system operating in parallel, is absolutely essential.

In-Service Support

System software must be subject to strict configuration control to prevent serious deterioration of software quality. In this vein, built-in fault indicators are very valuable for detecting internal faults, although they sometimes produce false alarms. Therefore, field engineers should be knowledgeable about built-in test devices.

Remedial actions to correct operational problems are difficult to implement: operational personnel are not technical specialists. Furthermore, troubleshooting tools are limited. And materials and processes involved in logistics support themselves constitute a complex system.

Major System Upgrades: Modernization

Logistics cost is a large part of system cost. Therefore, P^3I facilitates improvement of systems during major upgrades. Advanced features are defined during system development, and advanced planning permits minimum disruption to system operation.

Operational Factors in System Development

Possible sources of operational knowledge include operational and installation tests—by observing system operations within its environment. Of course, assistance from operational and maintenance personnel is invaluable.

PROBLEMS

15.1 Identify and discuss four potential problems associated with the installation and test of a complex navigation and communication system aboard a transoceanic cargo vessel. Assume that some of the subsystems have been integrated at land sites prior to shipment. Assume that a number of contractors are involved, as well as the shipping company and government inspectors.

15.2 Interface problems are usually difficult to diagnose and to correct during final system integration. Why is this so? What measures should be taken to minimize the impact of such problems?

15.3 Operational readiness testing is an important function for deployed systems. As a systems engineer who is familiar with the design and operation of a large complex system, describe how you would advise operational personnel to define and conduct this type of testing.

15.4 Many complex systems incorporate a built-in fault indicator subsystem. This subsystem can itself be complex, costly, and require specialized training and maintenance. List and discuss the key requirements and issues that must be considered in the overall design of a built-in test subsystem. What are the principal trade-offs that must be addressed?

15.5 An effective logistics support system is an essential part of successful system operational performance. While the support system is "outside" of the delivered system, discuss why the systems engineer should be involved in the design and definition of the support system. Discuss the functions of some of the characteristics that must be considered, such as the supply chain, spare parts, replaceable part level, training, and documentation.

15.6 Discuss the types of systems that are best suited for applying P^3I during the design phase. Describe the key elements in justifying the additional cost of a P^3I approach.

15.7 In maintaining an operational system, hardware faults are usually corrected by replacing the offending subcomponent by a spare. Software faults are typically coding errors and must be eliminated by correcting the code. In complex systems, software changes must be made with extreme care and must be validated. Discuss ways in which software faults can be handled in a controlled manner where the operating system is remote from the development organization.

FURTHER READING

B. Blanchard and W. Fabrycky. *System Engineering and Analysis*, Fourth Edition. Prentice Hall, 2006, Chapter 15.

Performance Based Logistics: A Program Manager's Product Support Guide. Defense Acquisition University (DAU Press), 2005.

N. B. Reilly. *Successful Systems for Engineers and Managers.* Van Nostrand Reinhold, 1993, Chapter 11.

Systems Engineering Fundamentals. Defense Acquisition University (DAU Press), 2001, Chapter 8.

INDEX

Systems Engineering Principles and Practice, Second Edition. Alexander Kossiakoff, William N. Sweet,
Samuel J. Seymour, and Steven M. Biemer
© 2011 by John Wiley & Sons, Inc. Published 2011 by John Wiley & Sons, Inc.

WILEY SERIES IN SYSTEMS ENGINEERING AND MANAGEMENT

Andrew P. Sage, Editor

ANDREW P. SAGE and JAMES D. PALMER
Software Systems Engineering

WILLIAM B. ROUSE
Design for Success: A Human-Centered Approach to Designing Successful Products and Systems

LEONARD ADELMAN
Evaluating Decision Support and Expert System Technology

ANDREW P. SAGE
Decision Support Systems Engineering

YEFIM FASSER and DONALD BRETTNER
Process Improvement in the Electronics Industry, Second Edition

WILLIAM B. ROUSE
Strategies for Innovation

ANDREW P. SAGE
Systems Engineering

HORST TEMPELMEIER and HEINRICH KUHN
Flexible Manufacturing Systems: Decision Support for Design and Operation

WILLIAM B. ROUSE
Catalysts for Change: Concepts and Principles for Enabling Innovation

LIPING FANG, KEITH W. HIPEL, and D. MARC KILGOUR
Interactive Decision Making: The Graph Model for Conflict Resolution

DAVID A. SCHUM
Evidential Foundations of Probabilistic Reasoning

JENS RASMUSSEN, ANNELISE MARK PEJTERSEN, and LEONARD P. GOODSTEIN
Cognitive Systems Engineering

ANDREW P. SAGE
Systems Management for Information Technology and Software Engineering

ALPHONSE CHAPANIS
Human Factors in Systems Engineering

JOHN E. GIBSON, WILLIAM T. SCHERER, and WILLAM F. GIBSON
How to Do Systems Analysis

WILLIAM F. CHRISTOPHER
Holistic Management: Managing What Matters for Company Success

WILLIAM B. ROUSE
People and Organizations: Explorations of Human-Centered Design

GREGORY S. PARNELL, PATRICK J. DRISCOLL, and DALE L. HENDERSON
Decision Making in Systems Engineering and Management

MO JAMSHIDI
System of Systems Engineering: Innovations for the Twenty-First Century

ANDREW P. SAGE and WILLIAM B. ROUSE
Handbook of Systems Engineering and Management, Second Edition

JOHN R. CLYMER
Simulation-Based Engineering of Complex Systems, Second Edition

KRAG BROTBY
Information Security Governance: A Practical Development and Implementation Approach

JULIAN TALBOT and MILES JAKEMAN
Security Risk Management Body of Knowledge

SCOTT JACKSON
Architecting Resilient Systems: Accident Avoidance and Survival and Recovery from Disruptions

JAMES A. GEORGE and JAMES A. RODGER
Smart Data: Enterprise Performance Optimization Strategy

YORAM KOREN
The Global Manufacturing Revolution: Product-Process-Business Integration and Reconfigurable Systems

AVNER ENGEL
Verification, Validation, and Testing of Engineered Systems

WILLIAM B. ROUSE (editor)
The Economics of Human Systems Integration: Valuation of Investments in People's Training and Education, Safety and Health, and Work Productivity

ALEXANDER KOSSIAKOFF, WILLIAM N. SWEET, SAM SEYMOUR, and STEVEN M. BIEMER
Systems Engineering Principles and Practice, Second Edition